Emptying Their Nets
Small Capital and Rural Industrialization in the Nova Scotia Fishing Industry

The day of the independent fisher is done, according to conventional wisdom. There is a widespread conviction today that the survival of small capital in the fishing industry is anachronistic or, at best, exceptional, and therefore relatively unimportant in the grander scheme of things.

Richard Apostle and Gene Barrett take a different view. They argue that the differentiated structure of the industry, while rooted in a number of resource-centred conditions, stems from a range of socio-economic factors which the fishery shares with other industrial sectors. It is the overall constellation of events and forces that continually regenerates a structural basis for small capital and frustrates the centralization and concentration of capital.

They suggest that a variety of types of small capital must be distinguished, each with particular internal structures – for example, relations to the harvesting sector, large-scale capital, and the market. While the coastal zone fishery is vital to the small capital processing sector, family-based organization, intermediate technology, surplus labour, and community structures are central to this structure as well.

They argue as well that a populist ideology rooted in individualism and free enterprise emanates from this structure and represents its major weak point in terms of concerted action during times of crisis.

The culmination of a six-year research project, this work incorporates primary research data collected through surveys of plant managers, fishers, plant workers and brokers, four community studies, and a financial analysis of fish plant operations. It identifies various strands and patterns that make up the fabric of rural community, revealing the complex interconnections among household, class, gender, and community and the structuring of economy and society.

RICHARD APOSTLE is Professor and Chair of the Department of Sociology and Social Anthropology, Dalhousie University. He is co-author of *The Anatomy of Racial Attitudes*.

GENE BARRETT teaches sociology at Saint Mary's University.

Emptying their nets

Small capital and
rural industrialization in the
Nova Scotia fishing industry

RICHARD APOSTLE and
GENE BARRETT

with contributions by

Pauline Barber

Kevin J. Barrett

Anthony Davis

Marie Giasson

Leonard Kasdan

Leigh Mazany

Lawrence Willett

UNIVERSITY OF TORONTO PRESS
Toronto Buffalo London

338,372
A64e

© University of Toronto Press 1992
Toronto Buffalo London
Printed in Canada

ISBN 0-8020-5894-9 (cloth)
ISBN 0-8020-6831-6 (paper)

Canadian Cataloguing in Publication Data

Apostle, Richard.
  Emptying their nets

  ISBN 0-8020-5894-9 (bound) – ISBN 0-8020-6831-6 (pbk.)

  1. Fisheries – Nova Scotia. I. Barrett, Gene.
  II. Title.

HD9464.C23N67 1991 338.3'72709716 C91-094624-8

This book has been published with the help of a grant from the
Social Science Federation of Canada, using funds provided by the
Social Sciences and Humanities Research Council of Canada,
and with the assistance of a generous gift to the
University of Toronto Press from Hollinger Inc.

To Faye and Thomas
and to Susan, Brynle, and Tessa

And makest men as the fishes of the sea, as the creeping things, that have no ruler over them? They take up all of them with the angle, they catch them in their drag: therefore they rejoice and are glad. Therefore they sacrifice with their net, and burn incense into their drag; because by them their portion is fat, and their meat plenteous. Should they therefore empty their net, and not spare continually to slay the nations?

Habakkuk 1: 14–17

Contents

Illustrative material

Preface

This volume is the culmination of six years of research conducted under the auspices of the Gorsebrook Research Institute for Atlantic Canada Studies at Saint Mary's University in Halifax. The Land and Sea Research Project began as a collaborative effort to investigate the persistence of small-scale production units in the fishing and fish processing industries of Nova Scotia. To tackle this problem, various approaches and designs were needed. The project's designers were impressed by the array of research methods used by a team of researchers on Canada's West Coast who were studying the many dimensions of that coast's fishing industry (Marchak, Guppy, and McMullan 1987; Apostle 1990).

In 1983 the research group – Richard Apostle, Gene Barrett, Anthony Davis, and Leonard Kasdan – mapped out a series of data-collection projects that included surveys of fish plant managers, fishing captains, and fish plant workers, as well as four case studies of selected port and labour markets (see Figure 1 and Map 1). In 1984 the researchers received funding for a multiphase project to investigate the structure of fish processing in Nova Scotia.

The group completed the survey of plant managers (chapter 4) and port market case study I (chapter 8) (Map 1, Site I) during 1984. Because of other commitments, Anthony Davis left at the end of the first year's activities. Apostle, Barrett, and Kasdan continued with the research plans and in 1986 undertook the survey of captains (used in chapters 7, 12, and 14) and hired Lawrence Willett, an anthropologist, to conduct case study II, of port markets (chapter 9) (Map 1, Site II). Leonard Kasdan left the project at the end of the second year. At this point, Apostle and Barrett assumed primary responsibility for organizing the research program and completing the manuscript. Apostle and Barrett conducted the survey of plant workers in 1987 (used in chapters 10, 12, and 14) and recruited two anthropologists to complete the two remaining labour market studies. Marie Giasson's work on Clare (Case Study III) (see Map 1, Site III) became chapter 11, and Pauline Barber's case study (no. IV) (Map 1, Site IV) of 'Northfield' became chapter 13.

Two further studies emerged in 1985 as 'spin-offs' from the initial management survey. Kevin Barrett was commissioned to analyse some detailed financial data on selected fish plant operations, which were collected in a follow-up to the survey of plant managers (chapter 5). And Richard Apostle and Gene Barrett collaborated with Leigh Mazany, an economist at Dalhousie University, on a study of key US fish brokers purchasing fish from the plants in our 1984 survey (chapter 6).

The research program was financially sustained by a series of grants from the Social Sciences and Humanities Research Council of Canada, as well as a grant from the Donner Canadian Foundation. Four grants from the Social Sciences and Humanities Research Council of Canada (SSHRC) funded much of the work for this book. In 1984, SSHRC funded a pilot study (Project No. 410-83-0978); this money was used to conduct the survey of fish plant managers, the first case study, and the follow-up financial survey of fish plants. In 1985, Apostle, Barrett, and Kasdan received funding from SSHRC (Project No. 410-85-0608) for the survey of fish brokers in the New England and New York areas, a survey of fishing boat captains, and the second case study. In 1986, Apostle and Barrett employed a SSHRC grant (Project No. 410-86-0198) to finance a survey of plant workers and the two remaining case studies of labour markets. In 1987, a final SSHRC grant (Project No. 410-87-0116) was used to complete analysis of data and the manuscript of this book. Especial gratitude is owed to Patrick Mates, of the Research Council, for his advice and understanding in administering the project.

Apostle and Barrett also received funding from the Donner Canadian Foundation to help finance the survey of boat captains. Leigh Mazany received separate support from the Donner Canadian Foundation for her work on the survey of fish brokers.

Most of the material in chapters 2 and 3 originated in Gene Barrett's doctoral thesis. Portions of chapter 4 originated in an unpublished project report that Apostle and Barrett wrote with Anthony Davis and Leonard Kasdan. Chapters 6, 7, and 10 represent substantially revised versions of previously published articles.

The authors would like to thank a number of people for comments and discussions during the past six years: Ralph Bannister, Manfred Bienefeld, Tom Bottomore, Tony Charles, Wallace Clement, Neil Guppy, Suzan Ilcan, Svein Jentoft, John Kearney, Bonnie McCay, R.D.S. MacDonald, Ian McKay, John McMullan, Pat Marchak, Leigh Mazany, Marguerite Overington, Lynn Pinkerton, Peter Sinclair, Victor Thiessen, Torben Vestergaard, Keith Warriner, and Fred Winsor. Virgil Duff, at the University of Toronto Press, and two anonymous reviewers of the full manuscript have also provided much-needed support and helpful comments at a time when our enthusiasm was beginning to wane. Last, we are deeply indebted to John Parry and Lorraine Ourom for magnificent editorial assistance on the final draft of the book.

Large research programs like this one inevitably require the co-operation and goodwill of a number of people. The authors are particularly grateful for the support received from Saint Mary's University, especially Kenneth Ozmon, president; Joseph Jabbra, former academic vice-president; and James Morrison, former dean of arts. They also appreciate the physical resources and services provided by the Department of Sociology and the Survey Centre at Saint Mary's and by the Gorsebrook Research Institute for Atlantic Canada Studies. Pat Connelly and Henry Veltmeyer, former chairs of the Sociology Department; Tony Winson, John Chamard, and Ken MacKinnon, successive directors of the Gorsebrook Institute; and Madine VanderPlaat, director of the Survey Centre, were unfailing in giving assistance. Anne Martel, Janice Raymond, and Paul Smith were helpful with technical aspects of survey work. Shirley Buckler suffered through indecipherable manuscripts with more than the usual amount of grace and good humour. The project would also like to thank Andrew Cochrane for his assistance in collecting financial data from fish plants, Claudia Kingston for translating Marie Giasson's paper, and Peter Lambly for bibliographic and editorial assistance.

We would like to thank above all the plant managers, fishers, and plant workers of Nova Scotia for taking the time to explain their activities to us. We owe a tremendous debt of appreciation to Luis Araujo, Bev Banks, Donald Blades, Paul Blades, Marcel Comeau, Clayton D'Entremont, Noel Despres, Ove Hjelkrem, Bud McLeod, Cornelius Mutsaers, Susan Peterson, Donald Smith, Roger Smith, Gerrit Vanderheyde, and Dewey Waybret.

MAP 1 Nova Scotia: counties and general locations of case studies
I, Digby Neck (chap. 8); II, 'Gangen Harbour' (chap. 9);
III, Clare (chap. 11); IV, 'Northfield' (chap. 13)

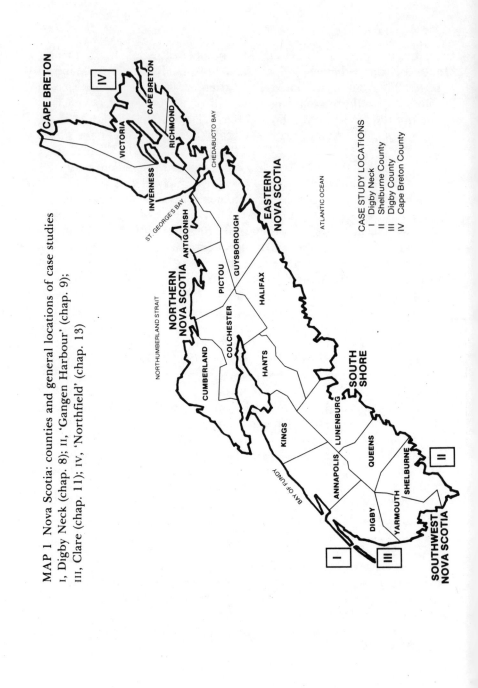

FIGURE 1
Land and Sea Research Project: surveys and case studies

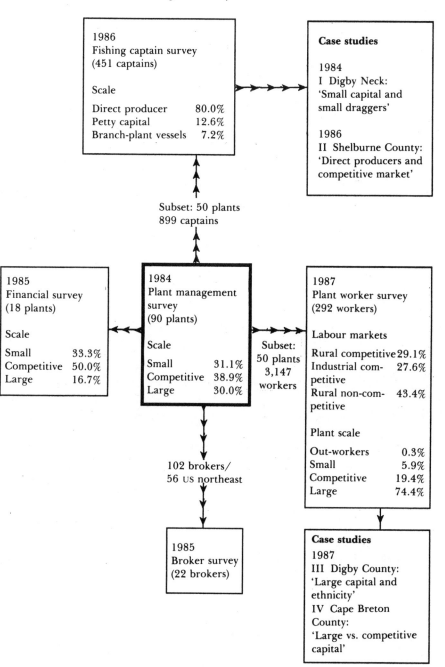

EMPTYING THEIR NETS

Introduction

It would be hard to overstate the importance of the fishing industry to the Atlantic region of Canada. This area is bounded by over 96,000 square kilometres of the world's richest fishing grounds (see Map 2). Aside from highly productive inshore waters, a continental shelf extends outward to a depth of 100 fathoms, in places up to 300 nautical miles from shore. A cold current running south from the Davis Strait helps make this environment rich in plankton and a fertile feeding ground for fish. There are approximately 30 significant species of fish pursued commercially in this area. The northwest Atlantic fishery, according to 1987 figures, accounted for over 3 per cent of the world's total catch (Northwest Atlantic Fisheries Organization 1989; United Nations, Food and Agricultural Organization 1987). While the Canadian share of the total northwest Atlantic catch was 44.6 per cent, this area contributed 80 per cent of the total Canadian catch. In 1988, Canada was the world's largest fish exporter and the Atlantic provinces exported fish valued at $1.8 billion – over two-thirds of the total (Canada, Fisheries and Oceans, 1988). A recent study of the economy of Atlantic Canada observed that fish processing is the single most important component of manufacturing in terms of either employment or output. While the harvesting sector employed 50,000 fishers in the early 1980s, 19,000 workers were employed onshore in the processing sector, and a further 47,000 jobs were indirectly dependent on the industry (Atlantic Provinces Economic Council 1987: 53, 57–8). Most of this impact is concentrated in the rural economy of the region. 'Fishing represents the lifeblood of more than 1,300 small communities in the region [58 per cent of which are in Nova Scotia] and a cornerstone of industry. It dominates economic activities outside big urban

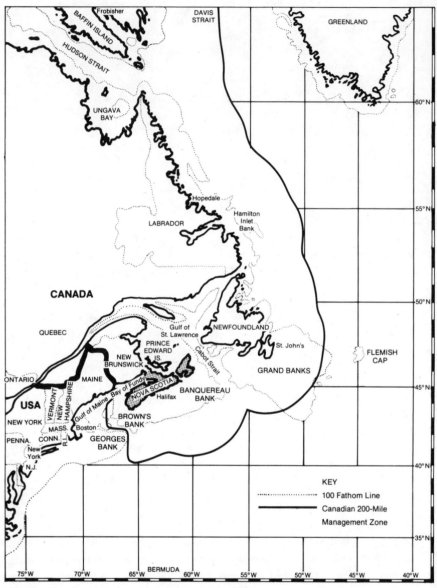

KEY

···················· 100 Fathom Line

————— Canadian 200-Mile
Management Zone

MAP 2 Nova Scotia and the northwest Atlantic fishery
Source: L.B. Jensen, *Fishermen of Nova Scotia* (Halifax: Petheric Press 1980),
113

areas, and influences the social fabric of many coastal rural areas' (Atlantic Provinces Economic Council 1987: 45).

At first glance, the fishing industry has, since the 1920s, looked like a shining example of capitalist development, to the point that Nova Scotia now claims to have the world's largest fishing company (*Sou'Wester* 1 May 1990). In this era of expanding scale and concentration, however, the overall industrial structure of the industry has been developing in the opposite direction. Since the Second World War, an independent coastal zone fishery has emerged to rival the industrial fleets of large companies, while in communities along the coast of Nova Scotia small plants have sprung up to revitalize rural economies impoverished for centuries by the ravages of debt dependence. How does one interpret these trends? New patterns of rural industrialization, characterized by productive decentralization and the rise of small capital, pose interesting problems for social science, since they challenge many orthodox assumptions about the development process. This is no less true for neoclassical economic theory than for Marxism. Industries relocating in rural areas represent one problem. Industrialism that springs from an indigenous rural base is quite another issue. The roots of rural industry and its structuring into larger regional and national economies have become the twin themes of a new industrial sociology.

This book examines industrial differentiation in the fishing industry. Prevailing views see it as anachronistic or, at best, exceptional and relatively unimportant to the grander scheme of things. We take a different position. Our thesis is that industrial differentiation in the fishing industry is the result of a constellation of factors, some unique to the resource and some shared with other sectors. Resource-centred factors stem from the nature of the fishery as common property – particularly the existence of an independent coastal-zone fishery – and from idiosyncratic, product-based factors that affect production processes and marketing relationships. More general factors, such as managerial style, family-based organization, flexible technologies, surplus labour, community structures, and inter-firm networking, are vital to the survival of small capital as well. It is the overall structure – resource, economic, and social – that accounts for the ongoing regeneration of small capital and the frustration of attempts by large capital to consolidate the industry. Lastly, we argue that a populist ideology rooted in individualism and free enterprise emanates from this structure and represents its major weak point in terms of concerted action during times of crisis.

Interest in pursuing this research program stemmed from two factors: dissatisfaction with existing research on the fishery and the need for more substantive directions in political economy and segmentation theory. Fishery social science had reached an impasse by the early 1980s. Prevailing perspectives were proving deficient, and specialized research was leaving too many issues outstanding. Durrenburger and Pálsson (1987) observed that anthropological research on fishing focused too narrowly on producers and indigenous management and economic research concentrated on econometric models of fisheries management. A more fruitful approach should 'focus on the national systems of fisheries production themselves, the whole systems, not just the fishermen and policy makers. Because economists ignore petty commodity production in fishing is no reason for anthropologists to ignore capitalist production and the capitalists' control of inland portions of industry' (1987: 519). At the same time, a number of paradigms in political economy, industrial sociology, and geography reached a conjuncture that necessitated renewed emphasis on research. Increasing theoretical impotence threatened the intellectual gains produced by a widespread critique of orthodox empiricism in the early 1970s. The interface among industrial geography, the sociology of organizations, and labour and management economics, which has been so fruitful in the United States and Europe, seems to have been largely absent in Canadian scholarship. Little of this scholarship has cross-fertilized with the logical area of interest – rural sociology.

We come from divergent intellectual backgrounds and wanted to reconcile aspects of segmentation and dependency theory in the analysis of rural development, generally, and the fishing industry, specifically. We wished to use the case of the fishery to examine the relevance of dualist and dependency theories of industrial structure in light of a new perspective in industrial sociology on small capital and flexible specialization. A number of questions informed this interest. For example, under what circumstances does capital decentralize production? When is capital unable either to centralize or to concentrate production? What industrial patterns emerge under these conditions? How are these trends related to conditions of rural development – to labour surplus, community life, plural activity, households and family life, and so on? In particular, how is industrial differentiation in the fishery affected by resource access and the competitive structure of the market?

What effect do these factors have on the social organization of the enterprise, the nature of the industry's 'labour problem,' and so on? Given recent emphasis in political economy on ideology and organization in the fisheries, we wondered what cultural patterns emerged from decentralized patterns of industrial development, especially where primary producers, workers, and communities were concerned.

In chapter 1, we review three perspectives on industrial structure: dualist theory, dependency theory, and differentiation theory. A differentiation perspective, which is emerging in industrial sociology on the basis of research on small capital in Europe, is seen to hold the greatest promise for our purposes. The chapter explores a number of relevant insights concerning the sources of decentralization, differentiation, and the internal dimensions of the new small-capital sector. Family business, paternalism, intermediate technology, and surplus labour are some topics examined. The resource sector and the fishery are then examined in detail to clarify the range of conceptual and substantive problems that need to be considered in assessing the particular case of the fishing industry. Property relations and the conditions that affect the survival or subsumption of primary producers are given particular attention.

Next, chapters 2–6 examine the historical and contemporary structure of the fishing industry. Chapters 2 and 3 offer an extended introduction by Gene Barrett, to the early and recent history of the industry in Nova Scotia. This material addresses issues that are largely absent from the differentiation literature, such as the connection between cycles in the accumulation process and industrial differentiation, the differences between 'new' and 'old' small capital, and the case for long-term trends toward either centralization or differentiation. Historical evidence is garnered to show that differentiation is closely related to 'stable' long cycles in the accumulation process, while attempts to centralize production, particularly in the industrial age, have been crisis-ridden and have precipitated state subsidization.

Chapter 4 examines the industrial and locational structure of the industry today. The original research team inquired about a number of dimensions, including the degree of contracting-out between large and small capital vis-à-vis an 'independent' small-capital sector, the geographical and social bases of locational differentiation, the artisanate origins of 'new' small capital, the nature of managerial decision-making, and so on. A ninety-case survey of fish plant managers in 1984 produced significant evidence for a tripartite plant-scale typology – small, competitive, and large – that distinguished among ownership and control dimensions, internal organization of management and labour processes, and external relationships to the harvesting sector and product markets. Post-war factors related to resource access, labour-market supply, and product-market expansion account for acute industrial and locational differentiation in the industry.

Chapters 5 and 6 offer detailed explorations into two facets of this structure. Based on a detailed financial survey of a subsample of fish plants, Kevin Barrett in Chapter 5 outlines the financial correlates of differentiation based on the scale typology developed in chapter 4. He has found evidence that suggests that small plants in the sample could be characterized as both comparatively profitable and efficient, that competitive plants were efficient but not particularly profitable, and that large plants were neither efficient nor profitable. Chapter 6 presents some collaborative research conducted with Leigh Mazany in 1985 concerning the fish product market in New England and New York. This study involved twenty-two in-depth interviews with leading fish brokers who purchased fish from the plants in the 1984 survey. Again, a high degree of correlation emerged between participation in key product markets and differentiation in the processing sector.

Chapters 7–9 concern the structure of the port market in Nova Scotia. The small-boat, coastal zone fishery is the backbone of Nova Scotia's 765 fishing settlements. Chapter 7 explores the extent to which this sector is the basis for the resilience of small capital, the ties that link fishers and buyers, and the distinction between competitive and monopsonistic port markets. The backward integration of the harvesting sector appears to be connected closely to the use of centralized processing technology, participation in key product-market segments, and state policy; further, the nature and function of informal economic ties between fishers and processors have changed since the post-war modernization of the coastal zone fishery. Close ties now reflect producers' bargaining power in getting capital to assume a portion of the costs and risks associated with the harvesting enterprise.

Chapters 8 and 9 are community-based studies of port market relations. In chapter 8, Anthony Davis and Leonard Kasdan explore the effect of post-war modernization on the coastal zone fishery in a peninsular-island region of Digby County. They look in detail at the state's policies encouraging industrialization of the fleet, the role of a fraction of small capital in developing inshore groundfish draggers, and the intrinsic connection between this technology and overfishing. In chapter 9, Lawrence Willett focuses on the familial and friendship ties that bind coastal zone fishers in a highly competitive port market in Shelburne County.

Chapters 10–13 deal with the labour market and community structure within which the fishing industry operates in Nova Scotia. Mean weekly income for fish plant workers in Nova Scotia in 1986 was $165.40 – 38 per cent of the mean weekly income in the manufacturing sector as a whole. Were conditions of labour surplus in rural Nova Scotia as undifferentiated as they seemed? Given perspectives about 'independent' in contrast to 'sub-contracting' small capital, it seemed useful to examine the relative importance of labour 'availability' as opposed to labour 'cost' and whether unionization has affected pay scales. Also, when does the labour force become segmented along gender or ethnic lines? – do men for example, represent as cheap a labour pool as women? The role of community-based considerations, such as patronage and clientism, in the structuring of local-level labour markets also seemed worth investigating.

Chapter 10 integrates findings from surveys of plant managers and of 292 fish plant workers and demonstrates that surplus labour is vital to the structure of today's fish processing sector, both large and small. However, plant location and plant scale stratify the work-force along gender, ethnic, pay-scale, and work lines. In chapter 11, based on a case study of women in fish plant work in the Acadian (French) district of Digby County, Marie Giasson reports on high levels of gender segregation and the central role that women play as captive labour in the operations of both a large plant and its small subcontracting operations. The chapter illustrates the central dilemma of fish plant workers between accepting jobs at low pay in fish plants and having to leave their communities to find employment.

Chapter 12 fuses community and household data from the surveys of captains and of plant workers. It explores the significant differences between fishers and plant workers in job satisfaction and community attachment, the primary role of women in reproduction of labour and community structures in rural Nova Scotia, and the double workday faced by women plant workers.

Chapter 13 is a community study by Pauline Barber of gender and work relations in family and fish plant settings. It explores paternalistic employer-worker relations in competitive-scale fish plants. These contrast sharply with the corporatist model employed in a large-scale plant and strongly affect women's predisposition to unionize.

Chapter 14 is an examination of class, ideology, and organization in the fishing industry. While differentiation theory focuses on the populist roots of ideology and class consciousness in the small-capital sector, political economy has remained captive to an orthodox Marxist model based on the capitalist centralization–proletarianization thesis. Material from the three major surveys provides evidence that shows 'proletarian' class consciousness and trade unionism to be highly location-specific and, in overall terms, weak. The chapter, offers an unconventional political economy of the industry which points out the relevance of populism and populist forms of political organization and expression in the industry today.

The concluding chapter is a recapsulization of the instructive conceptual points made in chapter 1 and the contributions that this study has made to the literature in a variety of contexts. The strength of this research stems from its historical depth and its multidimensional analysis of one industry. Various strands and patterns that comprise the fabric of the rural community are unravelled to reveal the complex interconnections among household, class, gender, and community in the structuring of economy and society. Finally, the current crisis in the fishing industry is examined, along with some obvious implications that this study has for explaining its origins as well as understanding the conflicting and contradictory responses that have been forthcoming from a number of quarters. While this chapter offers no 'grand plan' for the restructuring of the industry, it proposes a self-sustaining approach based on decentralized development.

1   A theoretical overview

Richard Apostle
Gene Barrett

Despite predictions to the contrary, small capital has not disappeared
in the modern age. In fact, since the mid-1970s, the growth of small
business in a number of economic sectors has been remarkably widespread in many Western countries (Storey 1983, 1985). While Japan
and Italy have traditionally had a large and vibrant small-firm economy
(Shinohara 1968; Weiss 1988), Bannock (1981) reports that growth in
small-enterprise employment in the 1970s outstripped generation of
employment by large firms in the United States and the United Kingdom (Granovetter 1984). Bechhofer and Elliot (1985) point out that
such change is not just a consequence of the growth of the service
sector in industrialized countries. In Canada, Switzerland, the United
Kingdom, and the United States small firms have increased their proportional share of manufacturing employment.[1] Bechhofer and Elliott
conclude that 'petits bourgeois enterprises have, all things considered,
shown remarkable resilience in the face of major changes – not only
those of general booms and slumps, but of major shifts in the technical
bases of our economies and substantial alterations in political and social
environments. It is the adaptability of the stratum and its role as a
dependent but integral part of modern capitalist economies that most
[impress] us' (192).

Social science has responded to these changes with renewed interest
in the industrial structure of modern capitalism. The size and significance of the small-capital sector, underlying bases of growth, internal
dynamics concerning management structure and labour relations, and
relationships to large capital, state policy, and regional development
have informed a diverse body of scholarship in the sociology of organizations, industrial and regional geography, management, and la-

bour economics. The literature has passed through three phases corresponding roughly to the three decades – the 1960s, 1970s, and 1980s – that have witnessed these dramatic changes. The first perspective was of industrial dualism. In spite of dramatically opposed ideological underpinnings, this view seems to have been based on a convergence theory shared by neoclassical economic theory and orthodox Marxist theory alike. A second schema emerged as a radical critique of both strains of the first approach and was based on a novel dependency theory of industrial structure. A third perspective – differentiation theory – has two different streams. The first is a rather ideologically motivated trend that goes back to the early 1970s and originated in a populist critique of big business by economists and ecologists. The second is a more balanced assessment of the small-capital sector that emerged in the late 1970s and the 1980s. The divergence theory that underlies this latter approach represents a deliberate epistemological break with some fundamental tenets of the Marxist theory of capital accumulation and exploitation.

In the first part of this chapter these three perspectives – dualism, dependency, and differentiation – are reviewed in detail, with particular reference to conceptualizations of the structure and dynamics of the small-capital sector. While the literature has gone through three phases, there may be no necessary antipathy between the dependency and differentiation theories, if sacred doctrines can be assessed in an open-minded fashion. A useful model may be the Italian literature, which attempts to develop typologies of small capital. The second and third sections of the chapter develop a differentiation perspective on the fishing industry as a particular case in point. While idiosyncratic features of the fishery are discussed, factors shared with other resource industries or branches of manufacturing illustrate how this perspective can help explain uneven industrial development. Given recent debate on the subject in the Canadian context, these sections consider relations between capital accumulation and property rights and factors underlying the survival of petty commodity production in the fishery.

Theories of industrial structure

Dualism

The dualistic theory of industrial structure reached its zenith in neoclassical economics and Marxism in the 1960s on the basis of the flowering of two distinct bodies of economic theory.[2] For each it seemed as if all the predictions had finally come true. For neoclassical economics, the post-industrial corporate economy had brought rationality and efficiency to an imperfect market. For Marxist economics, monopoly capital had finally fulfilled its diabolical mission of socializing production on a world scale. Paradoxically, each looked at the industrial structure of the Western world and saw largely the same thing: the leading economic sectors, in terms of any standard measures, were characterized by modern, large-scale firms, while the backward sectors were the last bastions of the small-scale, competitive firms of a bygone era. The industrial structure was conceptualized in dualistic terms – modern versus traditional, large-scale versus small-scale, monopoly versus competitive. The loci of dynamism, change, and growth were in the hands of modern, large-scale capital and would necessarily lead to the gradual shrinkage of the traditional sector: either its increasing marginalization or its subsumption through an inexorable process of capitalist concentration.

The champions of the modern corporation were Schumpeter (1950), Galbraith (1956, 1967), and Servan-Schreiber (1968). As a form of economic organization, the modern corporation emerged as a harbinger of progress and innovation. It represented the best solution to date for the two key economic problems in any society: efficient allocation of resources and the optimization of social welfare (Blair 1972: 112; Dewey 1974: 3–4; Bannock 1982: 80–2).[3] The modern sector was dominated by 'core' firms which were large-scale in organization and activity, were technologically advanced, and employed skilled or well-educated workers. These firms initiated technological change, captured new markets, and controlled whole industrial sectors (Gordon, Edwards, and Reich 1982: 190; Goldthorpe 1984). The 'central work world' characterized the employment milieu generated by these firms. It has recently been described as follows:

Central work world establishments ... offer (or are required to provide) high wages, extensive fringe benefits, interval career ladders, and job security provisions. They are characterized by the interfacing of powerful actors on both the employer and employee sides, and include the public sector as well as establishments in the large-scale, highly capitalized, private sector ... Well-organized self-employed occupational groups such as the professions and the highly-skilled craft guilds are also included in the central work world category, with the organization of the occupation providing the structure that is provided for other workers by their establishment. (Clairmont, MacDonald, and Wien 1980: 290)

The traditional sector was characterized as backward: productive activity was overwhelmingly geared toward traditional crafts, subsistence agriculture or fishing, or service activities. Enterprises were small, individual- or family-based, employing few workers, who engaged in manual labour or used archaic technology. Productive surpluses and incomes were small. Economists characterized labour use in this sector as inefficient. The productivity of labour was low and declining, and economic growth was stagnant. The traditional sector was dominated by 'peripheral' firms:

These enterprises are the ones usually dominated by a single individual or a family. The firm's sales are realized in restricted markets. Profits and retained earnings are commonly below those in the center; long-term borrowing is difficult. Economic crises often result in bankruptcy or severe financial retrenchment. Techniques of production and marketing are rarely as up-to-date as those in the center. These firms are often, though not always, technological followers, sometimes trailing at some distance behind the industry leaders. (Averitt 1968: 7)

This sector was termed the 'marginal work world' and was characterized by low wages, low educational achievement, irregular work, and poverty.

Paradoxically, the Marxist theory of monopoly capitalism shared a dualistic perspective of industrial structure with neoclassical theory. The classical statements of the 1960s by Baran and Sweezy (1966) and Mandel (1968) restated the basic tenets of Marx's theory of capital accumulation (see also Cowling 1982; Norton 1988). Marx conceptualized two laws – combined development and uneven development – which captured the central dynamic of the system as well as a number of exceptions and contradictions to the main trend. The former law was based on two premises. First, the logic of capital accumulation led to the sectoral concentration of capital in fewer and fewer hands under conditions of increasing concentration of productive scale (Wheelock 1983). Second, spatial agglomeration emerged as a reciprocal factor in the organization of society as a whole. While monopoly capital was the chief illustration of the first trend, urbanization was the major consequence of the second.

A number of tendencies and contradictory processes in capital accumulation accelerated the law of combined development while propelling individual capitals in quite different directions, toward spatial and sectoral decentralization. This latter process was Marx's law of uneven development (Mandel 1968: 371-3). Under certain conditions capital decentralizes a variety of functions to maximize its rate of profit. In so doing, it preserves and exploits regional conditions of backwardness.

Additionally, Marx discussed at length the heterogeneous forms that individual capitals, or capital in entire branches or regions, might take because of obstacles to accumulation, periodic crises, and so on. Such anomalies might include resistance to the transformation of manual to machine production,[4] cheap labour as a basis for 'special branches of capitalist exploitation,' domestic industry and domestic 'out'-departments of factories, small capital operating on the 'basis of old methods of production,' capital in the resource sector with distinctive natural advantage, or capital with a temporary technological advantage (Barrett 1983). However, Marx saw the differentiated industrial structure as backward and associated with temporary obstacles to the inexorable process of capitalist concentration (Curran and Stanworth 1979b: 427; Semmler 1982: 110). For example, the small-capital sector has been termed a 'competitive fringe' by Wheelock (1983: 35) in her study of monopoly and competition in Marxist theory.[5]

While the polarity between traditional and modern sectors and its reconciliation were the subject of the early dualistic literature, a segmentation perspective emerged in the 1970s to account for the relative permanence of industrial dualism in the Western industrialized countries and the general failure of the modern sector to complete its historic mission.[6] Although the theory of structural transformation was amended, the segmental approach to industrial structure was not[7] (see Stolzenberg 1978; Tolbert, Horan, and Beck 1980; Aldrich and Weiss 1981; Kalleberg, Wallace, and Althauser 1981). Segmentation theory argued that barriers to the mobility of both capital and labour slowed the transformation process. At the level of the firm, barriers to entry associated with technology, access to raw materials, market structures, advertising costs, and economies of scale are commonly discussed in the study of sectoral dualism in industrial structure (Bain 1956; Mann

1966; Blair 1972; Storey 1982).[8] Locationally these factors spawned an agglomeration tendency in modern industrialism most commonly associated with urbanization and growth poles (Perroux 1955; Myrdal 1957; Holland 1976).

As for the traditional sector, small marginal firms required maximum flexibility in their operations and therefore could not operate under the rigid job classifications of a union contract. Small firms could not afford to pay union-scale wages. Workers were often trapped in a cycle of marginal employment by their lack of skills, education, or experience (Apostle, Clairmont, and Osberg 1985). Theorists of the dual labour market added an ascriptive dimension, arguing that women and particular ethnic and racial groups suffered differentially from these disadvantages through institutional discrimination. Locational isolation further inhibited modernization. Other things being equal, the marginalization of entire localities tended to be self-perpetuating – the 'backwash' or 'internal colony' effect (Blauner 1969; Bonacich 1972; Harris 1972).[9]

Dependency

Writing in 1978, Enzo Mingione exclaimed, 'We cannot ignore it any more – whereas Marx could' (214). Two things that could no longer be ignored were the tendency of capital to decentralize production (and the related diseconomies of large scale in certain sectors) and the interconnections between large-scale and small-scale capital. A new dependency perspective posited the integral relationship between capitalist and pre-capitalist sectors. This relationship was seen to be fundamentally unequal and exploitative, although it might have benign phases, depending on the business cycle. For some theorists (Rubery and Wilkinson 1981: 116; Gordon, Edwards, and Reich 1982), this perspective emerged out of a critique of orthodox segmentation theory; for others (Rainnie 1984) it was a sympathetic revision of Marxism; and a third strand relied heavily on points made in the dependency critique of orthodox development economics (Massey 1984).[10]

The tendency to capitalist decentralization, both spatially and organizationally, and the increasing effectiveness of intermediate-scale economies were seen to have undermined large-scale production. The industrial technology that gave rise to large-scale productive technique and vertical integration evolved to free capital from many inefficiencies, waste, and the limitations of mass production technology (Mingione 1978: 216; Brusco 1982: 172–3; Gordon, Edwards, and Reich 1982: 170–1, 203; Murray 1983: 77, 79).[11] The changes were felt in all aspects of industrial structure but most notably in production, circulation, and organization.

It was argued that the new technologies increased the use of synthetic materials, reduced costs, and increased availability and locational flexibility. Automation expanded prospects for constituent component assembly and multiple-input sourcing. Storper and Walker (1984: 21) observed: 'Rigid, mechanically integrated systems have been replaced by electronic controls, permitting greater versatility and smaller production facilities (for the same or greater output).' Brusco (1982: 178) noted the immense flexibility of numerically controlled machine tools as opposed to single-purpose ones. Transportation and communication technology reduced the costs and time associated with circulation and facilitated standardization and internationalization of inputs over a wide space. The communications revolution rationalized these calculations and improved the efficiencies associated with centralized organization. Computer technology allowed capital to reduce idle capacity, increase inventory and information control, and institute low-cost 'dry run' modelling (Murray 1983). Murray (1983: 88) concluded: 'The introduction of computer-assisted management is a potentially valuable weapon for capital because it can increase management's control over all aspects of production, firstly through the further expropriation of workers' knowledge (mental labour) and secondly, through an "objectification" of control over labour that ensures the maximum saturation and co-ordination of labour time.' The multi-plant firm, with centralized administration and decentralized production facilities, became the latest incarnation of monopoly capital. The inevitability of capitalist concentration was, however, still unquestioned (Rainnie 1985a: 153, 158; see Norton 1988).

The precise contours of capitalist decentralization were complex.[12] Some industries, by virtue of their production lines, could not only relocate their central physical plant but decentralize operations altogether. Others could not decentralize and relied on firm-level restructuring and contracting out to independent capital. In a detailed study of the semiconductor industry in southeast Asia, A.J. Scott (1987) found that some aspects of decentralized production were reliant on intrafirm controls and, therefore, more likely to remain closely integrated through branch-plant organization. Other aspects of the industry lent themselves more readily to divestment. Production embodying 'firm-

specific' capital and managerial and technical control of intra-firm production schedules tended to be 'especially susceptible to ... functional
integration combined with spatial disintegration' (A.J. Scott 1987:
153–4). Independent assembly operations emerged around activities
that were free from such 'technological complementarities.' Their markets were more volatile and unpredictable, and their survival was more
dependent on close control of costs and the pooling of contracts. 'Lo-
cally-owned plants cut back on quantities of fixed capital per worker
as a way of minimizing the deleterious effects of this volatility, for
employment levels can be raised or lowered much more easily than
levels of fixed capital investment' (A.J. Scott 1987: 152).

Subcontracting was, for dependency theory, a central trend in the
monopoly era (Rubin 1978; Rubery and Wilkinson 1982; Mattera 1985).
Metropolitan capital subcontracted to small capital on an extended basis
in order to transfer risk in highly competitive markets (Murray 1983:
81; Goldthorpe 1984: 333). In so doing, it lowered costs and increased
flexibility, circumventing union contracts (Brusco 1982: 171; Lane
1982: 11), avoiding production that was difficult to standardize (Gordon, Edwards, and Reich 1982: 191). This process has been termed
'vertical disintegration' in the context of new patterns of rural industrialization in southern Europe (Hudson and Lewis 1984; Lewis and
Williams 1986). In Japan, subcontracting is the basis for a large 'split
level economy' (Curran and Burrows 1986: 274; Patrick and Rohlen
1987).

Three issues stand out in the dependency perspective on small capital: the underlying forces behind decentralization; the inherent backwardness of the small-capital sector, and the question of subsumption
– the real independence of this sector and the prospects for development based on small business.

Two interrelated causes of the post-1970 trend toward decentralization are generally discussed in the dependency literature. The 1970s
are seen to have witnessed a dramatic economic crisis for capital. First,
the increasing power of organized labour was felt through wage demands but also as constraints on managerial efforts to restructure mass
production plants (Mingione 1978; Brusco 1982; Murray 1983; Mattera 1984). The competitive drive to lower wage costs and increase
labour flexibility is seen as a primary force behind decentralization.
Small capital has a low level of unionization, is able to pay only low
wages, imposes irregular or intermittent work schedules, and varies
the numbers of workers and type of work required (Rainnie 1985a:

154). Murray (1983: 83) cites the example of an engineering firm in Bologna:

In 1972 46% of production work was put-out of the firm, employing indirectly the equivalent of 570 full-time workers in small firms and workshops, whereas in 1969 only 10% had been put-out. In 1974–5 production fell rapidly, and work put-out dropped to almost nothing, resulting in the loss of approximately 550 jobs. That is, while the level of employment in the firms working for the company went through a massive fluctuation, employment in the company itself was relatively stable. The company putting the work out did not then pay a penny of redundancy money and nor was there any disruptive and socially embarrassing struggle over job losses. This illustrates clearly the flexibility putting-out can provide. In this instance the reason for putting-out was not so much the exploitation of wage differentials as the minimisation of costs and conflict over job losses with the union.

However, the same firm does also put-out work for savings on wages, where the outworkers are paid up to 50% less than their counterparts in the factory. The work put-out here is not mechanical work but wiring and circuit board assembly and involves women working in small firms and sweatshops where they have no legal or union protection. (See also Brusco 1982: 170.)[13]

New technology, it was argued, revolutionized the social organization of capital, making locational diversity and flexibility a crucial feature in the drive for capital accumulation. Labour costs became the central determining factor in locational decisions (Massey 1979; Storper and Walker 1984: 22; A.J. Scott 1987: 144). While restructured capital had highly diverse labour requirements, capital was fundamentally more mobile than labour in resolving these 'problems.' Labour, as a production input, was highly 'idiosyncratic' and location-specific, and analysis of the regional impact of productive decentralization stimulated a new direction in regional geography. Regional conditions of cheap labour were seen as being both exploited and preserved through capitalist decentralization (Massey 1977, 1984; Fothergill and Gudgin 1982: chap. 5).[14] Lane (1982: 9) pointed out the significant problems that new rural plants represent for the urban-centred mass-collective trade union (also Murray 1983: 94–6).

A second issue that distinguishes the dependency perspective on small capital produced less consensus. The concern with labour as the central driving force behind decentralization led to the inescapable conclusion for many that the small-capital sector was still fundamentally backward and, if not pre-capitalist, then certainly archaic (Weiss 1988; 5). On the one hand there was an a priori emphasis on the 'grim reality' of small capital: of sweatshops, homeworkers, instability, and marginal profits (Mattera 1985; Rainnie 1985b). On the other hand, there was grudging acknowledgment that a proportion of new firm expansion had been in the area of high technology. Rainnie (1985a: 147, 156) provided the most doctrinaire interpretation of this trend, however, arguing that it was not as significant as it was made out to be in terms of job creation or technological innovation and, in any case, was still subject to the indirect domination of 'giant' capital.

This raises the third issue that characterizes the dependency perspective: how independent can small capital be? The dependency perspective also described two further types of small capital – apart from branch plants and subcontractors – as interstitial cases. The first was the one just mentioned: firms that were independent but ultimately subject to the market power of large capital in all important areas (Mingione 1978: 217–18; Bechhofer and Elliott 1985; Rainnie 1985a: 152). The small-capital sector was also seen as encompassing, second, a variety of activities inherently or temporarily unprofitable to capital. Petty producers and small capital survived in these niches because they were left alone and because internal labour processes, based largely on self- and family-exploitation, underwrote low profit rates. Mann and Dickinson (1978) felt this to be especially true of agriculture, where the discrepancy between labour time and production time posed an inherent obstacle to capital accumulation on an extended scale. The provision of cheap wage goods was another example: it was crucial to the reproduction of surplus labour but of little direct interest to metropolitan capital (Mingione 1978: 218; Gerry 1979: 135; Mingione 1983).[15] Capitalist hegemony was omnipotent, since the institutional and industrial structures with which small capital and petty producers dealt at every stage – most notably, in provisions of inputs – were as powerful as ever. Ultimately, the small-capital sector was still encapsulated and constrained. Gordon, Edwards, and Reich (1982: 191) emphasized that 'the periphery survives because the central corporations find the remaining periphery unprofitable for direct investment; at the same time, the existence of a periphery enhances the profitability of the corporations at the center.'

The question of the relative independence of the small-capital sector relates directly to a final issue: the nature of industrial development policy based on small business. The dependency perspective is unequivocal on this issue. In fact, the critique of small-business studies is usually begun with a diatribe against Thatcherism, Reaganomics, and the ideology of the new right. Most of the revival in small-business studies is seen as a thinly veiled attack on government social services and the labour movement (Rainnie 1985a: 145–6). At the very least, it is argued, the individualist ideology of a self-sufficient small-business economy serves to mystify and deflect individual attention away from the domination of monopoly capital and the need for collectivist strategies for economic reconstruction.[16]

Differentiation

As mentioned above, a third perspective on industrial structure generally, and small capital in particular, emerged in the 1980s. It has two very different strands. The first is ideologically based; the second is a diverse social science of small capital, with contributions from sociologists, historians, economists, and geographers. While the first strand tends to make broad and sweeping statements largely on faith, the latter has emerged out of detailed case material on management, labour relations, technology-use, production strategies, and family enterprise.

Neoclassical economics has always had a fringe anti-monopolist group that has argued against the evils of corporate concentration – inefficiency, but especially monopoly pricing – and the strengths of free competition (Simons 1948; Machlup 1952; Drilam and Kahn 1954). State anti-trust policy was strongly supported by this group, and its influence reached a peak in the 1960s, with the deliberations of the US Senate's subcommittee on anti-trust and monopoly (Blair 1972; Goldschmid, Mann, and Weston 1974). The crisis of the late 1970s and the rise of neoconservatism in Great Britain and the United States gave these ideas new respectability. The things they had been saying in opposition to Schumpeter and Galbraith about entrepreneurship, innovation and the small firm, the diseconomies of large scale, and the inflationary effect of monopoly pricing – ideas stemming from the 'perfect competition' wing of micro-economics – dovetailed with a new anti-Keynesian, monetarist macroeconomics. Quoting Haberler (1976: 17–18), Adams and Brock (1986) identified this new synthesis as follows: 'I agree with William Fellner, Fredrick A. Von Hayek, and Fredrick Lutz, who are of the opinion that a tight monetary and fiscal policy must be supplemented by measures designed to make the economy more competitive. In the jargon of economics, macroeconomic measures aiming at overall guidance of demand must be accompanied by microeconomic measures designed to promote competition.'

A related and probably more influential position is represented by E.F. Schumacher (1973), Gershuny (1978), and Dahrendorf (1975). This work presents a humanistic critique of industrialism that argues against the wasteful and alienating effects of large-scale industrial society in favour of an improvement in the quality of life through small-scale enterprise. While essentially a philosophical critique, with a strong 'futurist' bias, this perspective has had tremendous influence in terms of the 'green' movement. Marxists have seen this trend as a modern reincarnation of the nineteenth-century anarchist position and have strongly criticized it for being romantic, anti-labour, and fundamentally 'counter-revolutionary' (Kitching 1982; Rainnie 1985b).

A significant epistemological break with Marxist orthodoxy seems to have occurred in the sociology of industrial structure in the 1980s. Numerous case studies focused on industrial, regional, or social problems contributed to a growing body of substantive evidence that either contradicted or fell outside the prevailing dependency perspective. This has contributed to the post-modernist critique of many of the leading assumptions of the Marxist model in general. Two major points have been made: first, that the capitalist mode of production is heterogeneous and is characterized by distinctive and coexistent forms and, second, that the interconnections between these forms are equally diverse and cannot be derived a priori from theoretical propositions about how things must necessarily function. Methodologically, therefore, this perspective stresses that questions about the social and economic organization of small capital and its position in the larger economy need to be addressed case by case through careful empirical study (Curran and Stanworth 1979a: 338; Bechhofer and Elliott 1985: 189; Brusco and Sabel 1981: 99; Goffee and Scase 1982: 108; Solinas 1982; Weiss 1988: 1–8).

This line of inquiry is more than simply a populist critique of Marxist theory. It offers three important ideas: first, that the tendencies to capitalist centralization and concentration can no longer be assumed to be reciprocal – and in fact have to be assessed through careful analysis of specific industrial sectors over time; second, that the internal structure of small capital should not be assumed to be pre-capitalist, or uniform, but should be seen as highly diverse and resilient; and, third, that the development potential of this sector is unique and cannot simply be dismissed as of marginal consequence.

A good place to begin the discussion is with the likelihood of industrial differentiation. Storper and Walker (1984: 26) advise, as a rule of thumb:

If industries converged, their materials, labor, and marketing needs would then have purely idiosyncratic origins ... An entirely better starting point for an explanation of locational patterns is the obvious difference in the specific physical and social properties of products to be produced, transported, and consumed. The common forces of competition, class struggle, and concentration of capital have led industries down different evolutionary paths because each product sector faces fundamentally different sets of possibilities and limits in marketing, production technology, and organization. (also Rubery and Wilkinson 1981: 119–20)

Early anti-monopolists such as Blair (1972) and Mann (1966) identified the product cycle as a central variable in the concentration process. They identified 'centrifugal' factors that inhibit concentration or that continually undermine corporate control once it is established. The key to modern corporate economies is regularized, predictable flows of both raw materials and markets. The dependence of many sectors of production on raw materials whose harvest or production is highly seasonal makes problematic the organization of production by all industries downstream. The fishing industry, agriculture, the fur industry, and the wool trade illustrate this structural imbalance. Where capital has been able to stockpile inventories, the highs and lows in supply can be controlled. Many consumer markets are also inherently unstable and subject to annual seasonal variations in demand or cyclical oscillations. An example of this trend is the garment industry. In a recent study of the Chinese garment industry in New York City, Wong (1987: 127) observed:

New York's garment industry is labor-intensive and seasonal. As a center of fashion, New York has 120,000 jobs in the garment industry. Of these three-quarters are in garment manufacturing and one-quarter in designing. Almost all jobs are available through garment contractors. Virtually all garment factory owners [500 in Chinatown alone] experience a shortage of manpower during the busy season (January–April), but during the rest of the year business is quite slow. However, even during the slow season owners cannot abandon their seamstresses, so it is a challenge for many owners to keep their workers. Many factories remain open every day so employees can come to work.

The balance between seasonal demand and overhead costs is also to be found in tourism, various segments of retail trade, fishing, and construction. Cyclical variations in demand represent chronic problems in industrial sectors tied to intermediate markets, such as the machine tools industry: 'Having first arisen during the nineteenth century, machine tools firms were never able to consolidate. The demand for the products of the machine tools industry is highly cyclically sensitive, because the demand for the capital goods grows rapidly during upswings and plummets during recessions' (Gordon, Edwards, and Reich 1982: 191).

The essential advantage of small capital in these contexts has to do with flexibility and adaptation. This is the most recurrent theme in the differentiation literature: management has to be able to seize new opportunities and make unrestrained decisions at a moment's notice, labour has to be adroit and continuously available for work, and technology has to be inexpensive and flexible in its applications (Lawson 1981: 59–60; Brusco 1982: 179; Goffee and Scase 1985: 57; Dore 1986; Weiss 1988: 200–3).

The situation tends to be self-reinforcing, as low entry costs foster competition. A good example is the textile industry: 'Because textiles cannot be automated much more at this time and the markets are large relative to any one factory's output, there is a continual opening for new competitors entering the field; hence the industry has a low level of concentration which, coupled with steady but unspectacular sales growth, translates into modest profits' (Storper and Walker 1984: 27). While the high fixed costs of technology limit entry into the petrochemical, steel, or auto industries, such is not the case in textiles, woodworking, furniture manufacturing, sawmilling, food and beverage manufacturing, the retail trade, or the tourist industry (Edwards 1975). It is the inability of corporate capital to integrate these sectors horizontally and thereby capture 'monopoly economies' that is of particular concern to differentiation theory.

Differentiation theorists argue that one of the most intriguing trends underlying the 'new firm' phenomenon is not subcontracting but innovation. In contrast to dependency theory, technological and product innovation are seen as central features undermining corporate concentration. The post-war technological revolution has seen the emergence of the 'leading-edge' firm. Gordon, Edwards, and Reich (1982: 20) stress that 'the tendency toward concentration of capital may be periodically arrested and even turned back by the appearance of new technology, the opening of new markets or sources of raw materials supply or the entrance of new firms seeking to expand from successful operations in other industries.' Contrary to what Galbraith and Schumpeter thought, technological innovation is not monopolized by large corporations with specialized research and development branches (Blair 1972: 199–200). Since the Second World War, there has developed an 'independent primary [core] sector' of professionals and technicians who become more mobile as the competition in their sector intensifies (Gordon, Edwards, and Reich 1982: 202). Such has been the case in the semiconductor industry: 'In semi-conductors ... high growth rates and low entry barriers attract new capital and encourage technical innovators to go into business for themselves, whereupon their new products further revolutionize the industry, opening up more markets, continued rapid growth, and so on' (Storper and Walker 1984: 27; also Oakey 1985: 135).

Product innovation also offers small capital a market niche. Brusco and Sabel (1981) argue that this is a necessary condition for 'artisanal' production:

The central feature of the independent small firm ... is its capacity to innovate. Where the traditional artisan accepts the definition of demand given by the local market and the dependent small firm the demand mediated by the large firm, the independent small firm defines its own demand: it tells the customer what it really wants. The independent small firm, in other words, invents new needs and satisfies them at the same time. The secret of this trick lies in the particulars of the firm's internal organisation, its close relations with its clients and its collaboration with other firms in the sector. (106)

While most traditional studies of entrepreneurship have focused on 'demand-side' factors that explain risk-taking, innovation, and the like, the differentiation literature has rejuvenated a deeper sociological interest in the 'supply-side' factors affecting entrepreneurship (Storey 1982). A strong interest in ethnicity and 'middlemen' minorities has stimulated research in the United States and elsewhere (Light 1979; Wong 1987), most interestingly in the study of social class. Storey (1982: 78–9), for example, cites Swedish research on the 'livelihood principle': entrepreneurs coming from skilled sections of the working class who found firms in the same industry and within the same locality in which they previously worked, often as an alternative to unemployment. They are not the classical 'risk takers,' as their key objective is to maintain a standard of living. (By contrast, upper-middle-class, well-educated individuals are seen to be behind the growth in new high-technology firms.) The particular class-related features of an independent 'artisanate' are discussed in more dynamic terms by Solinas (1982: 342) and by Brusco (1982: 178):

The number of artisans or even major entrepreneurs previously employed as workers is very high, particularly as foremen, maintenance workers, and co-ordinators of putting-out networks. For each of these groups, their knowledge of some part of the productive process facilitates their passage to independent work. Even easier in some sectors, particularly that of garments, is the transition from subcontracting to direct contact with the market. Many subcontractors through their relations with their customers learn how to prepare samples, come into contact with the network of distribution, and eventually reach the point where they can circulate samples on their own. If these are well received they will produce a few copies within the firm and will put out the rest. At the same time they will continue to work as subcontractors, thus avoiding undue risks. The system therefore operates as a 'forcing' ground for entrepreneurship.

The internal organization of small capital is essential to its survival and success. Three principal features have been stressed in the literature: the family enterprise, the paternalistic nature of employer-employee relations, and ties to the larger community. Kinship is often more important than either product cycle or technology. The family is used as a 'shock' for absorbing the twin problems of irregular product demand and irregular labour supply. Wong (1987: 124–5) observes: 'When business is slow, family members do everything themselves and thus cut down on outside help. In adverse situations family members can simply stop their salaries from the factory or reduce the profit margin for every garment. Low profit margins and reduced production costs in the family firm environment have enabled the survival of many Chinese garment factories during slow seasons.' Wong observes also that even when small firms are able to establish enough continuity of demand to establish sectional (assembly-line) production, the family is vital to its management structure. A core of loyal, dedicated family members act as supervisors, controlling quality and working long, irregular hours for comparatively little remuneration (Benedict 1968: 3–5; Wong 1987: 125–6).

Benedict (1968: 9–11) also argues that the family structure offers key advantages in terms of risk-taking through the use of family revenues which enable quick response to opportunities. Bechhofer and Elliott (1985: 200) situate kinship at the very core of the petit bourgeois class:

For the petits bourgeois, work dominates their lives. In the struggle to keep the business afloat, husbands, wives, children, and often a wide network of kin and friends are drawn into the routines of the workplace. The insulation of work from home, family from job – so familiar to most people in a modern capitalist society – is neither possible nor, in many instances, desired. Their work is their life. The sense of self-hood, the day-to-day domestic relationships take their character from the rhythms and routines of the shop, the bakery, or the small farm. The relationships between the owners of the productive resources and those who work alongside them are seldom simply those of boss and worker. Commonly the worker is also a spouse, or a son or daughter, for the greater number of petites bourgeois businesses are family businesses ... Such enterprises are not simply run by families, they are fun for families.

In 1984, Goldthorpe reiterated a dualistic view of employer-employee relations in small business, arguing that they were essentially pre-capitalist and 'non-economistic expressive.' The stable, paternal relations in small firms led workers to be essentially self-selecting. In the 1970s, Newby (1977, 1979) and Curran and Stanworth (1979a, 1979b) questioned this view, arguing that the exploitative link between workers and employers in this sector was as acute as elsewhere and that workers found themselves employed in small enterprises for reasons other than choice. Differentiation theory has broken the typecast nature of these two conceptualizations. For example, Granovetter (1984) and Bielby and Baron (1983) argue against the assumption that employee-employer relations in small firms are necessarily either paternal-particularistic or contractual-exploitative (see also Bechhofer and Elliott 1985: 199–200).

The family defines much of the labour process in the small-capital sector, particularly in rural communities. Anthropologists have drawn attention to the hidden reciprocity in small communities and enterprises that accounts for the persistence of seemingly unfair and exploitative relationships. In exchange for loyalty and dedication, workers may receive such non-monetary benefits as work preference for family or friends (Solinas, 1982: 338), flexible hours, variable job tasks, and a personable and friendly work environment (see Murray 1983: 94).

Often a sense of community is reproduced within the workplace. Manager-owners, in exchange for their consideration and indulgence, get tremendous flexibility in how they organize production and often a cheap work-force (Wong 1987: 125).

For the worker this trade-off represents a fundamental dilemma. While these kinds of ties are generally painted in paternalistic-patriarchical terms (Lawson 1981: 48), some observers have linked the class background of the owner-entrepreneur and the nature of employer-employee relations, especially in the artisanate. In these situations, relations are non-hierarchical and fraternal. While not lacking in conflicts, they reflect long-standing personal and familial ties between owner-tradespeople and both the industry and other workers (Goffee and Scase 1982: 112). Curran and Burrows (1986: 272) observe that fraternalism reflects 'the inability of employers to control the market and their dependence on the indispensable skilled tradesmen. Employers work alongside their employees, keeping management tasks to a minimum or carrying them out outside normal working hours.'

The role of community has also been interpreted in two ways. Batstone (1975: 127–8) provides a classic study of paternal relations rooted in the 'ethos of small town capitalism.' Community ties reinforce primordial loyalties and clientism (see Alavi 1973; Long 1975), ultimately reinforcing a traditional social structure (Curran and Stanworth 1979a; Curran 1981). Weiss (1988) and Piore and Sabel (1984), in contrast, resurrect Tonnies's approach to the organically based rural community. Far from reproducing a competitive, cutthroat economy and society, the 'Emilian model' of small-capital development in Italy is based, according to Weiss (1988), on community networking and a spirit of collectivism, even with trade unions (Perulli 1990). Weiss points out: 'The small firm sector works best when competition is restrained by cooperation. As Piore and Sabel observe, in so far as the latter is vital to its success, "flexible specialization works by violating one of the assumptions of classical political economy: that the economy is separate from society"' (1988: 275).

A number of authors find the small-capital sector unique in its growth potential for regional/community development. Goffee and Scase (1985) speak of unique 'managerial' restraints to firm growth. Family-based capitals are run on the basis of 'quasi-organic' management systems: managers often sacrifice growth and revenue in favour of preserving personal control. They often have a well-founded fear of the lethargy, inflexibility, and waste associated with bureaucratic organization. Given the structural basis of small capital, Goffee and Scase (1985: 57) argue that 'the "organic" firm, then, tends to be more appropriate where flexibility and adaptability are required, fluctuating market conditions prevail, and specialist products are made in "short runs."[17]

A number of observations, however, contradict this latent stagnation view. For example, Bannock (1981) cites data on the high levels of new firms being formed that are interdependent. Regions with sizeable small-business economies experience self-reinforcing growth, while regions with large-scale or branch plant economies do not (97–8). Weiss (1988: 199) and Solinas (1987: 342) point to the 'diffused industrialization' among the independent artisanate in the 'Third' Italy. Goffee and Scase (1985: 54) term this a 'culture of entrepreneurship.' Schmitz (1989) points out that this is the heart of the flexible specialization model.

In summary, the obvious question that begs an answer is whether Weiss (1988) is correct in claiming that there can be 'more than one form of capitalism.' As Murray (1983) and Rainnie (1985a) point out, the issue rests on two questions concerning the internal structure and organization of small capital: 'Has the exploitative basis of the relationship between the labour and capital been abolished?' and 'Is small capital immune from the inexorable logic of capital accumulation?' On the first score, it would seem that the answer is no. The differentiation perspective does not claim that exploitation has ended but argues that it is mediated by a number of considerations related to the nature of work – e.g. artisanal versus assembly – and in particular to the degree of linkage between small capital and the larger economy in terms of either community (rural versus metropolitan) or contracting out (franchise versus independent firm). On the second score, the differentiation perspective would seem to argue more strongly in the affirmative, distinguishing between accumulation and the raw profit motive, on the one hand, and a technological imperative, on the other hand. Small capital is motivated primarily by efforts to maintain economies of intermediate scale (defined by product and structurally determined niches); that is, it tries to control costs while maximizing managerial control and productive flexibility. Profit maximization may be important, but it is pursued in quite specific ways.

From our point of view, therefore, the differentiation perspective offers a useful approach to the study of industrial structure. It is based on an analysis of capital, labour, and community structures that situates the small firm not in 'precapitalist' structures but in familial and kinship structures that are 'pre'-capitalist only in so far as they have existed a lot longer than the social contract in the modern sense.

Industrial structure and the resource sector

The resource sector has a variety of technical and natural conditions that affect the productive enterprise. The great promise of the industrial era has yet to be realized completely in natural resource production. Human knowledge and technology have neither replaced all natural sources of raw materials nor removed all natural obstacles – such as location, time, distance, and environment – to resource extraction (R. Murray 1977, 1978). These factors help determine the nature of the extractive or harvesting process, as well as downstream manufacturing and realization processes. From its beginnings, capital has simultaneously attempted to use these factors to its advantage and been prevented from doing so completely.

Four factors make the resource sector special and distinguish a variety of subtypes. First, at the most general level, the origin and age of the resources determine whether they are 'renewable' or 'non-renewable.' Things that can be husbanded, such as terrestrial and aquatic plants and animals, soil, water, and air, are considered renewable. Minerals and metals are examples of non-renewable resources: large-scale production is considered 'accidental' in nature and takes a comparatively long time. This division tends also to spell the difference between extractive industries, such as mining and drilling, and harvesting industries, such as silviculture and agriculture. Second, another fundamental division is between resources with open access and those where access can be closed. With the emergence of horticultural society, terrestrial or subterraneous resources have been subject to restricted access, while aquatic resources, which are far more difficult to exploit, have generally defied attempts at enclosure (Andersen 1974; McCay 1987).

Renewable-resource industries are subject to two further distinctions. Third, an important division still separates plants and animals that are cultivated from those that generally are not. Agriculture, horticulture, silviculture, and aquaculture represent some of the science and practice that humankind has devised to husband nature in the 'artificial' reproduction of resources. However, many activities still rely entirely on nature for reproduction: much of the fishing, forest, and trapping industries. In some cases, this reliance is based simply on considerations of cost and efficiency; in others, the 'wild' condition of natural reproduction cannot be artificially reproduced at all. The major reason for this condition is another distinguishing feature: fourth, sedentary resources – all plants and many 'domesticated' animals – are far easier to cultivate than resources such as many species of fish, game, and birds that migrate over vast areas as part of their reproductive cycles. The migratory nature of many fish stocks has, more than any other factor, limited the ability of particular groups to enclose them.

A number of secondary characteristics constrain capitalist reproduction in harvesting and processing. The most obvious is the specificity of location. Human activity is dictated by nature, by availability of resources, and by the environment within which they exist. Location is often remote from concentrations of human settlement, so that distance remains a significant cost of production. The temporal availability of many resources is conditioned by natural processes of regeneration and growth, fecundity, fertility, and so on. Seasonal fluctuations in input supply therefore inhibit capitalist rationalization of throughput. Many organic resources are perishable or deteriorate over time, generating strong reliance on preservation and storage technology and adding to locational/distance cost considerations. The fragility of products has also limited application of advanced forms of conversion and transfer technology to mass production in processing (Storper and Walker 1984: 34–7).

The fishing industry and capitalist differentiation

Few cases could better illustrate the technical peculiarities of production than the fishing industry. Scholars have developed an avid interest in this case, particularly in the Canadian context. Their studies represent central points of reference in a growing interdisciplinary field of fishery social science (Gordon 1954; Innis 1954; Andersen 1979b), and three important books have been published in the last few years (Sinclair 1985b; Clement 1986; Marchak, Guppy, and McMullan 1987). It would seem useful to focus on factors that affect production in both harvesting and processing sectors. Such an approach permits us to address two theoretical issues in recent political economy research: the relationship between capital accumulation and property rights, and the factors underlying the survival of small capital and independent commodity production.

The relation between capital accumulation and property rights is affected by the unique resource ecology of fishing: most fisheries are open-sea, uncultivated resources. Biologists have for the last one hundred years been concerned with the impact of human harvesting on the resource (Bannister 1989). While evidence indicates that a certain degree of fishing will increase the state and size of fish stocks, scientists disagree as to the point at which harvesting will damage natural reproduction (Schaefer 1957; Crutchfield 1975). Relevant factors are the nature of the harvesting technology – vessel size and catch capacity; vessel mobility; and type of fishing gear and the numbers of vessels and the amount of fishing gear being deployed. Since the turn of the century it has become abundantly clear that not all technology is equal in its effects on fish stocks.[18] Some fishing gear – for example, hook-and-line technology – is selective of certain species and year classes (size) of fish, while other gear is non-discriminatory – for example, most net technology.[19] Some fishing gear is fixed in certain locations – such as trap technology and some hook-and-line and net gear – while others are deployed on a mobile basis, such as some net technology. Basically, fixed gear waits for fish to come to it, while mobile gear chases the fish. Also, some types of gear are non-disruptive – e.g. gear set in mid-water – while others, such as those dragged along the ocean floor, destroy fish habitat.

Productivity is always raised by economists as a measure of the gross efficiency of various fishing technologies. The more non-discriminating and mobile the gear, the greater will be the catch. Fish, being renewable and uncultivated, represent a highly productive but fragile resource that petty producers and capital alike can easily disrupt and destroy. While humans do not have to expend capital to cultivate the resource, their investments in harvesting technology are potentially far more critical than in agriculture. Anthropologists have drawn attention to different 'management' goals and their relative effects on the resource ecology itself: between production imperatives based on increasing short-term catches and those based on sustaining long-term yields. McCay (1981) and others argue that producer-based systems are far less likely to upset ecological balances than large-scale, industrial systems. Capital, they argue, is inherently prone to discount future rewards in the interest of immediate gain, since it can withdraw from the fishery altogether when the resource becomes depleted (Clark 1973; McCay 1981; A. Davis 1984). In addition, large capitalist enterprises are better able to finance the high capital costs associated with an industrial model and are therefore more likely to develop it. It is argued below that industrial technology deployed in an unrestrained manner destabilizes ecological balances and leads to the continual undoing of long-term interests, necessitating the intervention of the state as 'resource manager' (Marchak 1987). In contrast, producer-based selective and intermediate technologies that equilibrate short-term benefits with long-term yields are much more stable and thrive under laissez-faire conditions.

Most fisheries are located in an open marine environment and are composed of migratory stocks.[20] Since the emergence of sedentary human communities this openness has posed a problem. How does one group enclose for itself the sea and marine resources in order to cut off access by other groups? And, since the emergence of privatized wealth, attitudes toward the hunt – and the catch – have compounded the problem for humans: 'One person's gain is seen to be another person's loss.' Scholars have pointed to the 'tragedy of the commons' in the fishery (Hardin 1968): 'If no one owns the resource, and everyone is out for as much as they can get, then they have a real incentive to catch the last fish before someone else does.' While economists concentrate on the utilitarian problem (H.S. Gordon 1954; A. Scott 1955; Munro 1979, 1980), Marxists often cavalierly dismiss it as 'the common tragedy of virtually all economists' (Clement 1986: 69).

Since the emergence of agrarianism, the issue of tenure has been fundamental to the social organization of resource production in society. The fishery resists privatized control and has been 'successfully' exploited only where management systems have created the illusion of ownership. Two such ownership conventions currently prevail: customary management and state regulation. Each is based on management of conflict and competition among harvesters and on tacit cooperation by participants, under the threat of exclusion, coercion, or worse. The former is based on the rule of custom; the latter, on the rule of law. Each regulates access to property – the productive marine environment.

The classic confrontation of capital and landed property illuminates the tenure issue in the fishery. Landlordism initially slowed application of capital to agriculture, since landowners demanded payment in the form of rent for providing capital with access to land. Ground rent was, therefore, a cost to capital. As capital privatized access to land, this cost became a source of surplus profit, as capital exploited a resource that gave it a productive advantage over its competitors. Capital realized profit as a return on its capital investment in technology and surplus profit – differential rent – as a return on its ownership of land (Marx 1967b; Barrett 1983). For the sea fishery, common property – the sea – inhibits the extended application of capital. Capital neither confronts private seaholders demanding rent nor encloses the sea to privatize access. Fish, however, continue to be an attractive resource, since no capital has to be invested in their cultivation. Harvesting and transportation technology – the means to capture and realize the value of the resource – assume exaggerated importance, since capital investment can yield both profit and differential rent (Breton 1977).

A number of observations follow from this analysis. First, common property and open access have allowed artisanal fishers autonomy and independence unparalleled in other sectors. Low entry costs and capital's inability to enclose the sea have insulated these producers from what some see as the 'normal' tendency toward capitalist subsumption. Common property therefore inhibits the extended reproduction of capital in the harvesting sector of the fishing industry.

Second, dependence on technology has made direct producers highly vulnerable to external agents if they wish to operate beyond subsistence levels. The social conditions that give rise to dependence have been the subject of much attention in recent literature, particularly in Atlantic Canada.[21] These discussions are taken up in later chapters. The direct producer's vulnerability is strongest where isolation and weak competition characterize both input and output markets. Local-level monopolies are both a necessary and a sufficient condition for the 'proletarianization' of artisanal fishers. The nature and evolution of these 'land-based' social conditions, and the circumstances under which producers are made dependent by capital or retain their independence, are the major foci of this book, particularly in chapter 7.

Third, there is risk associated with harvesting a resource that is migratory and uncultivated, under uncontrollable environmental conditions that can determine success or failure.[22] Coupled with the producer's inability to secure private tenure and limit competition, these vagaries have made the risks of direct investment high. This has tended to make artisans 'conservative' investors, especially since fishers have been subsumed by capital through capital debt rather than through expropriation. Operating debt, however, is a necessary cost of production in any enterprise. Fishers will logically attempt to relieve this risk through credit arrangements with external agents, notably those who want to buy the product after it is caught. Many recent scholars have interpreted this situation as an advantage to capital and one key factor in the subsumption of fishers by fish-processing capital. Such a conclusion, however, may well be overstated. When capital provides loans, loan guarantees, outfitting supplies, or services, these represent real costs, unless prices can be manipulated to cover them. Capital will prefer other, less costly ways of inducing dependence. It is argued in chapter 3 that in Nova Scotia a change in the nature of debt dependence transformed 'port' market relations between capital and artisanal fishers after the Second World War.

Insecurity of returns has made direct investments unattractive for capital. The tendency to let fishers assume most of the capital costs and risks of investment is therefore increased, except where extrinsic variables offset the added risk – for example, where the state underwrites capital costs or guarantees resource quotas, where the output market is such that high prices or captive niches outweigh the risks of harvesting investment, or where the risks and costs of such investment can be subsumed in the consolidated accounts of a diversified corporate entity. In chapter 4 these factors are seen to separate multi-branch fish processors from single-plant processors, and large- from small-scale operations, in ways that make the larger establishments more likely to need to own fishing vessels and the smaller operations less likely to want to own fishing boats.

The processing sector of fishing is the classic case of a manufacturing industry constrained by its close ties to the harvesting sector and by the unique qualities of its product. Three factors influence the patterns of capitalist differentiation there. First, the location of fish plants is a function of proximity to fishing grounds, in general, and harbours providing safe haven and good vessel berthing sites, in particular.[23] Second, the supply of fish – as capital's one essential input – is a function of predictable seasonal variations and unpredictable fluctuations in catch and product size. These factors wreak havoc with capital's ability to rationalize production schedules and have led to three dominant production patterns: specialized, multi-plant operations that rationalize throughput by means of consolidated operations; single-location, multiple-species operations that spread production over the entire year; and specialized, but highly seasonal, small-plant operations. While the latter two types adapt to natural supply patterns, the first is based on the backward integration of an industrial harvesting technology that can produce large supplies at precise locations throughout the year. This production pattern is therefore the most capital-intensive type as well. A technological imperative to reduce costs of idle plant machinery drives plant operators to seek large, regular volumes of fish. The contradiction between the resource ecology and large-scale capital production seems to have its real impetus here. The resulting production cycle undermines the resource's natural ability to reproduce itself. In chapters 2 and 3 we shall see how this arrangement leads to the continuing instability of large-scale capital in the fishing industry of Atlantic Canada. Capital established on an intermediate scale – in terms of both technology and organization – occupies a resource-derived niche which has contributed to relative stability and success over the long term.

Third, the perishability of fish affects fish processing at all levels. In particular, it influences the production process and the ability of capital to market the product. The labour process in fish processing is defined by the fragility, irregularity, and perishability of the raw material. As is argued below, this makes production in even the largest plants labour-intensive and resistant to mechanization based on either conversion or transfer technology (Storper and Walker 1984: 34–7). The greater the effort to produce a high-quality product that can be marketed in a fresh state, the more production methods resemble craft organization. The greater the effort is to produce a product that is preserved from deterioration or spoilage, the more likely production will be mechanized. In either case, however, the 'labour problem' is the same: skills can be learned relatively quickly, work will be intermittent and seasonally variable, and working conditions will be physically unattractive. The dominant need, therefore, is for a captive workforce with few employment alternatives.

Fish production for human consumption defies sophisticated technological or product innovation. Entry costs are as low or lower in fish processing than in the harvesting sector generally and are constrained only by distance from consumers and an abundant supply of labour. As with the entire manufacturing sector, capitalist differentiation in fish processing is highest in low-capital-entry production – such as fresh fish – which is near a market that allows the product to be sold in optimal condition. We argue that low entry costs, coupled with close proximity to a luxury product market, can generate substantial unit-scale profits. In competitive port and labour markets, this can translate

into high fish prices and wages. As is argued in chapter 10, the availability of labour, not its cheapness, is often more important for capital.

Conclusion

In the face of what has been said in this chapter, any degree of capitalist
centralization and concentration in the fishing industry is surprising.
Two categories of factors converge to undermine the stable, extended
reproduction of capital: those that the industry shares with other sectors
and a number of resource-related factors. In common with other industries, such as textiles, fish processing has low entry costs, a reliance
on unskilled labour, and a highly competitive product market. As in
other 'low-tech' industries, such as subcontract semiconductor assembly
operations, overhead is kept low and labour is made to bear the burden
of irregular markets. Along with many 'leading-edge' industries, however, new opportunities in fishing often emerge through market and
product innovation. As in garment manufacturing, seasonal fluctuations in production are an intrinsic part of operations and result often
in an organizational structure based on managerial flexibility, low capital costs, versatile labour, and the family firm. Resource-specific factors
in common with some aspects of forestry, agriculture, and trapping
create an ecological niche in the harvesting sector that enhances the
adaptation of intermediate technology. Locational isolation, in common with resource industries such as mining, often makes settlement
patterns and community structures important in labour-market considerations. Product perishability, as in agriculture, constrains technological development and raises transportation costs.

In addition, three idiosyncratic factors inhibit capitalist consolidation
in the fishery. First, the low-entry, migratory nature of the resource
poses problems for privatized property rights. This situation has given
the artisanate unparalleled independence. Subsumption of fishers by
capital has been a function of indirect control over inputs and outputs.
Second, the high risks associated with an open-sea fishery have increased the independence of fishers, who are left to assume operational
risks. This factor also strengthens the position of producers and small
capital organizationally adapted to the vagaries of the fishery. Third,
the ecological basis of the resource continually undermines the technological basis of large-scale capitalist organization, making it unstable
and prone to crisis. This factor accelerates modern tendencies to cap-

italist decentralization, although sub-contracting is only one pattern of capitalist organization. The technological revolution in communications may increase the likelihood of capitalist concentration based on intrafirm integration and intermediate scale. However, the circumstances under which contracting out takes place have to be explored further. The fishery offers a particularly interesting opportunity to examine the nature of obstacles to extended capitalist development. Capital can neither privatize access nor consolidate production in the harvesting sector in a manner that is consistent with the basic principles and organizational methods of industry elsewhere.

2 Mercantile and industrial development to 1945

Gene Barrett

Chapters 2 and 3 attempt to trace the various patterns in the accumulation process in the fishing industry from its beginnings in 1500
to the present. While the industrial history of the fishery is here being
interpreted in light of the accumulation process, no 'necessary' logic
toward increasing centralization and concentration of capital is uncovered (Norton 1988). Each historical period is characterized by a wide
range of efforts to protect profitability – in some cases to increase it,
in others simply to stay competitive – which only incidentally involved
attempts to increase the scale and concentration of economic power.
The obstacles to such developments – the uncharted consequences,
crises, and conflicts – explain emergent patterns. To the extent that
any inner necessity is uncovered, it appears that a 'technological' imperative associated with large-scale production explains a great deal
about capital, the origins of crises, and the role of the state.

As a recent bibliography suggests (Lamson and Reade 1987), there
is no dearth of scholarly interest in Canada's Atlantic fishery. Historical
problems of interest concern the causes of mercantile stagnation and
the general failure of the twentieth-century transition to industrialism.
Contemporary concerns have turned to the causes of poverty and underdevelopment in the Atlantic fishery, the chronic crises of monopoly
capital, and the role of the state in managing the industry.

Two theoretical perspectives seem to inform much of this research.
The first is a variant of the dualist theory of industrial development:
the 'staples' school that has emanated from the work of Harold Innis
(Watkins 1963, 1980; Drache 1976). Ommer's (1981) study of the
efforts of Jersey merchants to wrest control over the fishery in Atlantic
Canada is one of the clearest illustrations of this perspective. Revising

the classical Innisian argument (Innis 1954), Ommer blames the retardation of industrialism in the fishery not simply on idiosyncratic resource features but on mercantile attempts at control. This led to the impoverishment of fishers, market fragmentation, and a general failure to generate the kinds of backward and forward linkages that would stimulate economic growth. Alexander (1977, 1980) argues, in a similar vein, that the stagnation of the Newfoundland fishery was a consequence of mercantile weakness in the face of foreign interests. Consignment buying led, in particular, to low prices and the starvation of the family fishery. The entire structure was starved of capital and failed to stimulate even the barest basis for economic development – agriculture.

Writing in the 1960s and 1970s, Wadel (1969), Brox (1972), and Copes (1972) saw the legacy of traditionalism as the major problem facing the modern fishery. While their perspectives on change varied, the essential problem was the same: how to modernize the dualistic economy and society of Newfoundland (Barrett 1980). While this theory fell into academic disrepute in the late 1970s and the 1980s, the perspective lives on in government policy. A number of recent studies have focused on fragmentation and duplication in both the harvesting and processing sectors (Canada, Economic Council of Canada, 1980; Shrank et al. 1980; Roy et al. 1981), while a more tempered welfare-economic perspective has led other policy-makers to attempt to reconcile the needs of a rural fishery with the 'immanent' laws of rational economic efficiency (Newfoundland and Labrador 1981; Canada, Task Force on the Atlantic Fisheries, 1982; M. Kirby 1990).

The second perspective is a variant of dependency theory as applied to the regional question in general, and the fishery in particular (S. Antler 1979; Barrett 1979; Sacouman 1979; Neis 1981). Taking Marx's analysis of merchants' capital as a lead (Marx 1967b; Kay 1979); this perspective situates the stagnation of the fishery in the way merchants earned profit, through an unequal exchange with fishers and their families. The transformation to industrialism occurs in spite of mercantile strength rather than 'because of it.' Industrial capital is seen to have been conditioned by regional underdevelopment. Archaic technique and the partial subsumption of independent commodity producers are a consequence of this feature (Sacouman 1980). Inshore fishers are viewed as nominally independent and profoundly dependent on capital as price-takers in a market dominated by monopoly capital

(Williams 1977; 1978; 1979; 1987b; Carter 1983). A chain of dependence is also seen to encapsulate communities and households in a web of super-exploitation (Connelly and Macdonald 1983). Fishers have had to fight against capital, the state, and underdevelopment to gain better conditions of life (Cameron 1977; Williams 1979, 1987c; Inglis 1985; Clement 1986). A more subtle strand in this literature has focused on the differentiation of fishers. The state is seen as a primary factor in both the resilience of independent commodity production and the growing ability of a fraction of fishers to accumulate capital on a petty basis (Sinclair 1985b).

This chapter, and chapter 3, present a perspective on the history of capitalism in the fishery that attempts to answer a number of questions that existing research leaves unanswered. First, if mercantilism was so weak, why did it last for so long? Second, if industrialization represented the pinnacle of rational efficiency, why has it been characterized by crises and state support? Third, why have small-capital and independent commodity production increased at a time when many consider capitalist concentration and the proletarianization of fishers to be dominant trends?

In what follows it is argued that capitalism in the fishery has gone through two relatively stable 'long swings' and three relatively unstable 'short swings.' During the 'long swings' resource, labour, and market relations evolved with supportive state, cultural, and community structures to facilitate slow, evolutionary capital accumulation. In this chapter, this is shown to be true of the mercantile–resident family fishery from approximately 1600 to 1880. In chapter 3, it is shown to be the case for part of the industry – a competitive capital–petty producer segment – that emerged after the Second World War.

The three short swings represent phases of capital accumulation when all or most efforts to control resources, labour, or markets eluded large-scale capital. Chronic crises precipitated external state intervention to attempt to re-establish a stable structure. The mercantile-ship fishery from 1500 to 1600, the nascent industrial period between 1880 and 1945, and the corporate segment of the fishing industry after 1945 illustrate these characteristics. The latter case is examined in detail in chapter 3.

It is argued that the key to understanding the industrial history of the fishery, and the difference between long swings and short swings, lies in capital's ability to adapt to the resource-related idiosyncrasies of

the fishery and of the attendant community, class, and family structures. This is most clearly exemplified in the tendencies toward differentiation during both long swings, as opposed to the largely
unsuccessful efforts toward capitalist centralization during the short
swings. Thus, while these chapters take issue with the benign view of
capital put forward by dualist theory, the dependency view is shown
to be too simplistic, in its failure to identify a number of distinct and
contradictory phases in the capital accumulation process. Nowhere is
this weakness more obvious than in the post-1945 era, which is perhaps
best described as a shift from dependency toward modernization.[1]

The mercantile ship fishery (1500–1600)

The mercantile ship fishery was the first European attempt to exploit
the northwest Atlantic fishery.[2] It dominated fishery development in
Nova Scotia for the first 100 years of European contact, from 1500
onward, and lasted approximately 250 years in Newfoundland. From
the beginning of the ship fishery, cod was the most important species.
The cod fishery had two essential branches: 'wet' and 'dry.' Because
of a shortage of salt, England specialized in the latter, sun-cured product; France developed a strong urban market for the heavy-sate, 'wet'
cod product. This division in turn strongly influenced the orientation
of each colonial power to the land near the fishing grounds.

For England, the fishing season began each spring with a race from
'West Country' ports to the New World for the best land bases. The
master who was first to arrive was declared 'admiral' for the season.
Each ship carried a complement of eighteen to twenty-five men, and
fishing was done by hook and line from small dories, which ventured
at most three miles from shore.[3] The fish were 'made' on the land,
where fishing stages and fish houses were constructed. The 'making'
process involved light salting, extensive sun-drying, and pressing of the
cod with heavy stones brought over on the ship as ballast. At the end
of the season, the dried fish were loaded and taken back for sale in
England or the Mediterranean countries. While the ship fishery may
have indirectly serviced the Royal Navy as a training ground for seamen, these ventures aided the primitive accumulation of capital in
Britain (Innis 1954: 43–60; MacNutt 1965: 3).

For the French fishery, land bases were important only indirectly,
for access to fresh water, wood, and food. With the improvement of
onship storage, the French developed the offshore 'bank' fishery some

fifty years after initial contact and were largely able to ignore land bases in North America (Innis 1954: 25–6). The wet fishery entailed less preparation: fishing took place at sea, and the 'making' process involved heavy salting of split cod in barrels on board the mother ship (Biggar 1963: 25). The French dry fishery was carried out by Basque and Breton fishers, who would land at optimal staging points along the coast to dry and salt their fish, in direct competition with the English (Innis 1954: 90).

The ship fishery tested the limits of ideal-typical mercantile organization, which depended on strong control over inputs and outputs and low capitalization and risk (Kay 1975). The ship fishery offered few controls – fishing operations and land bases were 3,000 miles distant, under the scrutiny of 'bye-boatkeepers' or 'captains' for most of the year. Fish production was unpredictable, vessel construction and outfitting costs were high, and fish markets saw intense competition. Merchants, especially in the 'West Country' of England, used two methods to exert some control: a highly diffuse commercial organization, to offset risk and uncertainty, and an imperial monopoly to prevent competition from permanent settlers.

The ship fishery in both England and France was based on the share system.[4] Merchant, owner, master, and crew each figured for a fixed percentage of the value of the catch (Innis 1954: 21). In England, for example, master and crew took one-third, owners one-third, and victuallers one-third. This system parcelled out the risks of production uncertainty while guaranteeing merchants a share of the surplus in the event of a good catch. A contemporary critic of West Country merchants outlined the mechanics of what amounted to a classic unequal exchange:

Taking a 100-ton ship with 40 men, he estimated the cost of provisioning and outfitting her for a year to be £420 1s. 4d. With 8 three-man fishing boats catching an average of 25,000 fish per boat, the season's total catch would be some 200,000 cod ... [The total value of this catch would be £2,250.] Of this the master and the ship's company were entitled to one third, allowing a 'small' sum for victualing, ship's expenses, and charges. The owners were entitled to one third after deducting for the master's allowance and for bonuses, over and above their share in the first third, to those men who had proven themselves much better fishermen than the other; and the victualers were entitled to one third. This ... would leave £750, after deducting £420 1s. 4d. for supplying the 40 men, or 'a profit of £331 11s.: During the years of high prices for salt, bread, and beer the victualers were allowed one half and the crew and ship one half. (Innis 1954: 57–60; also 171)

Despite mercantile domination, however, the fishery was the birth-place of free trade. Except where the navy contracted large amounts of fish at set prices, strong evidence suggests that merchants paid competitive prices for fish. It was common for French or English ships to leave straight from the fishing grounds for 'the most promising market ... Marseilles, Spain, Portugal, the Biscay ports, ... or in the British Isles' (Innis 1954: 61; also 31, 32). It was only through supplies and insurance that West Country merchants indirectly controlled the market.

The insecurity inherent in the merchant's position increased with colonization of the New World and, following the Reformation, the decline in the British market for fish. Fish became an increasingly attractive commodity in trade with the Roman Catholic countries to the south. Innis (1954: 52) observed:

A three-cornered trade from England to Newfoundland, Spain and the Mediterranean provided a basis for expansion, and gave England an industry with an abundance of shipping, and outlet for manufactured goods and provisions, a supply of semitropical products and specie, substantial profits, and ideal possibilities for the development of a mercantile policy. England was able, in part because of her relatively shorter distance from Newfoundland and in part because of the nature of fish as foodstuff, to secure a strong and continuous hold on a product by which she obtained a share of Spanish specie and the products of the Mediterranean. Cod from Newfoundland was the lever by which she wrested her share of the riches of the New World from Spain.

The strengthening of trading interests in London came into direct conflict with traditional fishing interests in the West Country. Larger companies sought a monopoly of trade in British fish through passage of Navigation Acts in the seventeenth century (Innis 1954: 50, 64–70). West Country merchants tried to prevent traders from gaining land and settlement rights to parts of North America. Trading interests sent 'sack' ships laden with supplies, 'planters,' and settlers to Newfoundland. On their return the following year, the traders would purchase fish from the independent 'planters' and make their way directly to the Mediterranean. The new fishery that emerged represented a significant threat to the ship fishery.

The conflict between fishing merchants and settlers waxed and waned for the next 140 years and culminated in Palliser's Act (1775), which 'attempted to put an end to colonization by penalizing the masters of ships who failed to bring back to Britain the seamen they had conveyed to Newfoundland. By offering bounties it strove to give fresh spur to the West Countrymen and to reverse the trend to the eighteenth century by which "the free fishery", the source of manpower for the navy, had been in decline' (MacNutt 1965: 73; see also Innis 1954: 68, 207–9). Steven Antler (1979: 190) argues, 'Britain's anticolonization policy toward Newfoundland amounted to nothing more than the creation of a monopoly regulating the rate of utilization of the Newfoundland fishery.' By the mid-eighteenth century however, West Country merchants had become indistinguishable from trading merchants: it had become more profitable to convey settlers and supplies to Newfoundland and rely on a local planter fishery for the finished product, which could be conveyed directly to market.

Capital's first attempt to consolidate the fishery therefore proved unstable except under exaggerated state protection. The share system had diffused the risks inherent in the open-sea fishery while preserving an unequal exchange relation with producers. However, the end of 'captive' market demand in England and the inability of merchants' capital to control 'fishing labour' undermined its position. The latter weakness was related to capital's inability to monopolize access to the world's richest fishing grounds.[5] The centralized mercantile system broke down in favour of a decentralized pattern, to accommodate the new realities of an open-access fishery in the context of frontier settlement.

The mercantile–resident family fishery (1600–1880)

A white-settler mercantile-resident family fishery moulded a 'long swing' in capital accumulation which lasted almost three hundred years. It was based on a metropolitan-satellite chain involving European importers and traders, town merchants, outport merchants, and fishing families. Each unit was implicated in a web of credit and debt based on personal obligation and dependence. Merchants cultivated a class of independent producers to harvest the catch and produce dried cod while they concentrated their efforts on trade. As with the ship fishery, capital's control over the resource remained indirect, through control over producers' supplies and markets. However, the producers' relation to the resource had changed. They now controlled the harvesting unit, and their families were responsible for production (E. Antler 1977b; 1981: 127–44). But their dependence on merchants' capital increased dramatically with their geographic isolation from the world market. Two variations to this pattern developed, based on indenture and on truck.[6] The first form presumed strong controls over labour and proved unworkable in the frontier context of abundant land and resources.[7] The second was based on indirect control of the resource and labour and proved highly resilient, lasting, in some places, for 200 years.

By the eighteenth century, a number of Jersey firms faced the transition from the ship to the family fishery with an iron-fisted approach to control over fishing families and settlements in the Gulf of St Lawrence, Gaspé, and Acadian districts of Cape Breton.[8] The Charles Robin Co. was by far the most dominant and long-lasting mercantile force in these areas. Its empire was unparalleled for its time. Ommer observes, 'Once the strategies for control of the staple had been devised and implemented, the fishery had achieved a concentration of power and control that was more intensive than that created in any other nineteenth-century staple. The firm became at the one time supplier, importer, distributor, producer, processor, collector, exporter, marketer and financier of the Jersey-Gaspé fishery' (1981: 116).

The firm preferred isolated, unsettled locales, where competition could be restricted. Control was exercised through land ownership and personal indenture. While efforts to establish seigneuries were largely unsuccessful, Jersey firms often owned the land with preferential access to landing and drying locations (Innis 1954: 188, 192; Hughes 1981: 10–13; Ommer 1981: 108–10). The resettlement of exiled Acadians after 1760 was instrumental in the establishment of personal indenture (MacNutt 1965: 69–79; also Wade 1975: 42–65): 'Fishing magnates from Jersey, of whom Jacques Robin was a conspicuous example, dressed them in French seamen's shop-clothing and brought them to their stations on the Nova Scotian coast where their skill and experience were put to good use. Canso and Isle Madame drew large numbers. But the largest proportion made its way to the Bay of Chaleur and settled about Caraquet, where the Robin firm kept them employed.' By 1776, Robin employed at Cheticamp some sixty families who used fifty shallops in the inshore fishery. However, Jersey expatriates seem to have been the major group imported under these terms (Ommer 1979: 185–91; Hughes 1981: 7).[9]

Immigration and settlement undermined land monopoly and indenture as effective modes of labour control, while fragmentation of merchant capital reduced the control enjoyed by companies such as Robin Jones and Whitman. By the nineteenth century, the truck system emerged as an adaptation. Merchants assumed the risk inherent in the outlay of operating capital (Ommer 1979: 191–9, 1981: 111); fishers were made to bear the brunt of low prices and the insecurities generated by a wildly unstable market (Wadel 1969: 16–25; Hughes 1981: 10–11). As long as prices remained low, fishers would need credit to outfit at the beginning of each season, and as long as fishers were in debt, they would remain impoverished and unable to increase productivity (Ommer 1989).

The English family fishery in Newfoundland and Nova Scotia was characterized by small outport merchants tied into a metropolis–hinterland chain of credit and debt centred in either St John's or Halifax. In Nova Scotia, this structure emerged under pre-Loyalist settlement in the 1760s. For example, Innis (1954: 186) points out that seventy Connecticut settlers arrived in Yarmouth, Barrington, and Liverpool, under the auspices of merchant Simeon Perkins. Perkins traded with the New England schooner fleet and also conducted a coasting trade with client boat fishers and inshore fishing communities along the south shore. Innis (1954: 186–7) observed that 'supplies were distributed to ports along the shore in exchange for fish' (also Brown 1978). Writing in 1829, T.C. Haliburton (1973: 252) noted that the monopoly of fishing in Cape Breton, long enjoyed by the Jersey merchants, had been lost through the influx of 'other adventures'.

The American Revolution gave a dramatic impetus to mercantile power in Nova Scotia. From being a client of New England, Nova Scotia came to rely heavily on imperial protection in Britain's colonial markets. Under the protective cover of the Restraining Act and the embargo of American ports, Nova Scotia schooners plied New England's trade routes to Newfoundland, the West Indies, and South America.[10] Halifax became the new Boston of British North America, and trade expanded 'on flexible New England lines; for instance, small amounts of capital were invested in shares in ships and cargoes' (Innis 1954: 207).

Supportive institutional structures helped maintain the mercantile-family system. The colonial state apparatus was based on patronage, mercantile-family compacts, and a standing militia (Sharp 1976; Robertson

1988). The absence of broad-based social institutions, such as education, health care, and judicial advocacy, exaggerated patronage-clientism. Geographical isolation and poor transportation and communication, coupled with illiteracy, poverty, and limited economic opportunities, allowed élite-controlled institutional structures to thrive.[11]

Cultural and customary traditions rooted in kinship and community emerged to help people survive (Sider 1976). Extended kin networks and the household comprised production and consumption units in a highly resilient labour process which absorbed the worst shocks the mercantile economy meted out. Plural activity in both subsistence and 'cash' work formed the basis of the household economy. Hard work, self-reliance, and loyalty to friends and family bolstered these traditions. The levelling effect of the truck system was crucial: since all were equally poor, what was the good of competing? The cultural 'image of the limited good' of individual initiative was based on the limited 'goods' available (Long 1977). Mutual obligation, reciprocity, and co-operation superseded any competitive predilections associated with the open access fishery (Wadel 1969; 1973; Faris 1972; Sider 1976).

The truck system only indirectly controlled labour and did not attempt to restrict producers' access to the sea. Market control was never achieved, and middlemen proliferated. Participants kept overhead costs low and insulated narrow profit margins through low prices. An elemental form of merchants' capital was ideally suited to these circumstances, and so it thrived (Marx 1967b: 327–34).

This accumulation process had two points of vulnerability: merchants' failure to exercise market control and the poverty that stemmed from low prices. The proliferation of merchant middlemen undermined individual accumulation and accelerated factional intraclass conflicts, particularly between metropolis and outports, a process that was exacerbated by worsening market conditions. The perpetual poverty generated by the truck system also undermined the old system. People could be made to bear only so much. Market slumps led to out-migration and populist revolt.[12]

Self-government in Nova Scotia reflected the efforts of a small colonial élite seated in Halifax attempting to protect its position in international trade against American competition while fending off competition from satellite merchants in the outports. State intervention in the fishery was started to aid the Halifax élite. After the American Revolution, a better-developed family fishery, from which merchants could appropriate products for the West Indian market, was more needed than ever. Yet high-cost British imports, and the impoverishing effects of the truck system, continually contradicted these efforts.[13] It was vital to gain access to greater state revenues to encourage rural settlement and production. As early as 1751, bounties had been offered for the production of dry and pickled fish, and yet Halifax merchants could not resist appropriating these funds (Innis 1954: 241; MacNutt 1965: 147).

Halifax merchants were reluctant to produce salt fish themselves. Poor and widely variable qualities of salt, pickled and dry fish were endemic to the family fishery. The plantation market and protected trade had cushioned merchants against competitive pressures to re-structure production. The abolition of slavery in the British West Indies in 1834 shattered this niche. The merchants, however, demanded greater state regulation of existing techniques, putting the burden for modernization on the family unit. Seven new fish inspection acts were passed in 1827 and 1828, culminating in 1833 in an act that provided for appointment of inspectors in all districts. In the end, state revenue was used to increase bounties for the production of higher-quality 'mer-chantable' fish (Innis 1954: 263–7).[14]

Nova Scotia's élite had to struggle continually against American rivals by the mid-nineteenth century. The Americans developed larger and faster schooners and pioneered the use of new fishing techniques to increase their productivity. At the same time – and mostly because of overfishing – they expanded north to Labrador and Newfoundland waters for cod and herring, northwest into the Gulf of St Lawrence for mackerel, and around the entire coast of Nova Scotia for herring, mackerel, halibut, and cod (Innis 1954: 323–31). These incursions undermined local merchants' monopolies and threatened fish stocks. Under the pretext of taking on provisions, American vessels would purchase licences to enter and clear Nova Scotian ports, where they would sell provisions 30 per cent cheaper than local merchants and purchase local fish in exchange. Prior to 1867, Nova Scotians paid higher prices for supplies and provisions than Americans. Fishers on board American schooners received half the value of the fish caught (the 'half-lay' system). Higher wages, a protected domestic market and rising demand were a strong gravitational pull for Nova Scotia labour. By the 1850s, about one-quarter of the crews and half of the masters on American vessels were Nova Scotians (Innis 1954: 326–34; Brooke 1976: 46–8). With the development of coal mining in Cape Breton in the 1860s, fishers were attracted to mines, railways, construction, and, later, the steel mills (Watt 1963).

The effect of American fishing efforts on Canadian stocks is difficult to measure. There was accelerating state response, from 1786, with appointment of the first superintendent of trade and fisheries, to formation of the dominion Department of Marine and Fisheries in 1867 (Innis 1954: 232, 334–47, 364; Tallman 1975: 66–78). Merchants' responses fluctuated. Free trade arrangements, under the Reciprocity Treaty of 1854 and the Treaty of Washington in 1873, compensated for their weakened position in the West Indies. When access to the American market was threatened, as during the Civil War and in the mid-1870s (because of a shift in consumers' preference toward fresh fish), merchants became fervent supporters of Canadian sovereignty. From this protectionist position, they become strong advocates of modernization.

The industrial capitalist fishery (1880–1945)

Industrialism came to Nova Scotia's fishery in the 1880s, but by the 1940s the process was in a state of virtual collapse. Attempts at concentration, capitalist centralization, and the application of new industrial technology to the harvesting and processing of fish were neither widespread nor growth-sustaining in this second 'short-swing' cycle. Industrialism failed to mature. This pattern contrasts with situations elsewhere (White 1956; Tunstall 1968; Muszynski 1987). Except in the lobster canning industry, surplus labour and state protection cushioned capital from the need to restructure. Moreover, accelerated capitalist accumulation in the lobster industry precipitated an overfishing crisis. The outmoded patterns of industrial production that persisted throughout this period in the salt, fresh, and frozen branches, while making them more vulnerable to fluctuations in product and labour markets, perhaps protected them from the kinds of resource-centred crises that plagued large-scale capital in the period after 1945.

The lobster industry

The lobster fishery grew in New England after the American Revolution (Innis 1954: 222), based initially on development of local markets: lobsters were shipped alive and therefore had to be consumed quickly. The development of canning allowed wider distribution and inventory control to offset fluctuations in markets and catch. Inland markets burgeoned, and coastal-zone production soared along the eastern US seaboard (DeWolfe 1974).

The North American lobster has a relatively small migratory range, and its breeding and maturation stages are easily upset. Voracious demand soon led to overfishing. The decline in supply, coupled with a relatively strong market, increased the price of lobster and made ventures further afield attractive. The third quarter of the nineteenth century witnessed a massive flow of American capital into the Maritime provinces to exploit abundant supplies (Innis 1954: 437).

The six largest American lobster companies established seventy-one canneries in the Maritimes by the 1880s. Subregional conditions varied widely with access to the fishery, transportation facilities, and local socio-economic structures. These relatively modern branch-plant operations were often conditioned by local labour surpluses.[15] However, introduction of competition and cash sales further undermined the traditional truck system in Nova Scotia (Canada, Parliament, 1910c: 501–8, 536–43).[16] Low prices and seasonal production lessened the benefits to fishers. Fishers testifying before a Lobster Commission in 1909 stressed that low prices were often the result of poor quality control by the canneries, which often shut down early if a market slump were anticipated (Canada 1910: 521–6, 536–43, 660–6).

Overfishing quickly became the key problem in the industry. W.F. Whitcher, dominion commissioner of fisheries, stated in 1873 that 'it seems that excessive fishing has exhausted the lobster fishery along the north-eastern coast of the United States; and that the enterprise which was embarked in the same has now been transferred to Canada. Such being the case, if the same indiscriminate fishing should be practiced on our coasts, similar results might occur' (Canada, Parliament, 1876: lxxxii). Confronted with opposition from canners, the government did little conservation work. In 1874 the value of the lobster catch fell by $546,950 (Canada, Parliament, 1876: xxxi–xxxii). In his report of 1875, the fisheries commissioner remonstrated: 'An alarming decrease in the lobster fishery is reported by the fishery officers. It is ascribed to over-production and wasteful capture of spawners and undersized lobsters' (Canada, Parliament, 1876: lxxxii).

Market prices remained high enough to justify production. Producers sought even harder to maintain productivity. By 1910, for example, a fisher from Port Hood Island reported that 1,000 traps were needed to catch barely one ton of lobster; twenty-five years earlier, 100 traps would have easily yielded the same poundage; trap sizes increased, and the spacing between slats and mesh size on the traps decreased (Canada, Parliament, 1910c: 536–43, 1910a: 3).

The Royal Commission Investigating the Fisheries in 1927 (the McLean Commission) estimated an illegal lobster catch that approximated 60 per cent of the legal catch in some districts of the Maritimes, despite a decline in the average catch of 25 per cent in Nova Scotia and 40 per cent in New Brunswick (Canada 1927a: 11–12). It noted in dismay: 'We were told, too, of surreptitious canning and packing in barns and kitchens and woods, of illegally caught lobsters. The impression left on our minds in certain districts was that there was an utter lack of observance of the existing lobster regulations and little individual or community sentiment in support of their enforcement' (1927a: 12).

The failure of government regulation was rooted in the contradiction between long-term resource conservation and the imperatives of short-term capitalist accumulation. As early as 1873, processors and fishers from southwestern Nova Scotia petitioned against imposition of regulations protecting small and berried lobster. In 1874 the regulations were practically eliminated (Canada, Parliament, 1910a; Canada 1927a: 12). In 1912–13, the Shell-Fish Commission's call for a closed season on lobster in the Northumberland Strait was totally rejected 'because of a strong representation from fishermen' (Canada 1927a: 11). The McLean Commission characterized this history as one of blackmail: foreign processors threatened to close their canneries and move elsewhere unless sufficient supplies were maintained. 'Canneries ... and dealers must share the responsibility for the present unsatisfactory situation. The fishermen cannot do business without a buyer and we cannot absolve the canner and the dealer from blame for their indifference, or for their purchasing illegally caught fish' (Canada 1972a: 15).[17]

The dominion Department of Fisheries realized as early as 1910 that improvements in transportation would generate a 'live' lobster trade (Canada, Parliament, 1910b: xc). This development signalled the deathknell of the canning industry on the east coast.[18] It also marked the beginning of the next long swing in capital accumulation in the fishing industry. Low entry costs facilitated the proliferation of small buyers, improving port market competition and prices and, most important, reducing fishers' dependence on canneries.[19]

The salt fish industry

Loss of a captive market in the British West Indies after the abolition of slavery signalled the first major blow to the salt fish industry. The second crisis came with the rise of the US fresh fish industry. By the 1880s, merchant's capital attempted to restructure the primary and secondary aspects of the industry, thereby killing the family fishery in many communities. Innis (1954: 427–8) observed:

> The advantages of large-scale organization in the dried-fish industry made itself felt in that supervision of curing which large-scale organization could offer in its effective control of the product until marketed and its ability to compete in wide variety of markets, with their varying grades ... Independent fishermen were able to give smaller quantities of fish better supervision, and to cure a better product; but differences of skill and capacity in the individual and fluctuations of weather and catch inevitably produced wide variations. The risks that lay alike in extended credit, the careless grading of fish, and the intense competition in foreign markets made for both the disappearance and the amalgamation of companies.

Surviving the transformation most successfully were the old Jersey firm of Robin, Jones and Whitman (Ommer 1981); Halifax Fisheries and A.M. Smith of Halifax; and W.C. Smith, Zwickers, and Adams and Knickle of Lunenburg, which came to dominate the export trade in salt fish between the 1880s and 1945. Production of heavy-cure dried fish was centralized in Halifax and Lunenburg through the development of mechanized dryers.[20]

Primary production was of two basic types. In one, a number of firms, adapting steamers and schooners to the coastal trade, maintained buying stations or ties to outport merchants to collect 'green' salt fish from inshore fishing families (Barrett 1976). The other system was more revolutionary. In 1881 a steamship service to the West Indies and Brazil was inaugurated (Innis 1954: 372). Schooners were now adapted to the offshore fishery by a number of salt fish companies concentrated primarily in Lunenburg. A burgeoning offshore bank fishery developed in Lunenburg County. According to a Lunenburg captain, a salt fish company would own a small portion of the shares in each vessel, 'investing [just] enough money in the boat for certain captains to get their outfits from them' (Canada 1937: 31). The 'managing owner' – a representative of the company that controlled the largest block of shares in a vessel – was empowered to purchase supplies and equipment, sell fish, settle with fishers, and pay debts (Barrett 1976).

Capital ensured its centralized processing apparatus a large volume of supply without incurring too many of the risks attendant on direct investment in the harvesting sector. Investment risks were shared, and managerial control determined access to the product. Fishers were paid on a share basis after the boat and owners received their share. If the venture were unsuccessful, crew and boat shouldered most of the burden. High outfitting costs and low prices transferred a large share of the surplus from the crew in the event of a successful fishing trip. The schooner fishery represented a fairly stable relation of production for approximately twenty years.[21]

The post-war recession and the American tariff on Canadian fish imports in 1922 rocked the industry. Following the war, Norway, Britain, and Iceland re-entered the salt fish market, which had been assumed by Newfoundland and Canada. Substantial government supports in most European countries allowed restructuring and effective competition (Nova Scotia 1944: 82). World-wide overproduction followed (Canada 1927a: 41–3; Nova Scotia 1944: 81–2; Watt 1963).[22] Foreign importers continued consignment buying.[23] Predictably, salt fish companies continued to use the truck system to divest their risks.[24] Despite strong calls from fishers for state control and reorganization, and some sympathy from the McLean Commission, little was done. Rural conditions of labour surplus continued to cushion capital from competitive pressures to modernize.

In 1930s, for the first time, the fresh fish and live lobster industries offered fishers in southwestern Nova Scotia a significant alternative to out-migration. Bates observed: 'It was the desperation of this situation that forced many Canadian fishermen to abandon the salting trade entirely, and, to find a living from lobsters, or even fresh fishing' (Nova Scotia 1944: 81). The situation precipitated a crisis for large-scale salt fish capital. In 1938, Ottawa appointed O.F. MacKenzie, president of Halifax Fisheries Ltd, and Homer Zwicker, of Zwicker and Co., to investigate the industry and make recommendations for its rehabilitation (Canada 1937: 17). Their report repeated that the crisis was induced externally by restrictions in traditional markets, contractions in consumption abroad, and foreign subsidies. In proposing a Salt Fish Board to dispense government support, the minister pointed out that the report 'proved' that 'no one in Canada is responsible for the existence of the present conditions in the salt fish industry' (Canada, Parliament, 1939: 3656). The new board subsidized low prices with payments amounting to 25 per cent of the total value of fish exported. While the state hoped for a 'trickle down' effect, the lion's share of the subsidies went to the companies. The war boom saw many firms convert to fish canning for war production.

From disparate origins in the mercantile trade, the salt fish industry experienced a truncated transition from merchant's to industrial capital. In Nova Scotia between 1880 and 1939, large-scale capital enjoyed a short-lived boom based on buoyant southern markets and cheap labour[25] and then stumbled along on the basis of the truck system and state supports during an extended recessionary spiral from 1919 onward. As with lobster, decentralization of production eliminated large-scale capital in the salt fishery after the Second World War.

Fresh and frozen fish industry

In the emergent fresh and frozen fish industry, technological imperatives drove capital to unprecedented lengths to wrest control over labour and the resource. Entry costs were so high and the technology so new that the perennial problems of market control that plagued the salt fish industry were largely resolved by capital, for the first time. In the face of high foreign tariffs, restrictive state regulations, and the Depression, capital retrenched, however, expanding solely through exploitation of surplus labour on land and at sea.

A number of technological innovations made in the United States during the 1920s revolutionized that country's fresh fish industry and provided two large companies – General Seafoods Corp. of New England and Atlantic Coast Fisheries Co. of Groton, Connecticut – with sizeable technological 'rents' (Barrett 1976). Development in 1922 of a 'filleting' machine facilitated mechanization and routinization of production, as well as standardization of production (Nova Scotia 1944: 76). An advanced technique for 'fast freezing' transformed inventory control and made possible expansion of new and distant markets (Nova Scotia 1944: 76; Innis 1954: 425; Watt 1963: 2). The latter development would have longer, more far-reaching effects, but transportation had to be modernized before it could be fully used. Refrigeration and cold storage proved an intermediate solution.

The fresh and frozen fish industry witnessed significant capital investment in cold storage and packing equipment, processing facilities, and transportation services. It also spurred on industrialization of the harvesting sector through development of steam trawlers. In theory, the trawlers could fish under widely varying weather conditions, catch large volumes of fish, and supply plants on a regular basis (Innis 1954: 432, 435). The new technology made large investments in onshore processing facilities feasible by reducing idle-capacity costs. As discussed in chapter 1, the harvesting technique was an industrial adaptation of gill-net technology and, while highly productive, destroyed fish habitat and spawning grounds and was non-selective of species or year-classes of fish. Given the large volumes landed, the technology itself accelerated the potential for overfishing. The high fixed and operating costs of these vessels also exaggerated land-based technological imperatives. Left idle or used below maximum capacity, the technology was not cost effective.

By 1908, the industry was supported by a government transportation subsidy and by duties on American imports (Nova Scotia 1938: 30; Watt 1963: 20). By 1910, a rail line to Mulgrave had been built, and in 1913 a government-subsidized refrigerated express to Montreal was inaugurated (Fisheries Council of Canada 1967: 65). Three Canadian companies dominated the new industry: National Fish Co., with plants in Halifax and Port Hawkesbury; Leonard Fisheries Ltd, of Saint John and North Sydney; and Maritime Fish Corp. of Canso and Digby (Nova Scotia 1938: 29; Innis 1954: 433). These three firms led the number of steam trawlers to grow from four to eight during the First World War and to ten by 1927 (Canada 1927a: 105). Scarcity of fish off New England and availability of cheap British trawlers for sale or lease drew the American company – Atlantic Coast Fisheries – to merge National Fish and Maritime Fish. A domestic fraction of capital emerged around a coalition of the Smith interests of Lunenburg and the Bell interests in Lockeport. While the Bells relied on a coastal-zone, small-boat fishery for their supply, the Smiths fitted company-owned schooners with diesel engines (Canada 1927a: 75; Barrett 1976).

The Depression witnessed further consolidation, into Canadian-owned Maritime-National Fish, in Halifax and Digby, and Smith Fisheries of Lunenburg, in North Sydney, Port Hawkesbury, Lockeport, Liverpool, and Lunenburg (Innis 1954: 434). By 1939, two-thirds of the industry's freezing capacity was concentrated in nine freezing plants between

Halifax and Shelburne. In addition, 52 per cent of the filleting trade was concentrated in Halifax, 16 per cent in Lockeport, 10 per cent in North Sydney, 10 per cent in Lunenburg, 4 per cent in Shelburne, and 3 per cent in Canso. Of the eleven main fishing centres landing over five million pounds of fish in 1939, nine lay on the south shore, between Halifax and Digby Neck. This region accounted for 75 per cent of all fish taken in Nova Scotia, 68 per cent of the total value of landings, and 57 per cent of the fishing equipment (Nova Scotia 1944: 48, 79).

By the end of the Second World War, large-scale capital had emerged in the fishing industry based on both concentration of ownership and centralization of production. National Sea Products (NSP) was formed in 1945 through merger of Maritime–National Fish and Smith Fisheries. Sponsored by Keynesian policies promoting industrial modernization of the fishery, NSP embarked on expansion of the offshore trawler fishery and the frozen-fish-products industry (Barrett 1984).

In spite of a seemingly classic example of monopoly capital, industrial capital in fresh and frozen fish in fact had a distorted profile. Two crises undermined 'normal' capital accumulation by the Second World War. A populist struggle against the trawler technology led to state regulation of its use, and high wartime wage levels brought a labour shortage.

Protest against trawlers flared up as soon as they were introduced after the turn of the century. Local companies wanted to deploy the vessels within the three-mile territorial limit and used the Halifax Board of Trade to lobby the government. While the dominion subsidized infrastructural development for the fresh and frozen fish industry, it did not generally support trawler expansion.[26] In 1911, as a result of protests from fishers, bounty payments to trawlermen were cut off. During the First World War, trawlers could not fish within a twelve mile coastal zone (Canada 1927a: 90–1). Continual protests from fishers finally resulted in appointment of the McLean Commission in 1927.

The proceedings of the commission reveal that inshore fishers objected to trawlers on three grounds: that the technology ruined fishing grounds and destroyed spawn (Canada 1927b: vols. 3–6, 8, 9), that the trawlers fished in localities where they had fishing gear set, and that trawlers affected landings on the port market. The companies concerned could have diverted trawler fish to the frozen market, purchasing better-quality inshore fish for the fresh market (Nova Scotia 1944: 42, 72). However, capital used its trawler fleets to depress port-market prices. Trawlers could be operated by companies often at a loss, which could be recovered through lower overall input costs.

The most decisive objection, however, came from that fraction of capital content with traditional techniques. Second-hand trawlers were being used to land fish 'from a half cent to one cent per pound cheaper on the average' (Bell 1930: 75). Winthrop Bell (1930: 75) argued: 'It is this, perhaps temporary, supply of cheap foreign trawlers which constitutes the menace to the very livelihood of our native fishing population. Our attitude toward them is precisely that of other parts of the country introductions of temporarily cheap foreign labour.' He closed his remarks by saying: 'We don't want to see our fishermen robbed of their living.'

The commission's recommendation that all steam trawlers be prohibited from landing fish in Canadian ports (Canada 1927a: 98–9) was declared ultra vires by the Imperial Privy Council. A tax was imposed by the dominion Department of Fisheries of one cent per pound on certain fish landed by foreign-built trawlers and two-thirds of a cent per pound on fish landed by Canadian-built trawlers (Bell 1930: 76). From a high of ten steam trawlers operating out of Canadian ports in 1927, the number dwindled to three by 1939.

These regulations protected a domestic fraction of capital that survived on the basis of the truck system. The 1920s and 1930s witnessed a repeat of the labour crises in the salt fish industry of the 1850s. While capital adapted traditional harvesting technology (underwritten by state protection from competition), labour could not shoulder the risks of market collapse and the costs of price cuts. As with salt fish, profit levels in fresh fish were maintained throughout the 1930s because labour could be made to bear the brunt of price cuts. Stewart Bates (Nova Scotia 1944) observed later:

The price reductions that had to be made to widen the American market reflected themselves in the low standard of living to which the Canadian fishermen became increasingly subject. In other words, the usual method by the industry in trying to widen its market, was to cut the export price. Practically no attention was given to any other possible way of achieving the same end ... The labour of fishermen and plant workers was too cheap to force the industry into such alternatives, and in the milieu that existed, labour could be made to bear the incidence of low prices (64).

The exodus of fishers to the United States in the 1920s mounted into the thousands. In 1920, only three of 1,240 fishers in Lunenburg were from Newfoundland; by 1927 the overall number had declined by 11 per cent and the number of Newfoundlanders had increased to 289 (Canada 1927b: 2824–5). Further, the trawlers had glutted the market and survived by using cheap foreign labour (Canada 1927b: 1107, 1146). During the Depression, many fishers returned from New England, bringing trade union traditions (Barrett 1979). A series of strikes, boycotts, and lockouts took place in Lunenburg, Halifax, and Lockeport between 1937 and 1939 (Barrett 1976; Calhoun 1983).

While the nascent labour coalition between inshore fishers, crew members on schooners and trawlers, and fish plant workers failed, war production boosted wages and employment. Fishing capital was faced with a major labour crisis (see Nova Scotia 1944: 100). Stewart Bates observed:

Beginning in 1941, several corporations, particularly in Halifax, employed labour scouts to induce labour on the shores to move into the fishing centers. At least one attempt was made to decentralize operations to use available labour on the shores for fresh fish processing. Attempts were made to extend mechanization where equipment and building could be found. Women were employed wherever possible. Living accommodation was provided for incoming workers in centers like Halifax. As local supplies dried up, plants workers were imported from Newfoundland. (100-1)

The labour-intensive fishing industry was very vulnerable, and it could not survive without surplus labour. The labour crisis therefore precipitated major restructuring. Ottawa and the province bailed out the industry, through war purchasing, price supports for salt fish, vessel construction assistance for trawlers over seventy-two feet, depreciation allowances, and like measures (Nova Scotia 1944: 58, 95).

The need and rationale for extensive state support were laid out in Stewart Bates's proposals for post-war rehabilitation (Nova Scotia 1944: 125–37). The state was, however, also instrumental in the emergence of a vital coastal-zone fishery after the war. These efforts, combined with other economic and social developments, led to differentiation of the fishing industry in quite new ways and a second long cycle in the capital accumulation process – the subject of an extended treatment in the next chapter.

Conclusion

Three successive phases in the process of capital accumulation in Canada's Atlantic fishery up to 1945 have been examined: two brief and unstable short swings (1500–1600 and 1880–1945) and one relatively stable long swing (1600–1880). The stability of the mercantile-family fishery, as the first long swing, stemmed from three factors: its decentralized structure; its adaptation to the open-access, high-risk nature of the fishery (and the non-destructive mode and deployment of the harvesting technology); and its port-level monopoly of control over inputs and outputs. While the first two factors were successful adaptations to intrinsic resource peculiarities, the third suited the mercantile nature of capital and kept prices and incomes low, undermining modernization. Fragmentation prevented co-operative efforts to rationalize marketing so as to counter foreign importers. The absence of concentration lowered export prices and threatened merchants' profit margins. In response, merchants at every level trimmed overhead and passed price cuts on to fishing families.

Industrial capital (1880–1945) failed to create a more rational system. The difference between the two phases was the nature of capital. In its efforts to revolutionize production and modernize society, industrial capital threatened ecological balances and undermined a central basis for long-swing accumulation. Taking advantage of desperate local conditions, capital in the lobster-canning segment encouraged overfishing. In the fresh and frozen fish industry, capital introduced destructive, non-selective, industrial technology driven by its own cost imperatives. It is arguable, given what happened, whether capital could have survived – or even wanted to – solely on the basis of this harvesting technique, since trawler landings were used mainly as a lever to control port-market prices. Fish caught by independent fishers was still too cheap to warrant wholesale application of trawler technology. However, the shortness of this cycle was primarily caused not by a biological crisis but by continuing dependence on cheap labour. When regional conditions of labour surplus ended – albeit temporarily – during the Second World War, large-scale capital lost its only crutch.

3 Post-war development

Gene Barrett

This chapter tells two contrasting stories about the Nova Scotia fishery after 1945 – one concerns the growth of corporate capital and its chronic inability to establish any stable basis for accumulation without state subsidization; the other is about the rise of a competitive sector composed of small fish processors and coastal-zone fishers and the re-emergence of a long-swing phase in capital accumulation.[1]

Neither development can be understood without reference to post-war state policy in the fishery. From a backward-looking, protectionist stance in the 1920s and 1930s, both provincial and federal levels of government adopted an aggressive developmental attitude from 1939 onward. The Report on the Atlantic Sea-Fishery by Stewart Bates of Dalhousie University, for Nova Scotia's Royal Commission on Provincial Development and Rehabilitation (Nova Scotia 1944), was a benchmark (Alexander 1977; Hanson and Lamson 1984). Bates argued for modernization with a dual thrust: encouraging industrialization of the harvesting and frozen fish–processing sector and upgrading and modernizing rural districts to provide for a decentralized, shore-based industry.

Bates premised his view of industrial development on fairly orthodox notions of the inherent economies of large-scale organization (Nova Scotia 1944: 118–19). He called for greater assistance for large dragger construction and subsidies for plant modernization,[2] retail and wholesale storage, and refrigeration. He saw the industry's future in the US convenience frozen-food market. Concerning the impoverished shore fishery, the prospects of militant trade unionism or a second populist revolt against the trawler were not very palatable to the provincial government. From a welfare point of view, modernization of the 'shore' fishery – fishers, communities, and small plants alike – was necessary. Bates stated (123):

Socially the shore areas represent a special type of economy, in which there is generally an integration of fishing, farming, and forestry pursuits. In certain seasons men move from many of these areas to work at lumbering, stevedoring, etc., elsewhere. From a social point of view, the rehabilitation of these areas is an inter-administrative task, involving as it does farming, forestry, fishing, public health, education, electrification, roads, and other 'amenity' capital.

Bates called for modernized, diversified, year-round small plants in outlying areas, to be based on a revitalized small-boat harvesting sector. Boat loans to inshore fishers and introduction of longline and seine catch technology were key features. Bates also argued for development of small draggers to service the smaller plants, mimicking the integrated operations of large plants.

While most of Bates's recommendations were followed, the effects have been the reverse of what he anticipated. The corporate sector has emerged as a perennial 'lame duck,' while the coastal fishery has grown, in parts of Nova Scotia, into a vital modern industry. This 'tale of two industries' is what follows.

The corporate sector: crises and short-swing accumulation[3]

The post–Second World War era has been marked by dramatic expansion of the US fresh and frozen market, the offshore trawler fleet, centralized factory-based fish processing, and the concentration of capital.[4] National Sea Products is the pre-eminent example of all these trends. However, its development cannot be separated from post-war state policy. The state found itself underwriting the accumulation of large-scale capital on an extended basis.

In 1942, the state made a concerted attempt to industrialize the fishery. Vessel construction assistance for trawlers over 72 feet in length and for conversion of schooners to trawl technology was offered to corporations. In 1943 the assistance was improved: companies could get a subsidy of $165 per ton and accelerated depreciation allowances (Nova Scotia 1944: 95). Trawler licensing restrictions were relaxed in 1949 and again in 1953 to allow companies to purchase foreign-built vessels. Licence-holders were restricted to companies or individuals 'affiliated with a plant capable of handling the catch' (Watt 1963: 40). All restrictions on foreign-built trawlers were lifted in 1967.

Modernization was characterized in the 1950s chiefly by retrospective criticism of traditional methods and in the 1960s by concern with the competitive position of the Canadian fleet. Between 1960 and 1965, the Soviet Union increased its effort in the Northwest Atlantic by four and one-half times, while the Canadian catch increased by only 6 per cent. The Canadian share in the northwest Atlantic groundfishery dropped from 29.2 per cent in 1960 to 20.2 per cent in 1965 (Canada, Environment Canada, 1976c: Table 17). With government subsidies for vessel construction, the Nova Scotia trawler fleet grew from 5 in 1950 to 19 in 1952 and to 37 in 1962 (Watt 1963: 41). The number of Canadian offshore vessels over 50 tons in the northwest Atlantic increased by 165 per cent, from 211 in 1959 to 558 in 1968. Average size increased as well: the increase in total tonnage over the same period was 320 per cent (Canada, Environment Canada, 1976c: Table 16).

While most observers saw that the Soviet fleet was facing an efficiency crisis with diminishing returns to effort,[5] the provincial Department of Fisheries, the Atlantic Development Board, and the Atlantic Provinces Economic Council (APEC) all called for more federal aid for offshore expansion: foreign fleets would gradually withdraw, leaving Canada to exploit its 'proximity to the grounds' (Watt 1963: 72; APEC 1968: 7–9, 19, 28–9; Canada, Atlantic Development Board, 1969: 49–50, 69–72). Hédard Robichaud, the New Brunswick–born federal minister of fisheries from 1963 to 1968, spearheaded federal-provincial attempts to industrialize. Programs such as East Coast herring experiments with Pacific purse seiners, the Newfoundland Resettlement Program, vessel and plant construction, and research and development grants were monuments to federal-provincial co-operation. Robichaud saw them as giving 'practical guidance in the changeover to modern ways and [hedging] the industry against loss of risk capital' (Robichaud 1964: 20; 1966: 16; 1967: 19–20). By 1976, 61 trawlers, 95 feet and over, were licensed and fishing out of Nova Scotia ports (Canada 1976a). Over 90 per cent of these trawlers were company-owned vessels (Canada 1978). Locationally they were concentrated in the traditional fresh- and frozen-fish processing centres. By 1961, for example, Halifax and Lunenburg accounted for 56 per cent of vessels 75 feet in length and over (Watt 1963: 49). This pattern continued into the 1970s and 1980s except for the dramatic decline of Halifax as a large-scale fish processing centre (see below).

State support for expansion of large-scale processing plants was significant. The earliest case was a joint US-Canadian venture in Louisburg. Late in the war, the Nova Scotia Fisheries Division was given the mandate of making 'large capital contributions for fish freezing plants' (Watt 1963: 37). Extensive infrastructural support and subsidies provided for development of large-scale multinational capital in Canso, Mulgrave, and Petite-de-Grat in eastern Nova Scotia in the 1950s and 1960s (Watt 1963; Cameron 1977). However, the most spectacular case was continued support to just one company: National Sea Products.

The recent history of National Sea Products Ltd. (NSP) has been marked by two periods of expansion, in the early 1960s and early 1970s. Paradoxically, each took place in the face of an overfishing crisis. By the late 1970s, an economic crisis predicated dramatic state intervention. In each instance, the history of this company reveals the difficulty that large-scale capital has in establishing a stable basis for capital accumulation on its own.

Despite an 11 per cent drop in landings between 1959 and 1961, market values soared. NSP responded with large-scale expansion (Young 1962: 23). Using state support totalling $3.5 million, the company built the largest fish processing plant in North America at Lunenburg in 1964. Employing 600 people during peak periods, the new plant could process over 80 million pounds of fish per year and produced fish products valued at $11 million, or 14 per cent of provincial production, in 1966 (Lee 1953; W.A. Smith 1963; NSP 1965). From 1953 to 1965, 11 trawlers were added to the seven already constructed; seven had a revolutionary stern ramp design and catch capacities of 300,000 pounds. With 44 trawlers and draggers by 1965, NSP had the largest fishing fleet in North America (NSP 1963; Smith 1964; Watt, Papers; *Financial Post* 27 June 1964).

With buoyant markets, sales increased 47 per cent in 1965 to $48 million, and net profit jumped 52 per cent to $1.1 million (NSP 1965). Improved harvesting capacity increased landings by 22 per cent in one year and gave the firm 11 per cent of the total Atlantic Canadian catch by 1965. Operations in Lockeport, Louisbourg, and Halifax, NS, Shippegan, NB, and Rockland, Maine, were modernized in the 1960s, making NSP a dominant force in many local port and labour markets. Although no breakdown of the company's groundfish catch is available, NSP accounted probably for a significant portion of the 10 per cent increase in the Canadian Atlantic groundfish catch over the period

1965–8 (Canada, Environment Canada, 1977: Table 21). The year 1968 marked the peak in regional landings for the most important groundfish species, such as cod and haddock. Total groundfish catches then began to decline: by 3 per cent in 1969, 8 per cent in 1970, 10 per cent in 1971, and 35 per cent in 1974. Levels were 9 per cent less in 1974 than in 1955 (Canada, Environment Canada, 1977: Table 11).

NSP responded to the initial crisis by increasing its fishing effort with the addition of six new stern trawlers – financed largely through low-interest loans from the Provincial Fishermen's Loan Board. Increased market values kept prices high enough to offset higher harvesting costs (NSP 1968). Landings continued to increase until the 1969–70 season, culminating in a peak of 329 million pounds – a 51 per cent increase over 1967 levels and 13 per cent of the entire Canadian Atlantic catch (NSP 1967–70). However, diminishing returns to effort finally caught up with the company. The year 1971 witnessed an overall decline of 12 per cent in landings, followed the next year by a reduction of 15 per cent. This left landings 25 per cent below 1970 levels (NSP 1970).

The company responded with a development program which relied on improved 'catchability' and new locations. In 1970, NSP commissioned architectural designs for a new trawler with improved fishing techniques and speed. The next year, it expanded into Newfoundland to exploit 'Northern' cod stocks off the East Coast, Labrador, and Greenland. It purchased a plant in St John's from the Ross Group of England and ordered six new trawlers equipped with mid-water trawl. More than any previous expansion, this one was largely underwritten by the state. DREE grants amounted to $2 million, federal and provincial assistance subsidized 50 per cent of vessel construction costs, and guaranteed loans from the province provided another 40 per cent of construction costs. The Newfoundland Industrial Development Corp. also provided a $2.4-million long-term mortgage (*Chronicle Herald* 6 April 1972; NSP 1971). In the first two years of operation, the new trawlers cut their trip turn-around times by 50 per cent, and landings increased by 23 per cent (NSP 1973).

The decline in regional groundfish catches accelerated throughout the period. However, a previously underutilized species – redfish – was targeted with increased vigour. Redfish landings jumped 44 per cent and market values increased 92 per cent in 1972–3. Movement into this fishery in the Gulf of St Lawrence helped launch a temporary recovery in NSP's fortunes. Profits increased by 112 per cent, two new trawlers were ordered, and plans for a freezer trawler program were considered. In an atmosphere reminiscent of the early 1960s, the com pany called for renewed modernization (NSP 1972; see also NSP 1973).

However, 1974 marked a general economic crisis in the industry. Regional market values for groundfish slumped by 13 per cent. NSP's landings tumbled by 10 per cent, and, with the oil crisis, costs doubled. NSP's American market slumped, leaving it with large inventories and few prospects. Net profits plummeted by 91 per cent, and the company admitted taking its first loss on fishing operations. Three older trawlers were retired, the freezer trawler program was mothballed, and a whole-sale outlet in Montreal was sold (NSP 1974). By 1976, NSP's entire fleet was banned from the Gulf of St Lawrence in a desperate attempt by the state to rehabilitate redfish stocks (*Globe and Mail* 4 July 1977; *Mail-Star* 5 March 1977).[6]

Large-scale capital's tendency to overexpand (as in the early 1960s and early 1970s) was also evident in the development of the herring-based fishmeal industry. Overfishing had brought West Coast herring stocks to the brink of collapse after 1963, and companies such as B.C. Packers moved a large portion of their purse-seining fleet to the East Coast in 1964 to exploit stocks in the Gulf and on the Atlantic seaboard (APEC 1968: 29–30; Barrett 1984: 101). The collapse of the Peruvian anchovy fishery also generated a worldwide shortage of fishmeal, and prices soared (Canada, Atlantic Development Board, 1969: 11; Roemer 1970; Smetherman and Smetherman 1973; Molinari 1977). The fed-eral and New Brunswick governments convened a conference in 1966 to develop an industrial expansion plan. Existing plants were modified to produce fishmeal, while new plant construction began in Shippegan, NB, Burgeo, Nfld, and Pubnico, NS. Introduction of offshore purse-seiners more than tripled the catch in four years, to a peak level, in 1968, of 524,000 metric tons (Regier and McCracken 1976: 20). This marked a 3.6-fold increase over five years.

NSP was at the forefront of this expansion. Existing plants at Lu-nenburg, Louisbourg, and Shippegan were upgraded, with the latter receiving $320,000 in federal assistance in 1966 and $420,000 in 1968. New plants came into service at Mink Cove, NS, and Burgeo, Nfld (NSP 1966, 1968–70). The Newfoundland plant was set up as a joint venture with the Lake Group (NSP 1974). Herring purchased for reduction increased from 2 million pounds in 1966 to 143 million pounds in 1970 – a 14 per cent share of the East Coast herring catch. Fishmeal production became a significant revenue source by 1969 and 1970, nearly over-shadowing groundfish.

Within six years, however, herring landings had decreased by 57 per cent and the value of fishmeal and oil on the world market dropped by 73 per cent. Prices in 1974 were 50 per cent lower than in 1973 (Kearney 1984b). In retrospect, Regier and McCracken (1976: 20) observed: 'The industry developed an excess potential for capturing herring, for processing herring, and continued to expand the fishery even though some evidence suggested that the stocks were being heavily fished.'[7] The crisis prompted NSP to purchase the Lake Group's equity in its Burgeo plant and enter another joint venture with British-owned Sealife Fisheries Ltd to produce 'food' herring (NSP 1971; 1972). Traditional food herring products, as well as fresh and frozen products, enjoyed relative stability and improved markets throughout the 1970s (Canada, Environment Canada, 1977: Tables 25–38). Subsequently, NSP divested the Burgeo operation. The Newfoundland government purchased a 50 per cent interest in the subsidiary, netting the company $1.9 million in 1976 (NSP 1976).[8]

On the eve of the declaration of Canada's 200-mile fishery management zone in 1977, NSP was taken over by a smaller but more aggressive local fish processing company: H.B. Nickerson (Barrett 1984: 91–6). This marked the brief but spectacular pinnacle of corporate power in the fishing industry of Nova Scotia.[9] The new conglomerate's size was impressive for both the region and the fishery. In 1977, it had sales exceeding $234 million. According to a Dun and Bradstreet listing in 1976, this made Nickerson the largest regional company of any kind and nearly ten times larger than its nearest regional fish rival, Connors Bros. Ltd of New Brunswick (*Financial Past* 22 October 1977; NSP 1977). This size gave Nickerson relative parity with the largest fish conglomerate in North America: the two Weston affiliates, B.C. Packers Ltd and Connors Bros. Ltd, which had combined sales of approximately $238 million in 1977 (B.C. Packers 1977; Dun and Bradstreet 1976).

The structure of NSP, even at its peak, was highly segmented by sector and location, and it remains so. For this reason it is difficult to speak of the firm as a monopoly. For instance, according to unpublished government data, by 1977 the company owned 55 per cent of vessels 101 feet and over but just 15 per cent of all vessels registered in Nova Scotia between 65 and 100 feet. While its direct participation in the pelagic fishery was nil, it owned 48 per cent of all vessels over 65 feet in the groundfishery in Nova Scotia, but only 28 per cent of all such vessels engaged in the shellfishery (offshore scallops) (Canada, Fisheries and Oceans, 1978). As Table 1 reveals, by 1977 the NSP/Nickerson complex owned 11 per cent of all fish plants, with a combined ground-fish capacity of 45 per cent of the total in the province; employed 41 per cent of the permanent fish-plant work-force; purchased, according to 1971 levels, 34 per cent of all fish landed by fishermen in the province; and, according to conservative estimates, had a market share in fish products of approximately 40 per cent.

More detailed consideration of the province reveals a segmented picture. The company, through its network of 20 plants, had a clear monopoly position in just one county, Lunenburg (see Map 3). (Plant employment, plant processing capacity, and fish purchases were all above 80 per cent of county totals.) In eastern Nova Scotia, particularly Cape Breton, and in Guysborough and Halifax counties, NSP/Nickerson had a dominant position: its 10 plants accounted for 57 per cent of plant employment, 60 per cent of plant capacity, and 48 per cent of fish purchases.[10] The comparative weakness of corporate capital, both sectorally and locationally, is indicated in Table 2. In southwestern Nova Scotia, a region with over 69 per cent of plants, 45 per cent of plant employment, 42 per cent of plant capacity, and 50 per cent of fish purchases, NSP/Nickerson had a very minor position; it has consistently been unable to dominate the coastal-zone groundfishery, the lobster fishery, or the herring fishery.

**The competitive sector: modernization and
long-swing accumulation**

The post-war years witnessed a proliferation of small- and intermediate-scale fish processors, particularly in southwestern Nova Scotia. The total number of fish processing plants in the province more than doubled between 1933 and 1950 (Table 2). While the east, the north, and Cape Breton experienced moderate recovery after a wartime low, the number of plants in the southwest increased by nearly two and one-half times. The eastern/northern/Cape Breton share dropped steadily from two-thirds of all fish plants during the Depression to under one-quarter by 1983. In contrast, the southwest share increased from 37 per cent in 1933 to 62 per cent in 1950 and 77 per cent in 1983. Eastern and northern Nova Scotia and Cape Breton experienced sig-

TABLE 1
Locational and sectoral segmentation of corporate capital in Nova Scotia fish processing (county totals in parentheses)

Activity of NSP/ H.B. Nickerson	Cape Breton and northern and eastern Nova Scotia	Lunenburg	Queens	Shelburne	Yarmouth	Digby/ Annapolis	Total
1971 fish purchases (000 lb)							
Percentage of groundfish	47.0 (152,746)	91.4 (78,812)	11.7 (16,821)	27.7 (64,913)	0.0 (15,303)	0.0 (25,203)	46.3 (353,798)
Percentage of pelagics	60.0 (21,167)	53.9 (11,134)	0.0 (1,183)	40.2 (2,490)	0.0 (107,187)	0.0 (23,389)	11.8 (166,550)
Percentage of shellfish	20.3 (4,967)	81.6 (3,677)	8.2 (1,362)	1.7 (6,362)	1.5 (12,633)	0.0 (3,518)	13.6 (32,519)
Percentage of total catch	47.8 (178,880)	86.5 (93,623)	10.7 (19,366)	25.9 (73,765)	0.1 (135,123)	0.0 (52,110)	34.0 (552,867)
1976 fish processing							
Plants as percentage of total plants	18.9 (53)	40.0 (5)	20.0 (5)	3.9 (51)	8.6 (35)	5.7 (35)	10.9 (184)
Employment as percentage of maximum employment	56.5 (2,944)	90.1 (1,137)	11.3 (265)	18.7 (1,121)	8.4 (961)	5.0 (1,103)	40.7 (7,531)
Groundfish processing capacity as percentage of total capacity	35.5 (338,304)	87.6 (137,000)	22.2 (36,000)	19.1 (135,000)	8.3 (36,100)	7.2 (138,500)	44.8 (820,904)

Note: NSP/H.B. Nickerson plant list was compiled from sources cited in Barrett (1984).
Sources: Canada, Environment Canada, Fisheries Intelligence Branch (1973, 1976b)

TABLE 2
Percentage of plants by region

Region	1933	1944	1950	1970	1976	1983
Cape Breton, eastern and northern Nova Scotia	63.3	41.1	38.1	30.3	28.4	23.0
Lunenburg	1.7	4.2	5.6	5.0	2.7	4.3
Queens	4.0	2.4	3.9	2.8	2.7	3.0
Shelburne	12.4	17.8	17.4	25.2	27.4	34.6
Yarmouth	7.3	16.1	18.2	17.9	20.0	13.0
Digby/Annapolis	11.3	18.4	16.8	18.8	18.8	22.1
(N)	(177)	(168)	(357)	(218)	(186)	(231)

Sources: Canada, Dominion Bureau of Statistics (1934, 1946); Canada, Environment Canada, Fisheries Intelligence Branch (1970, 1976b); Canada, Fisheries and Oceans (1983); Nova Scotia, Department of Trade and Industry (1950)

nificant decline in all types of fish plants (Table 3), while Lunenburg, Queens, and Shelburne counties had a shift away from large and competitive plants toward small-scale plants. Yarmouth, Digby, and Annapolis counties lost small and competitive-scale plants but gained independently owned large plants (see chapters 4 and 11).

These changes raise questions, particularly in light of the theoretical perspectives discussed in chapters 1 and 2. One has to deal with the proliferation of small- and competitive-scale capital, particularly in Shelburne County, and the rise of independent, large-scale capital in Yarmouth and Digby counties. The discussion that follows examines three closely interrelated factors in the re-emergence of a long swing in capital accumulation in the fishing industry: market diversification, the role of the state, and technological modernization.

Market diversification

Lobster. The rise of the live lobster trade with the United States was mentioned in chapter 2. Its impact from the 1930s onward was differentially important to the western, eastern, and northern parts of the province. Bates (Nova Scotia 1944: 52) observed: 'Almost half the districts on Cape Breton Island, the districts in Cumberland, Colchester, and Pictou on the North Shore, from Canso to Halifax city on the east shore, and from Shelburne round to Digby, all got more

MAP 3 Corporate and competitive sectors

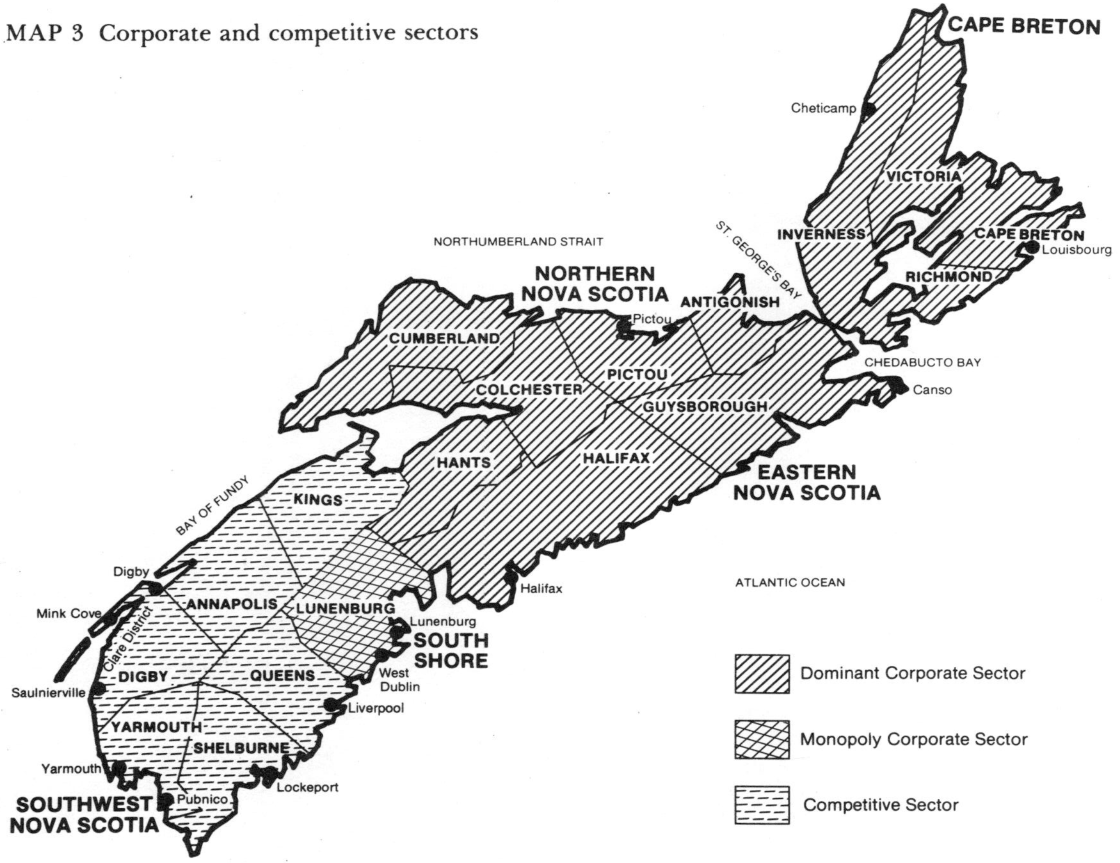

Dominant Corporate Sector

Monopoly Corporate Sector

Competitive Sector

CAPE BRETON

CAPE BRETON

VICTORIA

RICHMOND

INVERNESS

Cheticamp

Louisbourg

CHEDABUCTO BAY

Canso

EASTERN
NOVA SCOTIA

ATLANTIC OCEAN

ST. GEORGE'S BAY

ANTIGONISH

NORTHERN
NOVA SCOTIA

GUYSBOROUGH

Pictou

PICTOU

COLCHESTER

HALIFAX

Halifax

NORTHUMBERLAND STRAIT

CUMBERLAND

HANTS

SOUTH
SHORE

Lunenburg

LUNENBURG

West
Dublin

Liverpool

Lockeport

KINGS

QUEENS

SHELBURNE

ANNAPOLIS

BAY OF FUNDY

Digby

Clare District

DIGBY

YARMOUTH

Pubnico

Yarmouth

Mink Cove

Saulnierville

SOUTHWEST
NOVA SCOTIA

72 Emptying their nets

TABLE 3
Fish plants by scale of operations* and location over time

Region	Year	Small (%)	Competitive (%)	Large (%)	(N)
Cape Breton, eastern and northern Nova Scotia	1950	42.7	36.0	21.3	(136)
	1983	49.1	32.0	18.9	(53)
Lunenburg/Queens/ Shelburne	1950	36.5	42.7	20.8	(96)
	1983	63.9	22.7	13.4	(97)
Yarmouth	1950	66.1	27.7	6.2	(65)
	1983	43.3	26.7	30.0	(30)
Digby/Annapolis	1950	30.0	61.7	8.3	(60)
	1983	47.1	25.5	27.4	(51)
Total	1950	43.1	40.6	16.3	(357)
	1983	54.1	26.0	19.9	(231)

Sources: Calculated from: Canada, Fisheries and Oceans (1983); Nova Scotia, Department of Trade and Industry (1950)
* Since 1950 and 1983 data are not precisely comparable, an illustrative measure of scale of operations uses two reliable indices that have emerged from the detailed survey of fish plant managers (see chapter 4) – plant employment and production line. The 1950 data have been categorized using the following scale definitions: large plants have at least three product lines, one of which is frozen fish; competitive plants do not have frozen fish but have fresh fish and either salt fish or live lobster (or both); small plants incorporate plants that do not produce fresh or frozen fish but sell salt fish or live lobster or both. The 1983 data have been categorized using the following definitions: large, 51 employees and over; competitive, 21–50 employees; small, 1–20 employees.

than half their income from lobsters.' Lobster prices in the live trade ran three times higher than in the canning industry (Canada 1927a: 11). They were particularly high in the southwest, at $18.75 per hundredweight, in contrast to cod, which was $1.29 per hundredweight in 1939. This region had a relative advantage in the 'live' trade based on its proximity to the burgeoning US market and ice-free winter fishing (Nova Scotia 1944: 59, 66).

In the years 1926–30, 30 per cent of the total lobster catch went to the 'live' industry. This proportion increased by 1942 to 56 per cent, and by the 1950s, when stocks stabilized, to 83 per cent (Nova Scotia 1944: 66; Canada 1959: 255). The industry became the backbone of the small and competitive sectors of fish processing and of the independent coastal-zone fishery. While Clement (1986) tends to dismiss lobster in his analysis of the East Coast fishery, the attempt by corporate capital to consolidate control of this industry, and its complete failure to do so, are the real post-war story.

NSP moved into shellfish in 1964. It acquired the three largest lobster companies: Maritime Packers Ltd of Pictou, NS, and E. Paturel Ltd of Shediac, NB – recognized as the second- and third-largest frozen lobster meat producers in Canada – and Conleys Lobster Ltd of St Andrew's, NB – the largest live lobster shipper in the world (Watt, Papers). These takeovers gave the company plants in Pictou, NS; Grindstone on the Magdalen Islands; Morrell, PEI; and Shediac, Tormentine, and Shippegan, NB; as well as the largest lobster pound in the world, in St Andrews, and numerous other pounds in southwest Nova Scotia (NSP 1965). With its close proximity to the American market, NSP attempted to establish a monopoly in the Gulf shellfishery[11] and an effective monopsonistic position in parts of southwest Nova Scotia and the Bay of Fundy regions of New Brunswick and Nova Scotia.

NSP recognized that such control could not be garnered easily in southwest Nova Scotia. Lobster represented less than half of total receipts to fishers, and the proliferation of independent lobster pounds made it difficult to corner the market or generate a client labour force (Canada, Atlantic Development Board, 1969: 31). Consequently, from 1965 onward, the company promoted research efforts to develop an offshore lobster fishery through conversion of swordfishing vessels (NSP 1965).

By 1966, Atlantic lobster catches dropped dramatically in the first of a series of downward oscillations, and NSP moved to shift its newly acquired plants in Pictou and Shippegan to Atlantic Queen crab, converting company-owned side trawlers (NSP 1966; 1968). Both the offshore lobster and trawler-crab fishing experiments met with failure, as small-scale coastal-zone fishers proved more efficient operators. NSP admitted as much in its 1968 annual report: 'Our experiments with the "Fort Louisburg" in this fishery were not successful, and it is now apparent that smaller boats and crews can operate more efficiently and profitably than large units.'

In contrast, small and competitive capital created a stable niche for itself. By 1950, 48.2 per cent (or 171) of fish plants or buyers in Nova Scotia were involved in the live lobster trade. Seventy-four per cent of these establishments were new to the industry since 1922.[12] Twenty-six years later, 59.1 per cent of fish processors were engaged, to some extent, in the 'live' lobster trade. In the counties of Lunenburg, Queens, and Shelburne, the proportion was 65.6 per cent, and in eastern and northern Nova Scotia and Cape Breton Island, 69.8 per cent.[13]

Clearwater Fine Foods rose from a single lobster pound in 1976 to a large-scale corporate enterprise with eighteen plants in Canada, England, and the United States. It has developed new markets for live lobster – first Europe, then Asia – served by improved air cargo connections. A technological advance in lobster pound storage has improved quality. Recent estimates place Clearwater's share of total Canadian production at 12 per cent. In the mid-1980s it was the largest buyer in the province, but the market remains highly competitive (Kimber 1986a; Surette 1987). Since the general collapse of the fishery in the late 1980s, Clearwater's position in both the lobster and the groundfish sectors has declined significantly.

The lobster fishery has been crucial to direct producers since the war (Kearney 1984a). By 1959, 75 per cent of all fishers were engaged in it. The number increased to 76.4 per cent in 1965, and this fishery still engaged 73.0 per cent of fishers by 1974, when the federal government began to professionalize the fishery by eliminating occasional fishers (Canada, Environment Canada, 1977: 126). Estimates now place lobster receipts at 40 per cent of gross income for fishers in southwest Nova Scotia (Kearney 1986).

Fresh fish. As we saw in chapter 2, the growth of the fresh fish industry through development of ice-making equipment, refrigeration, and the US market was dominated initially by large-scale capital. By the 1930s, however, small firms in southwest Nova Scotia were making inroads into the US market. A reduction in the US tariff on fish imports in 1939 accelerated this trend. In 1938, 24 per cent of fresh and frozen groundfish exports went to the US market; in 1939, 43 per cent (Nova Scotia 1944: 70).

Development of paved highways, refrigerated truck transportation, and electrification led to proliferation of small fresh-fish plants in Nova Scotia (Blair 1972).[14] Bates observed:

In the future, as road transport improves, there may be a possibility that buyers will use trucking to a greater extent and so may widen the 'market' for those fishermen who, up to the present, have been compelled to salt their fish and to manufacture it themselves into pickled fish, boneless, or dried fish. A similar extension of the market usually occurs when refrigerated units are placed at a convenient number of shore points, as has been revealed in recent years in Quebec. The presence of these units and cold storage space, does allow opportunities for using the fish in an increased number of forms, and this in turn tends to raise the price to the fishermen, so long as markets are available. (Nova Scotia 1944: 53)

These factors made feasible intermediate-scale economies based on new technology. The truck in particular allowed small plants to market relatively small shipments quickly on flexible schedules and at low freight costs (Watt 1963: 65–6). Road transportation opened up the New England market to small capital, making redundant the Canadian Pacific rail line in southwest Nova Scotia by the late 1950s. The advent of year-round motor transport ferry service between Yarmouth and Maine greatly improved access to the New England market for fresh fish and live lobster (see chapter 4, below).

In 1933, 74 fish plants marketed $1.7 million in fresh fish – 43.1 per cent of total fish production (Canada, Dominion Bureau of Statistics, 1934: 172–3). By 1944, 107 fish plants marketed $10.7 million in fresh fish – 53.0 per cent of total production (Canada, Dominion Bureau of Statistics, 1946: 188–9). In 1950 the number of plants engaged in fresh fish production increased to 203, or 56.9 per cent of total plants. The highest concentrations were along the Atlantic coast in Cape Breton, on the Eastern Shore, and in the southwest, where small plants developed diversified business based on fresh fish, salt fish, and live lobster (Nova Scotia, Department of Trade and Industry, 1950). Twenty-six years later, though with fewer plants, fresh fish production was still a backbone of the industry, involving 50 per cent of all plants and 67.9 per cent and 60.0 per cent of plants in eastern Nova Scotia/ Cape Breton and Digby/Annapolis, respectively (Canada, Environment Canada, 1976b).[15]

Salt fish. A third post-war market shift was particularly dramatic for the emergence of small and intermediate fish production. It involved the practical elimination of all large-scale salt fish processors and exporters by the 1960s and restructuring based on small-scale units of production, where salt fish became an important but subsidiary product line. The roots of this change lay in an improved mechanical drying system developed before the war by the Atlantic Fisheries Experimental Station in Halifax (Watt 1963: 57–8). Adaptation of this prototype for small plants revolutionized production at a time when traditional West

Indian markets were giving way to a diverse, ethnic-based US market (Nova Scotia 1944: 63–4) and large-scale production organization was proving inefficient.

These developments created significant intermediate scale complementarities with fresh fish and live lobster production, and broke the hold of a few large exporting firms. By the late 1960s, all six large salt fish processors that dominated the industry up to the Second World War were either out of business or involved primarily in other lines of production. In 1950, 81.2 per cent (290) of fish plants in Nova Scotia produced salt fish (Nova Scotia, Department of Trade and Industry, 1950).[16] In the southwest, up to 93.8 per cent of plants handled salt fish. By 1976, the number of plants involved was substantially reduced, but 51.6 per cent of the province's plants were still active in the trade. In the southwest, 72.1 per cent of plants produced a salt product (Canada, Environment Canada, 1976b).

Herring. Other market developments helped to transform the fishing industry. Emergence of an ethnic delicatessen trade for vinegar-cured herring stimulated the fresh herring business, increased prices, and led to establishment of a number of US specialty product processors in southwest Nova Scotia from the Second World War onward (Nova Scotia 1944: 60).[17] As shown above, high entry costs limited involvement in the fishmeal boom of the 1960s to relatively few large corporations. Industrial over-expansion quickly brought this sector to the brink of collapse. Development of the Japanese herring roe market was done largely by independent large-scale capital based in the Clare district of Digby County. This segment remains highly seasonal and confined to Digby and Yarmouth counties but has become the basis for a rejuvenated coastal-zone herring fishery and an ethnically based corporate empire (see chapters 4 and 11).

These markets stimulated independent commodity production in the southwest. Stewart Bates (Nova Scotia 1944: 44) observed that chronic shortage of capital rendered two technologies conspicuously absent from the inshore fishery in the 1930s: longlining and seining. While Danish seining was used inshore to develop a flounder fishery for the fresh and frozen market, purse seining was highly regulated by government until the 1960s (Kearney 1984b). By 1962, 4,282 Nova Scotia fishers were engaged in the herring fishery – just over one-third of the province's registered fishers that year (Canada, Environment Canada, 1977: 126). Even after disintegration of the fishmeal market, the herring fishery still employed 3,514 of Nova Scotia's inshore fishers, or 33.8 per cent of the total (Canada, Environment Canada, 1976a). While most were involved in the seasonal herring gillnet fishery, twenty-three vessels still employed purse seine gear in 1976.

Scallops. Processors in New Bedford, Mass., developed the scallop industry during the Depression (Watt 1963: 30). Smaller dragger technology was introduced to southwest Nova Scotia, and two distinct scallop fisheries emerged in the 1950s and 1960s: a coastal-zone fishery centred in Digby and an offshore scallop fishery centred in Lunenburg County, Yarmouth, and Saulnierville. Scallop landings went from 1 million pounds in 1953 to 14 million in 1962 (Watt 1963: 50, 54). By 1976, the Atlantic catch was nearly 206 million pounds and, as a proportion of total landed value, had increased to 17.4 per cent, from 6.6 per cent in 1962 (Canada, Environment Canada, 1977: 49, 65–6). A buoyant luxury market gave capital and producers the highest profits and incomes in the fishing industry in the 1970s and 1980s.

The scallop industry has always been bifurcated into a vertically integrated corporate sector and an independent competitive-capital/coastal-zone fishery. High entry costs have restricted involvement in the offshore scallop industry. Independent companies based in Yarmouth and Digby counties pioneered conversion of wooden trawlers to scallop drag technology (Watt 1963: 51); by 1976, 939 offshore fishers were engaged on 100 vessels over 50 feet in length (Canada, Environment Canada, 1976a). NSP was involved in the small offshore scallop fishery from the early 1960s onward, but low prices and declining catches discouraged expansion. Rapid extension of the luxury market for scallops in 1973 led to increased effort and additions to the offshore fleet. By 1975 and 1976, NSP was converting many of its wooden side trawlers with scallop drag (NSP 1975; 1976). Within two years, however, the increased effort had led to diminishing average sizes – the classic symptom of impending stock crisis, which was clearly realized in 1980–1.

The Bay of Fundy scallop fishery is centred in Digby and has traditionally been pursued by fishers on independently owned boats under 50 feet in length. In 1976, 541 coastal-zone scallop fishers operated out of sixty-eight vessels under 50 feet in Nova Scotia, and the Department of Fisheries and Marine Science identified six small and intermediate processors in the Digby area that handled scallops (Canada, Environment Canada, 1976a, 1976b).

The role of the state

State support for the small and competitive sector can be traced to establishment of the Fishermen's Loan Board in 1936. Set up to provide fishers relief,[18] it was reorganized in 1944 under the Nova Scotia Fisheries Division and facilitated modernization of the coastal-zone fishery (Watt 1963: 37; Nova Scotia 1944: 39). After some research, a 50- to 60-foot-class longliner design, based on the 'Cape Island' lobster boat, was developed in 1950 by fishers, boatbuilders, and federal and provincial fisheries departments (Proskie 1959). The federal government adopted a new plan to subsidize fishers through the Loan Board by up to $165 per ton for draggers and longliners of 55 to 60 feet.[19] Later, the lower limit was reduced to 45 feet. Watt (1963: 39) recalled: 'From 1947 to June 1960, a total of 159 boats and vessels were built for Nova Scotian fishermen with the assistance of government subsidies and loans – 34 draggers; 125 longliners.'

The Loan Board displaced the merchant as the fisher's main source of credit for fixed capital investment[20] (Nova Scotia 1944: 30) and allowed many producers to become independent of capital (see chapter 1). Their vulnerability had stemmed from the central importance of technology in the fishery, and the Loan Board therefore assisted modernization.[21] Productivity increased, fishers' incomes rose, and a viable source of supply was established for a competitive, intermediate-scale processing sector (Watt 1963: 48). Chapter 8 (below) draws attention to the impact of the Loan Board's programs on the coastal-zone fishery on Digby Neck and the Islands:

By 1968–69, the federal vessel construction assistance programme had aided in the construction and acquisition of 285 vessels falling between 25 and 99.9 gross tons in Nova Scotia ... In addition, alterations in government loan and subsidy programmes during 1964–65 extended assistance to fishermen purchasing boats down to 35 feet in length and under 25 gross tons. For the first time, coastal zone fishermen were permitted ready access to financing programmes. Between 1965 and 1968, one hundred and ninety-two Nova Scotia boat purchases were assisted under this programme. (Mitchell and Frick 1970: 39)

The government extended unemployment insurance (UI) assistance to fishers in 1957, particularly important in areas with winter ice and restricted seasonal access to fish stocks. While UI is often interpreted as a hidden subsidy of low fish prices, prices can be said to have been perennially low in any case. UI provided cash that, along with other factors, contributed to fishers' and their families' independence.[22] Peter Sinclair (1985b: 109) recently observed: '[UI] is a vital part of the living expenses of the open-boat fishermen and, consequently, qualification for UI becomes the first financial priority of the fishing season.'

The state had been resource manager for some time in the lobster fishery. As we have seen, seasonal and size restrictions were undertaken to conserve stocks by the Second World War; limited entry based on licensing was introduced in 1967 (Sinclair 1985b: 112). After 1976 this method was extended to all other fisheries (Levellton 1973; R.D.S. MacDonald 1984; Sinclair 1985b: 113). After the stock collapse of 1974, the state tried to manage exploitation and reduce capacity (Barrett 1984). While declaration of the 200-mile management zone in 1977, and an upturn in the industry, reduced pressure on capacity, state-sanctioned licensing helped sustain independent commodity production.

Elimination of occasional and part-time fishers 'professionalized' the fishery. Fishers could no longer pursue an annual cycle of plural activity with the same freedom.[23] For some, the need to fish for at least a minimum period worsened conditions of life because of low prices in the groundfishery. For others, elimination of competition from occasional fishers seems to have increased productivity and incomes (Connelly and MacDonald 1983). Limited entry has, most importantly, helped solidify a petty commodity producer class in the fishing industry while aggravating, in certain localities, intraclass divisions through regulations on access to new species such as crab (Sinclair 1983; R. Matthews and Phyne 1988).[24]

State-based support for small and competitive capital in fish processing is difficult to document. While the new provincial Fisheries Division was 'to encourage ... modernization ... in such a manner that shore fishermen will share in the development' (Watt 1963: 37), financial assistance was restricted to large plants. At best, the government supported port-related infrastructural development, such as highways, power, water, sewage, and wharfage, although the federal Cold Storage Act encouraged public cold storage development, particularly fishers' co-operatives. In 1950, for example, fishers in Port Bickerton obtained a cold storage subsidy and a plant was built; in 1958 the Co-Op installed plate freezers and moved from the fresh to the frozen fish trade (Watt 1963: 48). In general, however, state-subsidized development has not characterized the small and competitive sector until very recently. The growth and vitality of this sector have been a function of a strong independent coastal-zone fishery; access to fresh, salt, and lobster markets in the United States; and the incidental provision of a decentralizing industrial infrastructure.

Technological modernization

Dependency theorists often fail to note the significance of post-war technological developments. They see technology as merely another technique used by capital to subsume direct producers. They cite most frequently vertical integration in agriculture and forestry and the use of new technology to manipulate independent producers (Williams 1987c: 196–7).[25] This interpretation perhaps misses the real significance of technological change from a historical perspective. While there are similarities between other resource sectors and the fishery – for example, producer debt (see chapter 8) – a number of post-war technological changes have released a large segment of fishers from their dependence on capital, rather than the other way around. This can be seen in changes in fishing and vessel technology and in the nature of the individual capitals controlling these technologies.

The 'Cape Island' longline design, stern trawlers and 65-foot groundfish draggers, large and intermediate scallop draggers, and purse seiners were developed after 1945. Longliners, seiners, and intermediate scallop draggers dramatically improved vessel mobility, productivity, efficiency, and, in the case of longliners, versatility. As chapter 8 shows, the 65-foot groundfish dragger class was a mixed blessing: while it was highly productive and cost-effective, construction costs and reduced landings tied fishers to processors through financial dependence by the 1980s. High costs for vessel construction, operation, and refit/maintenance have made the largest vessels – offshore scallopers and trawlers – subsidiary operations of processing capital (see chapter 7).

Innovations in fish-catching gear, electronic aides, navigational technology, onboard storage and hydraulics, and unloading equipment have also transformed post-war fishing. Patton (1981: 12) notes: 'Postwar equipment used by the fishery has included larger and more mechanized vessels (the largest single investment in the fisheries of developed countries), nets of synthetic rather than natural fibres, and the use of electronic gear. Mechanization through hydraulic systems for deck machinery and gear-handling apparatus has greatly increased the productivity of both individual fishermen and the large integrated companies.' The range of new equipment runs from marine diesel engines to hydraulic gear setting and hauling apparatus, radar and echosounders, UHF and CB radios, Loran C and plotters for navigation, and automatic pilots. These developments were accessible to coastal-zone fishers, who have become far more productive harvesters as a consequence.[26] These factors have accentuated the differences between selective and non-selective fishing gears mentioned in chapter 1: Levels of capital investment began sharply to differentiate fishers in Nova Scotia after 1970 and particularly after 1977. Availability of credit, limited-entry licensing, improved landings, and increasing prices have stratified fishers by scale, technology, and location (chapter 7; A. Davis and Kasdan 1984).[27]

Concerning recent government schemes to allocate 'fleet' quotas, the 'inshore fleet' has been deeply divided by conflicts of interest between small groundfish draggers and the longline fleet (*Sou'wester* May 1988). As we see in chapter 7, the class dimensions of these developments do not neatly follow the taxonomic categories of the 'proletarianization–petty capitalist' typology (Clement 1986: 63–5). The coastal-zone fishery has proliferated since the Second World War and, in becoming differentiated, has become far less dependent on capital. Modernization of technology, while increasing debt, has also decreased dependence. The effect of this class transformation on ideology and organization is taken up in chapter 14.

Post-war technological change has been carried out by agents, distributors, and branches of capital – local, national, and foreign. Marine supply companies, electronic communication distributors, banks and insurance agents, shipbuilders, and fuel suppliers began to establish regional service centres in the 1960s (Sinclair 1985b: 119–20). Fishers no longer had to go to the community merchant for everything they needed on credit. Cash accounts with various types of service capital became essential if a fishing enterprise were to be competitive. While Sinclair interprets this change in minimal terms, the technological revolution seems to have left corporate capital unable to consolidate significant backward linkages – apart from its own fleet servicing (Patton 1981). Fishers, for so long dependent at their most vulnerable point – technology – on the same players that controlled the outlet for their product, were freed from this constraint.

Conclusion

From a protracted infancy, Canadian corporate capital finally emerged in the fishing industry of Nova Scotia after the Second World War, but, despite the vision of many economists and government officials, it did not save the East Coast fishery. In spite of major attempts to centralize production, integrate the harvesting sector, and industrialize all aspects of the fishery, corporate capital has stumbled from one crisis to another. The post-war experience has only continued the saga begun in the 1920s.

The pre-war experience revealed the difficulty that industrial capital had in garnering direct control over the harvesting sector. It could not completely privatize the resource or displace independent producers, and deployment of industrial harvesting technology led to resource depletion. These problems continued after the war. Overexpansion in the groundfishery, the herring fishery, and the offshore scallop fishery has created a succession of resource crises for capital.

Before the war, the truck system and regional conditions of surplus labour created a cosy niche for large-scale capital, insulating it from competitive pressures to centralize production. A wartime labour shortage disturbed this position temporarily. While the role of surplus labour is discussed in chapter 10, cheap labour became less significant in absolute terms for corporate capital after the war. Seasonal variations in supply had always been the 'natural' basis for irregular labour demand. Large mechanized processing plants supplied on a regular basis by offshore trawler landings reduced the casual element of plant employment. However, unionization of plant labour since the 1950s has reduced the role of surplus labour in the largest plants. The increasing cost of labour has forced capital to become operationally more cost-efficient. In the 1980s, capital used production quotas, more part-time workers, and contracting out in thinly veiled attempts to recapture 'labour' rents to compensate for declining differential rents in the fishery.

Differentiation in the fish product market and growth of the US market have also helped corporate capital wrest control over key market segments. As is discussed in chapter 6, the key to NSP's market successes has been the US frozen fillet and block market, where it exerts a sizeable shared monopoly position. The luxury scallop market has also afforded corporate capital a dramatic margin of security in recent years, although a number of independent capitals in Yarmouth and Digby counties first developed the fishery.

The technological imperative behind capitalist expansion, in embryonic stage prior to the war, has been notable since then, with the dramatic centralization of capital. The need to minimize operating costs and maximize throughput drives these operations. Past a certain cost threshold, options open to managers during crises become limited, as was evident in NSP's successive expansionary responses to stock depletions. The separation of ownership and control may also exaggerate this inflexibility, since dividends assume primary importance. Were it not for post-war state policy, based on modernization and large-scale industrial development, corporate capital would probably not have assumed the form it took in Nova Scotia. The state-sponsored restructuring of corporate capital in the 1980s saw the gradual assumption of welfare economic policies in a continuing effort to underwrite corporate accumulation.

The post-war years have witnessed re-emergence of a long-swing trend in capital accumulation centred around a modernized coastal-zone fishery and a new small-capital sector. The proliferation of low-entry, competitive-product markets has been matched by state-supported modernization of the coastal-zone fishery where the versatile longliner has been the most revolutionary development. The factors that make the competitive sector distinctive and relatively stable in terms of capital accumulation seem to spring largely from its relationship to the harvesting sector.[28] Since the war, capital there has shown marked reluctance to invest in fishing vessels. Small capital's inherent need to keep overhead costs low has given it greater flexibility in adjusting to seasonal and market fluctuations and has left a major share of the harvesting sector to independent producers. As we saw in chapter 1, indigenous producer-based resource harvesting régimes are more likely to equilibrate effort and catch at levels that sustain long-term yields. This balance between capital's investment risk and indirect resource and labour control has re-created a competitive-based, long-swing accumulation process.

However, two contradictions continue to characterize this process. The most obvious is the competitive nature of the product market. As we see in chapters 4 and 6 below, strong competition has produced a highly volatile system, in which Canadian processors are placed at a substantial disadvantage in relation to American brokers. Such factors as the imposition of various US tariffs on Canadian fish over the years and fluctuations in exchange rates have exaggerated this problem. However, in a competitive market, consumer price elasticity is more likely to dictate price trends in each market segment. While the competitive structure may lead to fragmentation, low incomes, and reduced profit margins, as is discussed below, this is not necessarily the case. There is no guarantee of better prices to fishers under corporate consolidation. Union struggles for better wages and prices have pushed capital to restructure and control input and labour costs in other ways.

The other contradiction stems from irregular demand for labour. Seasonal fluctuations in supply and market demand, and vagaries in fish landings, lead to major highs and lows in plant employment. The success of the small and competitive sector is predicated on managerial flexibility. As we saw in chapter 1, this is a rule of small-business organization. While UI benefits have reduced the social impact of this problem in rural Nova Scotia, the dependence of the competitive sector on surplus labour remains a fundamental contradiction in the industry.

4 Small, competitive, and large: fish plants in the 1980s

Richard Apostle
Gene Barrett
Anthony Davis
Leonard Kasdan

The declaration of a 200-mile coastal management zone by Canada in 1977 signalled a remarkable expansion of the East Coast fishery. The Task Force on the Atlantic Fisheries (Canada, Task Force, 1982: 31) reported that between 1974 and 1981 the numbers of licensed fishers increased by 45 per cent while fish processing plants increased by 35 per cent: 'Provinces with no trawler fleets wanted them; provinces with trawlers wanted to add more and bigger vessels. Companies poised themselves for the growth in resources. Processing plants expanded; new ones were built. Fishermen who had left the industry since 1968 came back again; by 1980 fishermen were as numerous as they had been before. Banks loaned money with less than normal prudence' (19–20).

This phase in capitalist expansion proved to be the shortest in recent memory for large-scale capital. An economic crisis of unprecedented proportions brought the two largest companies – H.B. Nickerson/ National Sea and Fishery Products – to the brink of bankruptcy by the end of 1981.[1] To many, it seemed that the larger capital got, the more unstable it became. The Task Force was established to investigate the latest crisis and recommend plans for restructuring. Its report has been the subject of widespread review and criticism (Barrett and Davis 1984; Sinclair 1985a; Davis and Thiessen 1986, 1988; Williams 1987a).

Of particular interest here is its analysis of fish processing. For the first time a government report questioned the dualist assumptions that underlay government policy: 'Sometimes the inshore fishery is portrayed as the "social" fishery while the offshore is thought to be economically efficient. No such general statement can be made. There are many situations where the reverse is true. For example, the inshore-

based fishery in southwestern Nova Scotia is more successful economically than the trawler-based fishery on Newfoundland's south coast'
(Canada, Task Force, 1982: 34). A detailed financial survey of ninety-
nine fish processors by Woods Gordon pointed out the inherent adaptability of smaller-scale enterprise to an industry plagued by seasonal
fluctuations in raw material supply and market instability.[2] The survey,
however, excluded many small and intermediate salt and fresh-fish
processors and the entire lobster industry and failed to explore differentiation in a way that would identify strength and weakness in the
industry as a whole (Barrett and Davis 1984).

This chapter offers a detailed study of a representative cross-section
of the fish processing sector in Nova Scotia. The objective is to isolate
the social and economic factors that continually regenerate differentiation of capital and to underscore the short-swing and long-swing
cycles in capital accumulation. The focus is on the nature of 'smallness'
in the industry and the extent of subcontracting and dependence and
on the differences between large-scale independents and corporate capital.

The chapter begins with an overview of fish processing in the 1980s
and an outline of the survey sample design. It then assesses the results
in terms of four structural patterns: scale and social organization of
capital, port market, labour market, and product market. The chapter
concludes with a general discussion of the structure of the fish processing industry.

Fish processing in the 1980s

The Task Force on Atlantic Fisheries focused on the dramatic increase
in the numbers of fish plants entering the industry in the buoyant years
1974–81. It argued that this increase was one important aspect in the
development of overcapacity. This clear increase in participants, however, was not unprecedented. The crisis period 1974–6, the base years
for the Task Force, was an unusually low point in the industry. In
1970, for example, 218 fish plants were registered in Nova Scotia
(Canada, Environment Canada, 1970). By 1976 the numbers had declined by 14.7 per cent to 186 plants. The decline was particularly
severe (Table 4) in primary 'groundfish': Cape Breton and eastern and
northern Nova Scotia experienced a 19.7 per cent decrease, Lunen-
burg/Queens/Shelburne a 15.3 per cent decrease, and Digby/An-
napolis a 14.6 per cent decrease. The scallop and herring fisheries kept

TABLE 4
Total number of plants by location (percentages, with numbers in parentheses)

Region	1970		1976		1983	
Cape Breton, eastern and northern Nova Scotia	30.3	(66)	28.4	(53)	23.0	(53)
Lunenburg County	5.0	(11)	2.7	(5)	4.3	(10)
Queens County	2.8	(6)	2.7	(5)	3.0	(7)
Shelburne County	25.2	(55)	27.4	(51)	34.6	(80)
Yarmouth County	17.9	(39)	19.9	(37)	13.0	(30)
Digby/Annapolis	18.8	(41)	18.8	(35)	22.1	(51)
Total	100.0	(218)	100.0	(186)	100.0	(231)

Sources: Canada, Environment Canada, Fisheries Intelligence Branch (1970, 1976b); Canada, Fisheries and Oceans (1983)

declines in Yarmouth County to a moderate 5.1 per cent.

Post-1977 recovery had marked subregional variations. By 1983, 231 plants were in operation in Nova Scotia, a 24.2 per cent increase over 1976. As Table 5 reveals, however, the numbers of new firms were dramatically higher: only 58.1 per cent of the 1976 plants remained in 1983. Cape Breton and northern and eastern Nova Scotia registered no net increase, but 28 of 53 plants in 1983 were new; in Yarmouth County, with an 18.9 per cent decrease, 18 of 30 plants operating in 1983 were new since 1976. However, in Shelburne County, 41 plants in business in 1983 were new; in Digby/Annapolis, 30. Shelburne County's share of fish plants increased from one-quarter in 1970, to 27.4 per cent in 1976, to 34.6 per cent in 1983; southwestern Nova Scotia's from 69.7 per cent in 1970 to 77.1 per cent in 1983.

Most increases were among small companies, employing less than 21 workers (Table 6), particularly in Cape Breton and the east and north and in Shelburne and Digby/Annapolis counties. The numbers of large and competitive-scale companies increased substantially in Yarmouth County, which fact, coupled with the area's overall 1976–83 decrease, indicates a strong trend toward centralization. Lunenburg and Queens maintained the province's strongest concentrations of large capital.

Survey design

With establishments as the unit of analysis, interviews were conducted with 99 fish plant managers in 1984 (see Appendix). Sixty-one per cent

TABLE 5
1976–83 fish plant survival by location

Region	1976 Total plants	Fish plants still in operation in 1983 (%)	(N)	1983 Total plants
Cape Breton, eastern and northern Nova Scotia	53	47.2	(25)	53
Lunenburg County	5	100.0	(5)	10
Queens County	5	100.0	(5)	7
Shelburne County	51	78.4	(40)	80
Yarmouth County	37	32.4	(12)	30
Digby/Annapolis	35	60.0	(21)	51
Total	186	58.1	(108)	231

Sources: Canada, Environment Canada, Fisheries Intelligence Branch (1976b); Canada, Fisheries and Oceans (1983)

were plant managers; 33 per cent, owner-managers.[3] On average, firms covered had been in business for only eight years; these managers had had their positions five years on average. The median age was 42 years; 73 per cent were anglophone, and 20 per cent francophone; the average educational level achieved was grade 11; and the dominant religious affiliations were Baptist (34 per cent) and Roman Catholic (28 per cent).

Managers repeated the same theme: success required close understanding of the idiosyncrasies of the fishery. Firms with problems had owners or managers not familiar with the fishery. There were uncommonly strong ties between managers, communities, and the fishing industry. Despite little formal education and training in management, food technology, or mechanical trades, most managers had long experience in the industry.

Only 5 per cent had a father or father-in-law who was a fish plant manager; over one-third had fathers or fathers-in-law who were fishing-boat captains. Eighteen per cent were former fish plant workers or supervisors, 20 per cent were fishing-boat captains, and 22 per cent had managed fish plants elsewhere. The community connection was even stronger. Sixty-three per cent of managers lived within one mile of their plant, while 71 per cent were raised (up to age 16) within 15

TABLE 6
Fish plants by scale of operations* and location, 1976 and 1983 (percentages, with numbers in parentheses)

Region	Year	Small		Competitive		Large		Total	
Cape Breton, eastern and northern Nova Scotia	1976	24.5	(13)	47.2	(25)	28.3	(15)	100.0	(53)
	1983	49.1	(26)	32.0	(17)	18.9	(15)	100.0	(53)
Lunenburg County	1976	0.0	(0)	60.0	(3)	40.0	(2)	100.0	(5)
	1983	44.5	(5)	22.2	(2)	33.3	(3)	100.0	(10)
Queens County	1976	20.0	(1)	20.0	(1)	60.0	(3)	100.0	(5)
	1983	47.1	(4)	14.3	(1)	28.6	(2)	100.0	(7)
Shelburne County	1976	64.7	(33)	25.5	(13)	9.8	(5)	100.0	(51)
	1983	66.7	(53)	23.4	(19)	9.9	(8)	100.0	(80)
Yarmouth County	1976	75.7	(28)	13.5	(5)	10.8	(4)	100.0	(37)
	1983	43.3	(13)	26.7	(8)	30.0	(9)	100.0	(30)
Digby/Annapolis	1976	40.0	(14)	25.7	(9)	34.3	(12)	100.0	(35)
	1983	47.1	(24)	25.5	(13)	27.4	(14)	100.0	(51)
Total	1976	47.9	(89)	30.1	(56)	22.0	(41)	100.0	(186)
	1983	54.1	(125)	26.0	(60)	19.9	(46)	100.0	(231)

Sources: Calculated from: Canada, Environment Canada, Fisheries Intelligence Branch (1976b); Canada, Fisheries and Oceans (1983)
* Scale of operations is operationalized according to the following employment size categories: large, 51+ employees; competitive, 21–50; small, 1–20 (see discussion below for elaboration).

miles. Two-thirds had close relatives working in the industry: primarily brothers, brothers-in-law, or sons. Of 130 such relatives mentioned, 42 per cent were captains, 19 per cent plant managers, 18 per cent fishers–crew members, and 16 per cent plant supervisors. Clearly the great majority of managers had grown up in the industry, knew it well, and had a substantial personal stake in its operations and strong community attachment.

Scale and social organization of capital

Analysis of the interviews revealed the basis of capitalist differentiation in the fish processing sector and, in particular, an independent, intermediate stratum, between large and very small firms. A typology was therefore developed that grouped the plants into three broad sets according to scale of operations. A number of outworker operations were then separated out (these had the lowest scores on the scale measure).[4] Questions asked concerned number of workers employed, replacement value of fixed capital, and volume of sales – to construct an index for scale of operations – and distinguished three groups that tended to cluster together along this spectrum.[5]

The median values and ranges for plant types are presented in Table 7: 28 'small' plants, 35 'competitive'[6] ones, and 27 'large' ones. The median number of workers varied from 10 to 22 to 121 across the firm groupings, while replacement value of capital ranged from approximately $130,000 to $500,000 to $2 million, respectively. Median 1983 annual sales varied from approximately $360,000 to $1.5 million to $4 million, respectively.

Activity was heavily concentrated in rural southwest Nova Scotia (see Map 4). Eighty-six per cent of plants were located between Halifax and Annapolis County. Types of plants were distributed rather evenly (Table 8), except on the South Shore, which has proportionately fewer large plants.[7] The significance of industrial location is discussed in detail at the end of the chapter.

As in any industry, production line is a crucial defining feature and clearly differentiates plants in the typology. As is usually assumed, large plants were disproportionately heavily involved in production of frozen fish and scallops. As compared with 7 per cent of small plants and 15 per cent of competitive ones, 54 per cent of large plants were involved in frozen fish production. Eight of the 10 plants in scallops were large ones. While 64 per cent of small companies and 63 per cent of competitive ones were multi-product processors, 93 per cent of the large ones were. The 1980s boom in herring roe production was concentrated in large plants, particularly in Digby/Annapolis.

Small and competitive operations tended to specialize in fewer product lines and did not process frozen fish. Both groups were involved in salt and lobster production, but competitive plants were more likely (70 per cent v. 54 per cent) also to do fresh fish. The negative relationship between fresh and salt fish production among competitive operations suggests greater flexibility than in small plants in responding to market changes. This was particularly true of fresh fish processing in 1983–4.

Of 86 cases providing information on ownership status (see Table 9), 62.8 per cent were totally owned by one person or family, 30.2 per cent indicated outside participation, and 7.0 per cent were co-operatives. The connection between scale and ownership status is complex.

TABLE 7
Scale of production in fish processing

	Small (N = 28)		Competitive (N = 35)		Large (N = 27)	
	Median	Range	Median	Range	Median	Range
Plant employ-ment (number of workers)	9.83	1–46	22.00	1–120	121.00	25–875
Value of fixed capital (replace-ment cost)	$129,000	$4,000–$425,000	$498,714	$150,000–$5,000,000	$1,993,750	$500,000–$40,000,000
Volume of sales in 1983	$361,200	$25,000–$2,300,000	$1,499,906	$350,000–$6,000,000	$4,037,500	$2,000,000–$50,000,000

While large plants had more outside participation, some were 'independent' capitals and others were an integrated part of a larger organization. Ten of the twelve larger plants with outside participation operated as part of a multi-plant firm, while only 5 of the 11 family-owned firms had more than one plant. These two subgroups are distinguished further, later in this chapter. Of plants in the sample, 26.2 per cent were either subsidiary 'feeder' operations of large capital or informally tied to other fish companies through joint marketing or supply arrangements. The resulting horizontal integration particularly affected small and large fish companies. Competitive operations were noticeably more independent. The implications of such ties for labour are discussed in chapter 10.

Port market structure

Tables 10 and 11 profile the number of inshore and offshore fishing boats supplying the plants surveyed.[8] Seventy-one per cent of small-plant operators interviewed purchased fish from a total of 210 inshore boats, for a median number of eight boats each. By contrast, only 29 per cent of small plants purchased from any offshore vessels. Nearly two-thirds of competitive firms purchased inshore fish. This group of plants provided an outlet for 615 boats, for a median of 19 inshore boats per plant. The same number of competitive plants bought fish from 197 offshore boats. It emerged, contrary to prevailing assumptions, that 78 per cent of large firms purchased fish from inshore boats

MAP 4 Plant location areas

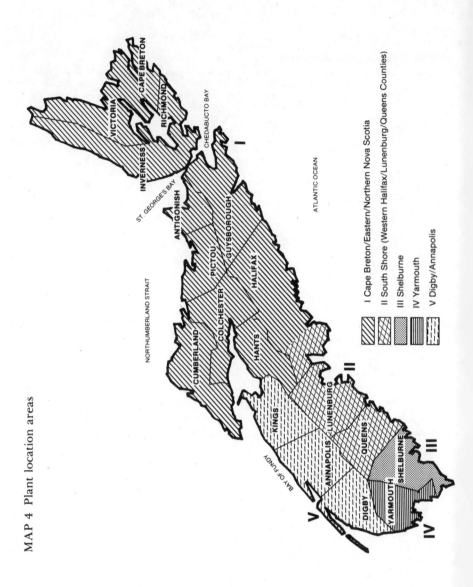

I Cape Breton/Eastern/Northern Nova Scotia
II South Shore (Western Halifax/Lunenburg/Queens Counties)
III Shelburne
IV Yarmouth
V Digby/Annapolis

TABLE 8
Sample fish plants by scale of operations and location (percentages, with numbers in parentheses)

Region	Small		Competitive		Large		Total	
Cape Breton, eastern and northern Nova Scotia	42.9	(6)	35.7	(5)	21.4	(3)	100.0	(14)
South Shore	33.3	(3)	33.3	(3)	33.3	(3)	100.0	(9)
Shelburne County	31.1	(9)	44.8	(13)	24.1	(7)	100.0	(29)
Yarmouth County	21.4	(3)	42.9	(6)	35.7	(5)	100.0	(14)
Digby/Annapolis	29.2	(7)	33.3	(8)	37.5	(9)	100.0	(24)
Total	31.1	(28)	38.9	(35)	30.0	(27)	100.0	(90)

– accounting for 443 boats altogether. Median inshore boat averages varied little among competitive and large plants. Not surprising, more offshore vessels sold to large plants than to competitive plants. Of greater interest, 70 per cent of small plants buying offshore fish did so from 'longliners,' compared to 40 per cent from 'draggers'; competitive plants tended to buy equally from longliners and draggers; and for large plants the division was 32 per cent from longliners and 79 per cent from draggers. Two-thirds of plants purchasing from offshore longliners were located in Shelburne County. 'Dragger-based' companies were disproportionately located in Digby/Annapolis and the South Shore.

Much of southwest Nova Scotia's offshore fleet is composed of locally designed wooden or fibreglassed hull vessels under 65 feet in length. Ice-free conditions and proximity to offshore banks have facilitated development of this fleet. By contrast, plants in Cape Breton and northern and eastern Nova Scotia are subject to supply interruptions caused by ice, as well as by migration of fish, and so they work with the largest offshore fishing vessels, which are less vulnerable to such conditions.

The plant typology helps to distinguish among sources and levels of supply. As indicated in Table 12, 25 per cent of small plants depended exclusively on other plants for supply, and 46.4 per cent relied on other plants and inshore boats. These plants represented a substantial proportion of the feeder plants in the sample. By contrast, many competitive and large plants had multiple sources of supply, with more of the larger plants (44.5 per cent v. 28.6 per cent) having access to all three major lines of supply.

TABLE 9
Ownership status by scale of operations (percentages)

Status	Small	Competitive	Large	Total
Outside participation*	22.2	27.3	46.2	30.2
Owned totally by individual or family	70.4	69.7	42.3	62.8
Co-operative	7.4	3.0	11.5	7.0
(N)	(27)	(33)	(26)	(86)

* 'Outside participation' is preferred to the more general classification 'public,' since some equity participation in private companies is quite common. It captures the real 'control' that such situations often involve.

Table 13 summarizes the proportions of total supply that came from combinations that involved only boats and ones that also involved plants. For plants that relied jointly on inshore and offshore boats, offshore boats provided by far the greater proportion of their supply. Competitive and large plants that received supply from various sources including other fish plants tended, unlike small plants, to have only minor amounts of fish coming from these other plants.[9]

The complex constituents of the 'port market' have been poorly understood (see chapter 7). This issue involves the dynamics that influence fish prices for fishers and the cost of raw material for processors. Plant managers were questioned about relations between fishers and plants. When asked about advantages derived from boat ownership, security of supply was frequently cited. Ownership guarantee processors access to at least some supply and gave them a way to regulate supply by controlling fishing effort – for example, by reducing it during summer, when independent boats maximized effort, and increasing it during winter when landings from independents were down.

Vessel ownership also provided some leverage in resource costs. By landing their own supply, fish plants could internalize costs. Fish prices therefore represented an internal transfer price and provided significant price control in the port market. At the formal level, 29 companies indicated that they owned fishing vessels. Fish companies usually invested in vessels only under certain circumstances. During periods of product buoyancy, they might expand into the harvesting sector, where supply was insecure and risk and costs were outweighed by supply volumes and/or price control. In the post-1981 economic climate, a

TABLE 10
Fish plants purchasing fish by source of supply and plant type

Plant type	Plants purchasing from boats (%)	(N)	Boats supplying plants Total	Median
Inshore supply				
Small	71.4	(28)	210	8.0
Competitive	65.7	(35)	615	19.0
Large	77.8	(27)	443	16.0
Total	64.5	(90)	1,268	
Offshore supply				
Small	28.6	(28)	58	7.5
Competitive	65.7	(35)	197	6.5
Large	77.8	(27)	228	11.0
Total	52.5	(90)	483	

number of companies divested themselves of vessels, passing on to fishers a greater share in the costs and risks of fishing.

Backward integration into the fishery tended to increase as scale of operations increased. Only 21 per cent of small operations owned any vessels, while 48 per cent of large plants did so. Eleven competitive plants owned vessels (8 inshore boats and 11 offshore boats), and 13 large firms owned 10 inshore and 90 offshore boats. Only 2 per cent of the 1,268 inshore boats selling to the plants in the survey were company-owned, while 21 per cent of the 483 offshore vessels were.

Locationally, 41 per cent of boat-owning plants were found in Digby Annapolis and accounted for 70 per cent of company-owned inshore boats and 25 per cent of company-owned offshore vessels. An occasional plant in the other designated areas might own an inshore boat, while the company-owned offshore fleet was concentrated in the South Shore and the Cape Breton eastern/northern region. Shelburne and Yarmouth counties were notably absent in any of these trends.

Formal ownership ties between harvesting and processing sectors are much less significant than informal ties. While only 7 per cent of plants had minority interests in fishing boats, factors such as credit-debt relations, personal obligation, and community or familial ties bound fishers to processors. When asked if they offered credit to fishers, 61.0 per cent of processors purchasing their fish from producers answered

TABLE 11
Fish plants purchasing fish by location and source of supply

Location	Plants purchasing from boats		Boats supplying plants	
	(%)	(N)	Total	Median
Inshore supply				
Yarmouth	57.1	(14)	102	13.0
Digby/Annapolis	64.0	(25)	247	12.0
Shelburne	56.8	(37)	357	13.5
South Shore	90.0	(10)	263	19.0
Cape Breton, eastern and northern Nova Scotia	76.9	(13)	299	18.0
Total	64.6	(99)	1,268	
Offshore supply				
Yarmouth	42.9	(14)	45	8.0
Digby/Annapolis	64.0	(25)	88	7.0
Shelburne	51.4	(37)	232	11.0
South Shore	50.0	(10)	40	10.0
Cape Breton, eastern and northern Nova Scotia	46.2	(13)	78	12.0
Total	52.5	(99)	483	

in the affirmative. Only 24.4 per cent of all buyers providing credit reported charging interest on accounts. This suggests the special role of credit, which is intimately connected with personal obligation. Managers expected that if they sold supplies to a fisher on credit, he/she would in turn sell them his/her fish.[10]

The impact of these ties on fish prices is more complex.[11] In ports with competition (e.g. Shelburne County), the obligation to sell would be honoured only while prices remained competitive. When fishers sold to another buyer offering higher prices, processors complained not so much about losing a dependable supply as about incurring uncollectable debts. In port markets without much competition (e.g. eastern Nova Scotia and Cape Breton), the obligation to sell would probably be honoured even if prices were low, as no alternative outlets existed. Substantial effort was being expended to exclude outside buyers.

Other resources, such as fishing supplies, fuel, bait, ice, and holding coolers/freezers, might also be made available to fishers in return for an understood obligation to sell to the processor. Seventy-two per cent

TABLE 12
Plant type by source of supply (percentages)

Source of supply	Small	Competitive	Large
Other plants only	25.0	14.3	7.4
Inshore boats only	0.0	5.7	11.1
Offshore boats only	7.1	11.4	11.1
Inshore and offshore boats	7.1	17.2	14.8
Other plants and inshore boats	46.4	11.4	3.7
Other plants and offshore boats	3.6	11.4	7.4
Other plants and inshore and offshore boats	10.7	28.6	44.5
(N)	(28)	(35)	(27)

of managers stated that fishers stored supplies in their freezers, 38.7 per cent sold fishing supplies to producers, and 33.3 per cent sold fuel to producers.

Many fish buyers assist fishers in other ways. They often act as intermediaries between fishers and the institutions controlling capital resources, such as governments and banks: 42.9 per cent of processors buying from fishers indicated that they help fishermen obtain boat loans or subsidies. Such external sources of capital are valuable to producer and processor. Often the fish buyer has both equity and experience in dealing with these institutions. Indirect economic relations are clearly a characteristic of the port market.

Labour market structure

The ninety plants surveyed reported employing more than 5,700 persons, half as regular employees and half as seasonal workers. Notably, females constituted a disproportionate share of seasonal workers (58 per cent) while males held a majority of the regular positions (58 per cent). Maximum employment usually occurred during the period from late spring through summer and early fall, when fish landings are at their yearly peak. This pattern was associated with the seasonal dimensions of fish plant operations. Fifty-eight per cent of the sample reported operating in some capacity throughout the year (Table 14). Only a little more than 14 per cent were seasonal plants, open for six months or less per year. Smaller plants were more seasonal than larger plants.

TABLE 13
Proportion of supply from various sources for plants with multiple sources of supply
(number of plants in parentheses)

Plant type	Percentage of total supply	Supply from inshore and offshore boats		Supply from other plants and inshore and offshore boats		
		Inshore boats	Offshore boats	Other plants	Inshore boats	Offshore boats
Small	1–25	100.0	0.0	29.4	25.0	33.4
	26–50	0.0	0.0	17.7	31.4	33.3
	51–75	0.0	0.0	35.3	31.3	0.0
	76–100	0.0	100.0	17.6	12.5	33.3
		(2)	(2)	(17)	(16)	(6)
Competitive	1–25	50.0	16.7	70.6	28.6	14.3
	26–50	33.3	0.0	17.6	42.8	14.3
	51–75	0.0	33.3	5.9	14.3	50.0
	76–100	16.7	50.0	5.9	14.3	21.4
		(6)	(6)	(17)	(14)	(14)
Large	1–25	75.0	0.0	64.3	46.2	23.1
	26–50	25.0	0.0	28.6	38.5	15.4
	51–75	0.0	25.0	0.0	0.0	23.1
	76–100	0.0	75.0	7.1	15.4	38.4
		(4)	(4)	(14)	(13)	(13)

Firms operating in Yarmouth and Shelburne counties, as well as the South Shore, are much less seasonal than those in Digby/Annapolis or the Cape Breton/eastern/northern region (see Table 15). Supply for fish plants in the latter region is often curtailed by ice and other winter factors. Winter conditions also affect plants in the southwest, especially those more dependent on inshore boats. The seasonal availability of certain species, such as herring, and the federal government's closing of offshore banks (e.g. Browns, LaHave, Roseway) to commercial fishing from March to June every year additionally constrain year-round operations. A significant number of plants in Yarmouth County and Digby-Annapolis specialize in production of items such as food herring, fishmeal, and herring roe and operate primarily while herring is available, particularly during summer. This explains the comparatively high percentage of plants there that operate for six months or less (see chapter 11).

Many plants that reported being open twelve months of the year actually were not. As one owner remarked, 'We're open year around

TABLE 14
Seasonal dimensions of fish plant operation by size of plants
(percentages)

Plant type	Length of time open			
	6 months or less	7–11 months	12 months	(N)
Small	25.0	28.6	46.4	(28)
Competitive	11.4	34.3	54.3	(35)
Large	7.4	18.5	74.1	(27)
Total	14.4	27.8	57.8	(90)

if there's fish. We can't keep going if there isn't any, though. Sometimes we truck some in from other places to make some work. But we have to close down when we can't get any.' These 'situational' closings, involving complete layoffs except for maintenance staff, occurred most commonly during winter, when fishing was intermittent.

These conditions make full-time, year-round employment difficult to attain. Regular employees, as well as seasonal workers, often experience periodic unemployment. Plant operators reported that only 5 per cent of their total labour force could be classified as trades and technical employees; 95 per cent had tasks and involved skills directly linked to fish processing, such as cutting, filleting, splitting, and trimming. Disruption in supply could result only in layoffs and plant shutdowns.

In different sizes of plants, fragmentation and generalization of tasks varied. When asked if they organized jobs so that workers could do different jobs at different times, 45 per cent of plant operators answered yes for all jobs, 29 per cent yes for most jobs, and 15 per cent yes for some jobs. However, over 50 per cent of the labour in small and competitive plants engaged in multiple jobs, compared to less than 20 per cent of those in large plants. Large operators reported that a plurality of their employees were light labour, engaged in trimming, machine feeding, weighing, packing, and so on.

Tables 16 and 17 profile the proportion of the total labour force comprised of men and women in each job classification by plant-size category and geographical location. First, women's proportion of total employees increased with the scale of firms – as work processes become mechanized and tasks specialized. Second, women were much more

TABLE 15
Seasonal dimensions of fish plant operation by location of plants (percentages)

Location	Length of time open			
	6 months or less	7–11 months	12 months	(N)
Yarmouth	25.0	8.3	66.7	(12)
Digby-Annapolis	20.0	36.0	44.0	(25)
Shelburne	8.1	29.7	62.2	(37)
South Shore	10.0	20.0	70.0	(10)
Cape Breton, eastern and northern Nova Scotia	15.4	38.5	46.1	(13)
Total	14.4	28.9	56.7	(97)

likely to be employed for specific light labour tasks. (Over 70 per cent of all light labour was performed by women.) Third, no women worked in any of the trade or technical categories. Fourth, while constituting almost 50 per cent of the total work-force, women were reported to hold 13 per cent of supervisory positions and 20 per cent of cutting and splitting jobs. Women received little or no encouragement to enter technical trades education or to 'cut' fish. Moreover, several operators indicated that the physical and uncomfortable nature of certain tasks, such as lifting boxes and working in the freezers and on the wharfs, predisposed them to hire men for these jobs. These factors do not, however, explain the extent of some of these distributions, especially in the foreman category, where women were systematically underrepresented.

This condition is underlined by the distribution of each gender across pay scales. Tables 18 and 19 profile the pay-scale distributions of men and women for each plant-size category and geographical location. Fish plant work pays substantially less than work in other industrial sectors. For instance, the Task Force on Atlantic Fisheries found that hourly rates in fish plants, while nominally lower than those in the food and beverage industry, were almost two dollars an hour less than the manufacturing average (Canada, Task Force, 1982: 68).

Women constituted 68 per cent of all persons earning $5.99 or less per hour, while men comprised 80 per cent of all persons earning $7.00 or more per hour. Moreover, 83 per cent of all persons earning less than $5.00 per hour were women, while 95 per cent of those earning

TABLE 16
Occupational categories by gender and plant type (percentages, with numbers in parentheses)

Plant type	Foremen			Trades/ technical*			Cutters/ splitters			Light labour†			Other labour‡			Multiple jobs			Total		
	M	F	(N)	M	F	(N)	M	F	(N)	M	F	(N)	M	F	(N)	M	F	(N)	M	F	(N)
Small (N = 28)	83.3	16.7	(12)	100.0	0.0	(9)	85.7	14.3	(35)	20.0	80.0	(65)	12.0	88.0	(25)	74.3	25.7	(191)	61.4	38.6	(337)
Competitive (N = 35)	90.2	9.8	(41)	100.0	0.0	(18)	70.2	29.8	(215)	31.3	68.7	(128)	75.0	25.0	(36)	64.5	35.5	(668)	63.7	36.3	(1106)
Large (N = 37)	86.1	13.9	(137)	100.0	0.0	(250)	85.8	14.2	(359)	27.7	72.3	(2432)	83.1	17.9	(236)	62.4	37.6	(834)	48.6	51.4	(4248)
Total (N = 90)	86.8	13.2	(190)	100.0	0.0	(277)	80.3	19.7	(609)	27.7	72.3	(2625)	76.1	23.9	(297)	64.6	35.4	(1693)	52.3	47.7	(5691)

* Includes plant and electrical maintenance and refrigeration and boiler room operators.
† Includes machine feeders, packing, trimming, and weighing.
‡ Includes clean-up, forklift operators, and machine operators.

TABLE 17

Occupational categories by gender and geographical location (percentages, with numbers in parentheses)

Location	Foremen			Trades/technical			Cutters/splitters			Light labour			Other labour			Multiple jobs			Total		
	M	F	(N)	M	F	(N)	M	F	(N)	M	F	(N)	M	F	(N)	M	F	(N)	M	F	(N)
Yarmouth (N = 14)	83.3	16.7	(6)	100.0	0.0	(12)	88.5	12.5	(26)	20.8	79.2	(106)	90.5	9.5	(21)	62.9	37.1	(383)	58.1	41.9	(554)
Digby/Annapolis (N = 25)	86.1	13.9	(36)	100.0	0.0	(35)	72.7	27.3	(231)	14.1	85.9	(587)	95.0	5.0	(40)	60.3	39.7	(771)	48.3	51.7	(1702)
Shelburne (N = 37)	80.0	20.0	(55)	100.0	0.0	(42)	90.6	9.4	(139)	17.8	82.2	(766)	66.7	33.3	(120)	80.8	19.2	(297)	47.3	52.7	(1435)
South Shore (N = 10)	85.7	14.3	(14)	100.0	0.0	(65)	79.0	21.0	(38)	19.4	80.6	(67)	83.0	17.0	(53)	61.6	38.4	(112)	66.8	33.2	(349)
Cape Breton, eastern and northern Nova Scotia (N = 13)	92.4	7.6	(79)	100.0	0.0	(123)	81.0	19.0	(175)	43.0	57.0	(1009)	71.4	28.6	(63)	60.0	40.0	(130)	55.9	44.1	(1669)
Total	86.2	13.2	(190)	100.0	0.0	(227)	80.3	19.7	(609)	27.7	72.3	(2625)	76.1	23.9	(297)	64.6	35.4	(1693)	52.3	47.7	(5691)

TABLE 18
Pay scales by gender and plant type (percentages, with numbers in parentheses)

Plant type	Less than $5.00			$5.00–$5.99			$6.00–$6.99			$7.00–7.99			$8.00 or more			Total		
	M	F	(N)	M	F	(N)	M	F	(N)	M	F	(N)	M	F	(N)	M	F	(N)
Small (N = 28)	9.1	90.0	(77)	37.1	62.9	(132)	90.9	9.1	(186)	81.5	18.5	(27)	0.0	100.0	(5)	57.8	42.2	(427)
Competitive (N = 35)	34.6	65.4	(110)	50.6	49.4	(399)	76.6	23.4	(243)	74.3	25.7	(222)	87.9	12.1	(58)	62.2	37.8	(1032)
Large (N = 37)	14.3	87.7	(488)	35.0	65.0	(938)	47.2	52.8	(2087)	76.2	23.8	(399)	100.0	0.0	(183)	45.7	54.3	(4095)
Total (N = 90)	17.0	83.0	(675)	39.4	60.6	(1469)	53.2	46.8	(2516)	75.8	24.2	(648)	95.1	4.9	(246)	49.7	50.3	(5554)

TABLE 19
Pay scales by gender and location (percentages, with numbers in parentheses)

Location	Less than $5.00			$5.00–$5.99			$6.00–$6.99			$7.00–7.99			$8.00 or more			Total		
	M	F	(N)	M	F	(N)	M	F	(N)	M	F	(N)	M	F	(N)	M	F	(N)
Yarmouth (N = 14)	12.7	87.3	(63)	41.2	58.8	(345)	100.0	0.0	(238)	100.0	0.0	(6)	0.0	0.0	(0)	60.4	39.6	(652)
Digby/Annapolis (N = 25)	16.5	83.5	(550)	41.8	58.2	(569)	75.7	24.3	(321)	76.8	23.2	(185)	91.4	8.6	(70)	45.9	54.1	(1697)
Shelburne (N = 37)	13.6	86.4	(22)	22.5	77.5	(227)	34.3	65.7	(811)	77.2	22.8	(215)	98.8	1.2	(83)	42.8	52.2	(1363)
South Shore (N = 10)	0.0	100.0	(1)	66.7	33.3	(18)	84.1	15.9	(207)	43.6	56.4	(62)	100.0	0.0	(31)	76.5	23.5	(319)
Cape Breton, eastern and northern Nova Scotia (N = 13)	33.3	66.7	(39)	43.9	56.1	(310)	43.3	56.7	(939)	83.3	16.7	(180)	91.9	8.1	(62)	49.9	50.1	(1530)
Total (N = 90)	17.0	83.0	(675)	39.4	60.6	(1469)	53.2	46.8	(2516)	75.7	24.3	(648)	95.1	4.9	(246)	49.7	50.3	(5554)

$8.00 or more per hour were men. In large fish plants, females constituted a higher percentage of those earning $5.99 or less per hour (72 per cent) than was the case in competitive fish firms (53 per cent). Conversely, males comprised a somewhat larger share of those earning $7.00 or more per hour in large plants (84 per cent) than they did in competitive plants (77 per cent). More developed task specialization in the larger plants linked pay rates to tasks arrayed on an axis of unskilled-to-skilled categories. Female labour was disproportionately located at the bottom end, male work at the upper end. When asked if they hired men and women for specific jobs, 76 per cent of all managers answered yes. In particular, they hired women for jobs such as packing and trimming. Clearly, female labour was viewed as having its 'place' in the production process.

A much higher proportion of both men and women paid $5.99 or less per hour were employed in Yarmouth County and Digby/Annapolis (916 of 1,471 positions). Almost 82 per cent of all women and 79 per cent of all men paid less than $5.00 an hour were employed in Digby/Annapolis alone. These data reflected the large amount of labour employed seasonally in processing low-value herring (e.g. roe, fishmeal, food herring), a type of work often subcontracted by one large independent plant to small feeder operations. The Acadian district of Clare has a large concentration of low-wage plant workers (see chapter 13). There is a high-wage labour market in Shelburne County, which has a competitive labour market despite almost total dependence on fishing (see chapter 10). The proliferation of competitive plants and independent coastal-zone fishers, and proximity to the lucrative New England market, particularly for lobster and fresh fish, account for this seeming anomaly. Competitive processors attempt to subcontract the most skilled aspects of salt fish processing – the splitting and boneless production – while devoting scarce plant floor space to more lucrative lines of production like fresh fish.

Fish plants survive on access not only to fish but to a captive labour force. Because of inherent seasonal fluctuations and competition, plant labour has to be intermittent and cheap, forcing all fish processing into rural locations, with few other employment opportunities. Segmentation of the labour market is defined by plant scale and location. The larger the plant, the more detailed the division of labour and the more likely segmentation by gender. Location influences labour-force characteristics, particularly the gender division of labour and wage scales. For instance, the extreme seasonality of herring roe production in

106 Emptying their nets

Digby/Annapolis is underwritten by gender and ethnic stratification, the high-wage competitive sector in Shelburne County makes large-scale fish processing unattractive, and high-wage industrial competition in Lunenburg/Queens has driven competitive and small plants out of business and forced large-scale capital into a gender segmented labour market (see chapter 10, below).

Product market structure

The importance of species/product niches in the structure of fish processing was established above, with large plants involved disportionately in frozen groundfish, scallops, and herring roe production. Competitive processors limited themselves primarily to three product lines: fresh fish, salt fish, and live lobster. Small processors tended to specialize in just salt fish and/or live lobster. The distribution of primary product lines is as follows: 18 per cent of processors shipped at least some portion of their product direct to the retail market; 5 per cent sold some to corporate consumers such as restaurants; 55 per cent marketed through brokers; 8 per cent had their own distributors; 41 per cent sold to other fish companies; and 2 per cent sold to government agencies such as the Canadian International Development Agency (CIDA). The relative significance of these six channels of distribution was measured by percentage of sales: brokers and other fish companies proved crucial. Of the 45 companies selling to brokers, 36 derived 75 per cent or more of total sales from this source, and of 41 companies selling to other fish companies, 33 did so. Twenty-five of the 54 firms selling through brokers dealt with ones either in Boston or in other parts of the US northeast (see chapter 6). Seventy-six per cent of companies selling to other fish companies sold to firms located in Atlantic Canada.

The channel of distribution through brokers was central to competitive-scale and independent large-scale plants. Roughly half of all competitive and large plants marketed over 75 per cent of their primary product through brokers. In sharp contrast, over 50 per cent of small plants, many of which were feeders for larger operations in Digby/Annapolis, used this avenue for marketing. (Approximately one-third of competitive and large plants accepted occasional contracts through other firms.) These trends indicate substantial bifurcation between small companies and all the others. The smallest companies often rely on other fish processors in the region for a marketing outlet. Large companies often use small plants for specialty or overload branch operations. Most competitive and large plants were much more willing and able to establish marketing links with independent brokers. Their products were made to their own specifications, and markets were sought for these, rather than the other way around.

Dependency theory would suggest substantial ownership ties between small and large fish plants, mirroring the 'contracting out' strategy. Sixty-six per cent of independent individual- or family-owned firms marketed fish through brokers, while only 46 per cent marketed through other fish companies. An equal proportion of non-independent plants sold to other fish companies and brokers. Multi-plant ownership therefore seems to have cemented a marketing connection between small and large plants, but it is not a necessary condition of this tie, since less formal arrangements also characterize the marketing of fish between fish companies.

The location of plants selling through either brokers or other fish companies reflected the overall distribution of plants in each case. Two-thirds to three-quarters of plants in the South Shore, Shelburne, and Yarmouth areas sold through brokers – pattern consistent with Shelburne County being highly competitive. In contrast, 50 per cent of plants in Digby/Annapolis and in Cape Breton/east/north tended to sell to other fish companies, reflecting the historical domination of large capital mentioned in chapter 3 and the increasing dominance of one firm in the herring roe industry of Digby County. In the former case, distance from the US market and environmental limitations on access to supply encouraged consolidation into a large-scale firm.

The survey of marketing strategies revealed also the importance of discrete species/product segments for the position of competitive plants. Plant managers dealt with three or more separate brokers during the year, and each broker specialized in only fresh fish or salt fish or lobster. Managers thereby maximized competitive advantage while participating in some highly concentrated market segments, such as salt fish (see chapter 6). This pattern complements processors' competitive relationship with the coastal-zone fishery and, arguably, their competitive position in the local labour market.

Conclusion

The survey of fish plant managers has served to pinpoint the structural basis for industrial differentiation in the fish processing sector. The three major facets of this process identified in chapter 3 continue to underlie this structure.

Small plants tend to conform most to the dependency perspective on small capital. They are closely associated with large capital in terms of inputs and outputs. This relationship can be formal or informal and largely services the cost or production needs of large capital. The survival of this sector rests on the continuing availability of surplus labour in the most highly seasonal enterprise in the industry. Most surprising, this sector is highly insulated from the independent producer fishery. (A central theme in the dependency literature in Atlantic Canada has been the importance of small capital supplying high-value fish from lobster and longline fishers to large capital.)

Competitive plants confirm the importance of a differentiation perspective on industrial structure. Thriving on independent access to fish – and independent producers who assume risks – this sector provides the basis for a significant level of competition in the port market. The structure of supply is diversified and flexible in a production structure that is well adapted to a multi-dimensional and seasonal harvesting sector. This pattern is clearest in the family-based management structure of competitive-scale plants, which are one step removed from 'artisanate origins'. Flexibility, versatility, personal obligation, and paternalism characterize worker-employer relations in these plants. The continuing strength of this sector stems from managers' ability to avoid over-commitment to one broker or one product market segment. Competitive advantage is maximized by 'playing the field.'

Large plants fall into two categories: single-plant independents and multi-plant corporations. The former group is quite similar to the competitive sector in management structure and connection to brokers. The groups differ in their large-scale investment in plant capital and, as a consequence, the backward integration of a significant proportion of their fishing boats. Multi-plant operations tend to be run by non-owning managers who make purchasing and production decisions in terms of the overall organization of the enterprise. These plants tend to specialize much more and get their fish almost totally from a centrally co-ordinated fleet and market through company-owned distributing houses on the basis of large contracts. Their operations thus tend to be much more isolated from the community within which they exist, except, of course, for their labour supply. The labour-market niche in which large-scale plants operate is the key to their operations. While wages are higher in unionized plants and work hours are more regular, working conditions and the pace of work are usually very unattractive. Turnover is high, and these plants have to expend major effort on recruitment. Large plants continue to rely on surplus labour that is made up increasingly of women.

This survey helps explain locational divisions that cut across industrial differentiation in fish processing. The effects of environment, distance from the US market, and historical legacy still divide the industry along geographical lines. Cape Breton and eastern and northern Nova Scotia and the South Shore still have high capitalist centralization and power, particularly in marketing. Shelburne, Yarmouth, and Digby counties are, in contrast, highly differentiated in types of plant, product specialization, and harvesting. While all three counties enjoy smaller proximity to the US market, each has followed a different path of development. Access to highly lucrative lobster and groundfish stocks has led in Shelburne County to specialization in fresh and salt fish and live lobster production – niches that are, of course, the strength of the competitive sector. Contrary to both dualist and dependency predictions, competition in the labour market among competitive-scale plants has made this area a high-wage sector. The region's industrial structure – particularly in Barrington Municipality – has promoted community and entrepreneurial growth patterns, within the fishery at least, that resemble the 'Emilian' model of central Italy.

Access to scallops and herring stocks has led plants in Yarmouth and Digby/Annapolis to develop these lines of production. The development of the herring roe market in Japan by one large-scale independent processor in Digby County has attenuated this process even further. Except in the inshore scallop fishery, these fisheries have required more capital investment, and plants have accordingly been larger and more likely to own their own vessels. This trend spilled over into groundfish plants as well. In this area, manager-owners have been more powerful and less accountable to the community than in Shelburne County. This insulation is bolstered further by the greater urbanization in Yarmouth and Digby. Lastly, the almost total absence of unionization in the large plants has made Yarmouth and Digby/Annapolis – in particular, the district of Clare – the cheap-labour capital of Nova Scotia.

5 Financial characteristics, 1974–1984

Kevin J. Barrett

Companies in any industry tend to have separate and distinct operating characteristics. Efforts to segregate groups of companies by these characteristics tend to follow groupings by scale of operations. It is assumed that companies with access to similar types and quality of human and natural resources exhibit similar management styles and operating philosophies and are affected in similar ways by changes in environmental factors. Accordingly, this chapter seeks to determine, by way of financial-ratio analysis, the extent to which these styles, philosophies, and effects differ between various stratifications (small, competitive, and large) of a representative sample of fish processors in Nova Scotia.

The report of the Task Force on Atlantic Fisheries (Canada, Task Force, 1982) probed the causes of the serious problems experienced by the Atlantic fisheries in the early 1980s. The chapter in that report devoted to economic conditions in the processing sector was based on analysis of a test sample comprised of companies from Quebec, New Brunswick, Nova Scotia, Prince Edward Island, and Newfoundland. Are that study's broad regional conclusions applicable to Nova Scotia? Tables 20 to 24 compare the operating characteristics of the test group used in the Task Force's report and those of an independent sample gathered from fish processing organizations in Nova Scotia (eighteen of the companies in this research project's plant survey: six small, nine competitive, and three large independent firms). The approach used here 'mirrors' the Task Force's study for the critical period of this crisis (1978–81) and, by way of financial-ratio analysis, shows that the characteristics of the region are reflected in this province.

Is Nova Scotia typical?: Task force v. independent sample, 1978–81

Risk

No evaluation or comparison of companies or industries is complete without a discussion of risk. This factor is implicitly weighed in any investment or credit decision. Risk levels in fish processing tend to be higher than in other, more stable industries. Factors contributing to this include: seasonal operations, cyclical profitability levels (high variability in returns), and dependence on favourable foreign exchange rates to maintain export demand. Returns taken over a period of 'good' and 'bad' years provide investors with inadequate returns to merit investment, reducing the availability of equity funds and forcing reliance on debt to finance capital expansion and operating losses. Because of the seasonal nature of the industry, higher levels of debt-service payments have introduced still more risk: payments must be made whether the company is in a peak period or not. Management has no control over foreign exchange rates. If foreign money has more buying power in terms of Canadian goods, demand will be high, and vice versa.

The need to attract equity funds has been identified as a principal challenge in the Task Force's report. Perceptions of risk levels must be changed to reflect more favourably on investment conditions. Improvements that are within the control of operators (labour productivity, cost-effective production, finding off-season supplies of fish, and reduction of reliance on debt) are key to the correcting the situation. Similar conditions exist among the independent (Nova Scotia) sample group, although the risk levels tend to be lower than those for the Atlantic region as a whole, because of more stable and profitable operations and less reliance on debt to finance expansion and operating losses throughout the crisis period.

Net current position

Current ratio. Analysis of the results of the Task Force's study show a declining current ratio during the period 1978–81 (from 1.18 to 0.75, a decrease of 36 per cent). This result is indicative of declining working capital levels and weakened cash flow caused by poor earnings performance, leading to over-reliance on short-term debt to finance inventory and receivable levels. Such a situation sends to the investing and credit communities a message of instability. Insufficient collateral exists to cover loans, and earnings are inadequate to finance operating assets. Risk must be compensated before investors will be attracted. An industry target level of 1.30 has been selected, as continued operations below 1.00 are impossible to sustain. Long-term improvement will require re-establishment of adequate internally generated cash flow to keep short-term debt more manageable.

The independent sample exhibited a slight upward trend, from 0.78 to 0.87, an increase of 12 per cent, though analysis shows that while working capital increased by 1 per cent profitability declined by 41 per cent. Coupled with the fact that long-term debt increased by 441 per cent while net fixed assets increased by only 128 per cent, the implication is that long-term debt has been used extensively to finance current assets. Such a strategy is born of cash-flow deficiencies and is as dangerous as allowing the current ratio to fall significantly below 1.00; in both cases credit limitations will be reached and cash flows generated will be insufficient to meet debt-load service requirements. With the use of long-term debt, the assets acquired are generally exhausted long before debt requirements are fully repaid. This 'dead weight' debt tends to be self-perpetuating, as ever-increasing interest costs eat into profit performance, furthering reliance on debt to get through cash-flow shortages. Improvement of this situation will require adequate internally generated cash flows to finance current asset increases to remove the need for reliance on debt.

Loan ratio. The Task Force's results show an increasing trend to a level well above the industry's target levels (from 0.50 to 1.13, an increase of 126 per cent). As noted above, over-reliance on short-term debt was caused by the poor cash flow generated from operations. Corrective measures similar to those noted above will be necessary to restore acceptable operating levels for creditors and investors alike.

The independent sample results showed a slight decreasing trend (from 1.38 to 1.29, a decrease of 7 per cent). While reliance on short-term debt is much less than that for the Task Force's sample (debt increased 69 per cent, the receivable/inventory ratio increased 80 per cent) long-term debt has been used to finance these current assets instead. If the Task Force's target ratio of 0.40 is accepted, the entire Atlantic fishery was in crisis during this period. Continued financial viability rests on the ability to cover current liabilities with current assets and for assets to be sufficiently liquid to meet obligations as they come due. This was clearly not the case in either the Task Force or Target Sample groups.

Average collection period (days). The Task Force's sample results are very steady (40 to 39 days) over 1978–81. Thus, when compared to the target level of 35 days, receivables management was not a critical problem during the crisis. The Test Sample group exhibited an upward trend from 21.7 to 25.5 days. Common levels were much lower than the target for the industry, in part because of a higher ratio of cash-sale–oriented to credit-sale–oriented companies and more conservative credit-granting policies.

However, companies in the independent sample exhibit more efficient and effective credit collection. Receivables management does not appear critical during this period, despite the upward trend. More comprehensive evaluation between various groups using this ratio would probably not provide significant information because of the impact that various credit policies can have on receivables balances.

Days sales in inventory. The Task Force's sample results show a declining trend for the period (72 to 64 days) caused by a decline in levels of inventory carried proportional to volume of sales. Dramatic increases in the cost of debt no doubt increased the efficiency with which management handled products on hand. Proximity to the industry target indicates an absence of significant problems in inventory management in the Task Force sample during this period.

The decreasing trend so evident in the independent sample almost exactly mirrors that found by the Task Force (44.0 to 38.7 days). Higher money costs forced better management of stock levels, and companies were generally more efficient managers of stock levels carried. Perhaps not coincident with this are lower levels of interest cost and greater profitability. Less money was tied up in inventories, meaning lower cash flow could be tolerated before resorting to debt and the attendant interest costs, which eroded profit margins so greatly during this period.

Equity position

Long-term debt ratio. The Task Force sample showed a dramatic increase over the 1978–81 period (from 1.23 and 8.75, an increase of 611 per cent). The declines experienced in the equity base through operating losses and lack of available investors, coupled with an extreme increase in non-current debt to finance capital assets, were the primary causes for the increase. The capital structure had been totally distorted by these factors, again sending out warning signals to the investment community. Debt requirements must be met before dividends to investors can be paid, and in an industry characterized by seasonal operations, vulnerability to uncontrollable factors (foreign exchange), and declining profitability, attracting outside investment can be very difficult. The industry target level of 0.80 indicates the seriousness of the situation. Access to equity funds must be kept open to ensure long-term viability. Far greater flexibility exists if available credit sources are not exhausted. Correction of this problem remains a challenge, which will be met only after continued profitability margins are established and existing debt loads brought back to more realistic levels.

The experience of the independent sample was similar, if not as severe (from 0.41 to 1.70, an increase of 314 per cent), causing similar problems. A possible explanation, apart from more stable profitability, is a more reasonable historical ratio level. The Task Force ratio level in comparison to target was already well out of line before the crisis. That for the independent sample was far lower than even the target at the same point. The South Shore thus had a more secure investment environment than the rest of the Atlantic region before and after the crisis. This is not to say, however, that the situation among the Test Sample group was ideal. Sustained profitability of operations is essential to bring debt levels back to more acceptable levels.

Equity as percentage of net fixed assets. The Task Force sample showed a declining trend over the period (from 0.63 to 0.10, a decrease of 84 per cent), brought on by erosion of the equity base through operating losses, over-reliance on debt financing (altering the capital structure of the industry for the region), the difficulty in attracting new equity capital to the industry, and general over-expansion of net fixed assets in an attempt to absorb government funds. A ratio level of less than 1.00 indicates inadequate collateral for long-term debt, further harming risk assessments. The ratio was weak compared to the targeted level of 1.00 prior to the crisis, for reasons similar to those noted for the long-term debt ratio. A more attractive debt-equity ratio is essential to overcoming the inability to attract equity funds. Sustained profitability is also essential if the equity base is to be rebuilt.

The results of the independent sample were similar, if not as severe (from 0.68 to 0.39, a decrease of 43 per cent), and indicate the same problems. The factors causing this decline are similar, although, for the same reasons noted in the diagnosis of long-term debt ratio, not as great.

Results of operations

Return on equity. The Task Force sample exhibited a downward trend (from 0.34 to a negative value), as compared to an industry target of 0.19. The levels reached in 1978 were artificially high, influenced by an already inadequate equity base and relatively stable demand and prices. The negative returns reached by 1981 reflected critically low profitability, caused by problems in labour productivity, depressed foreign demand for exports, high interest costs, and a 'flat' market price in general for products. The deficient equity base most definitely did not improve throughout the period, as shown by the analysis of equity position (above). Perceptions of operating risk are influenced by many factors, including variability of returns. The seasonal nature of the industry has created a situation where 'good' years must be adequate to offset 'bad' years and provide an acceptable rate of return to investors. Obviously, this has not been the case for the Task Force group as a whole and can only project the most negative possible image of the industry. Recovery to the target level for the industry will be possible only when internal problems related to productivity and debt costs are solved and environmental factors such as demand and prices rise to former levels.

The independent sample group experienced a similar downward trend in return on equity (from 0.24 to 0.11, a decrease of 54 per cent), again attributable to the decline in profitability experienced in the industry in this period. Causes for the decline are similar to those experienced by the Task Force group but have had less effect on profitability because of lower reliance on debt financing and therefore lower interest costs. Despite this, return on equity was still below target by 1981, causing a redoubled effort by management to solve environmental problems. Many of these factors are beyond the control of individual companies (market price, export demand, and the like). Therefore, streamlining of operations in areas within their control is essential as a long-term strategy.

Fixed asset turnover. The Task Force group showed a decreasing trend in this period (from 4.90 to 3.30, a decrease of 33 per cent), because of cyclical flattening of sales prices, slackening of export demand caused by exchange rate fluctuations, and the dramatic increase in investment in plant assets. The trend indicates lessening in productivity in sales per dollar invested in capital assets and indicates excess capacity in relation to sales levels. This excess is especially apparent when viewed in conjunction with interest and depreciation expense as a percentage of sales. Excess capacity is a historical problem in the industry caused primarily by the seasonal nature of the supply of fish. A change in government policy in the test period produced far greater government financing and many companies expanded beyond present or future ability to create demand. Achieving the industry target level of 4.00 (trawler plant) or 6.00 (inshore plant) will require rebound of selling prices to previous levels and recovery in the export market, accompanied by significant disinvestment of capital assets or the procurement of off-season supplies of fish to use some of the excess operating capacity.

The independent sample exhibited similar tendences (5.89 to 4.68, a decline of 21 per cent), with similar causes (sales volume increased 81 per cent, net fixed assets 128 per cent). Excess capacity appears not as significant in Nova Scotia as in the rest of the Atlantic fishery, as shown by the generally higher ratios and the smaller decline during the test period. Despite the decrease, Nova Scotia does not appear to be in a critical stage in this area, as shown by comparison to target levels for trawler and inshore operations.

Comparative income statement. The Task Force group exhibited a wide swing in profitability as a percentage of sales (from 4.50 to −6.10) for the test period. Approximately 50 per cent of the total decline was attributable to cyclical flattening in sales prices and slackening of export demand (as noted above) and to a general increase in cost of sales, caused primarily by problems in labour productivity. Approximately 40 per cent of the total decline is attributable to increased interest and depreciation changes caused by three factors: a drastic increase in general short/long-term interest rates (1978: 8.6–8.8 per cent; 1981: 11.3–19.8 per cent); over-expansion of plant assets relative to sales volume increases; and the proliferation of 'dead weight debt' (non–income generating, i.e. to finance cash-flow deficiencies from operating losses and inventory increases). The remaining 10 per cent is attributable to relatively higher selling and administrative expenses.

Inability to maintain profit levels during the crisis adds to the already lengthy list of items which increase the operating risk of investors. The level of increase (relative to sales) of interest and depreciation expense suggests excess capacity. If use and productivity levels were relatively consistent, percentage relative to sales would remain fairly constant as well. Many organizations have been pushed to the brink of financial unviability in attempts to meet debt-service requirements as they fall due. Long-term recovery is possible but, because of the industry's lack of control over key areas (interest rates, sale prices of products, and export demand), will not occur overnight or without significant improvements in internal operations and the management decision process.

The independent sample group (analysed below) tends to mirror the conditions and results exhibited (see Tables 20–23) by the Task Force group over the 1978–81 period, if to a lesser degree. (1) Both groups display high levels of operating risk caused by variable and declining profitability levels, over-reliance on debt as a source of financing, and susceptibility to fluctuations in environmental factors (interest and foreign exchange rates, export demand, price levels). (2) Problems or potential problems were apparent in the ability to meet debt-service requirements from operational cash flows or to provide adequate collateral for debt. Companies in the independent sample exhibited good ratio performance, but underlying causes did not create hope for the future. (3) The equity base of the samples was in serious trouble, mainly because of over-reliance on debt for financing expansion and historically high perceptions of risk, as noted above. (4) Profitability levels were insufficient to provide adequate returns to investors or even to cover debt-service requirements. Reasons for decline include depressed market prices and export demand, inflated interest rates, and excessive depreciation changes. Over-expansion of fixed assets created excess production capacity and added to already distended debt levels.

Differences between the groups were minor and centred generally around efficiency of operations and degrees to which the problems noted above affected the general level of operations. It would appear, then, that Nova Scotia is a representative subset of the Atlantic region.

Analysis of independent sample, 1974–84

In the independent sample, throughout the period 1974–84 for all types of plants, problems – including over-reliance on debt in the capital

TABLE 20
Analytic ratios, independent sample v. task force, 1978 and 1981

Net current position	Sample		Task force		Target	
	1978	1981	1978	1981	Trawler	Inshore
Current ratio	0.78	0.87	1.18	0.75	1.30	1.30
Loan ratio	1.38	1.29	0.50	1.13	0.40	0.40
Average collection period (days)	21.70	25.50	40.00	39.00	35.00	35.00
Days sales in inventory	44.00	38.70	72.00	64.00	70.00	70.00

TABLE 21
Analytic ratios, independent sample v. task force, 1978 and 1981

Equity position	Sample		Task force		Target	
	1978	1981	1978	1981	Trawler	Inshore
Long-term debt ratio	0.41	1.70	1.23	8.75	0.80	0.80
Equity as % of net fixed assets	0.68	0.39	0.63	0.10	1.00	1.00

TABLE 22
Analytic ratios, independent sample v. task force, 1978 and 1981

Results of operations	Sample		Task force		Target	
	1978	1981	1978	1981	Trawler	Inshore
Return on equity	0.24	0.70	0.34	neg	0.19	0.16
Fixed asset turnover	5.89	4.68	4.90	3.30	4.00	6.00
Profit margin (% of sales)	2.70	[0.50]	4.50	[6.10]	4.40	2.95

structure, potential shortages in collateral for debt, the seasonal nature of operations, variable operating results (cyclical sale prices and fluctuating export demand), and low labour productivity – all combine to reflect negatively on the industry.

TABLE 23
Comparative income statement: independent sample v. task force, 1978 and 1981

	Sample		Task force		Target	
	1978	1981	1978	1981	Trawler	Inshore
Sales	100.0	100.0	100.0	100.0	100.0	100.0
Cost of sales	79.3	84.0	80.9	88.0	79.2	82.0
Gross margin	20.7	16.0	19.1	12.0	20.8	18.0
General/administrative	11.7	7.2	8.5	9.9	8.0	9.0
Interest	2.6	4.4	2.2	7.6	3.7	2.9
Depreciation	3.4	3.9	2.2	2.8	2.5	1.7
Other	0.3	1.0	1.9	[2.2]	2.2	1.5
Operating expenses	18.0	16.5	14.6	18.1	16.4	15.1
Net income	2.7	[0.5]	4.5	[6.1]	4.4	3.0

Net current position (Table 24)

Current ratio/loan ratio. During the period 1974–84, the competitive group was more able to cover current debts as they come due with current assets, while the large group was least able. Profitability, however, tended to be lowest for the competitive group; if proportionate reliance on short-term debt is lower and internally generated cash flow is not greater than for the other groups, long-term debt is being relied on to finance current assets. This is not a viable long-term financing strategy and will leave to financial collapse.

The large group tends to rely significantly more on both short- and long-term debt (hence the poor net current position), probably because of higher overall investment in capital assets. Small firms, reflecting their size, tended to be less consistent in year-to-year results, although over time 'good' years offset 'bad' years to produce a fairly solid ratio.

All three classes exceeded or approximated the desired level. The small group varied around the target level, indicating a degree of risk, while competitive plants rarely dropped below and the large plants rarely rose above it.

From 1978 to 1981, responses to shrinking profit margins and declining internal cash flow were fairly consistent. The industry's difficulty in attracting investment led to extensive use of debt to finance

TABLE 24
Analytic ratios in test sample, 1978, 1981, 1974–84

Net current position	Small			Competitive			Large		
	1978	1981	1974–84	1978	1981	1974–84	1978	1981	1974–84
Current ratio	0.51	2.10	1.00	2.01	1.55	1.43	0.72	0.72	0.83
Loan ratio	1.89	1.26	1.44	0.94	0.73	0.98	1.38	1.39	1.29
Working capital (% total assets)	[0.21]	0.06	0.00	0.10	0.22	0.16	[0.16]	[0.12]	[0.08]
Average collection period (days)	4.45	14.76	14.47	19.69	21.74	17.89	25.24	26.60	30.91
Days sales in inventory	62.48	26.51	42.88	32.26	41.97	32.98	45.77	38.63	33.08

increases in current assets. A stable high ratio for the competitive group implies reliance on long-term debt in the face of shrinking internal cash flow. Large companies maintained a relatively consistent level of negative working capital, indicating use of short-term debt to finance current asset accumulation and operating losses: they used all available long-term debt to finance capital asset accumulation. While the higher ratios in the competitive group may indicate financial stability, the level of reliance on debt for financing was a serious problem for both competitive and large plants. The small group exhibited a rising ratio, probably because of the decreased reliance on debt associated with the rising cash flows generated by increased profitability.

Consistent with the foregoing, the competitive ratio tended to decline in the subsequent period, 1981–4, to below targeted levels (0.99 current ratio in 1984). Because of reliance on long-term debt to finance cash-flow deficiencies, requirements of debt-service agreements continued to mature long after the expiration of assets financed by the original debt. Long-term debt is needed in other, non-current areas, and profitability declines further. The small group also suffered significant declines after the crisis (current ratio of 0.98 in 1984), primarily because of lower profitability. The large group showed slightly better working-capital performance, despite adjustments to cope with low profitability, excess capacity, and crushing debt loads.

Long-term recovery rests on re-establishment of adequate internal cash flow to reduce reliance on short- and long-term debt. 'Dead weight' debt tends to create more cash-flow problems than it solves, and the quicker it can be removed, the sooner economic recovery can be achieved.

Working capital as percentage of total assets. Analysis of these ratios over the years 1974–84 indicates polarization of performance by size of operation. Most large companies had negative working capital, influenced heavily by dependence on short-term loans and the high opportunity and holding costs of tying up funds in current assets. Small companies tended to maintain a 'break even' level of working capital, indicating substantial non-current assets. Adequate collateral probably exists for both long- and short-term debt, and protection of investors' interests favourably affects risk assessment. Competitive organizations exhibited the strongest ratios, indicating positive working capital levels, although they apparently relied on long-term debt to finance working capital and operations.

Results for competitive and large plants for 1978–81 were consistent with 1974–84. The increase in ratio performance shown by competitive indicated improved cash flow, despite decline in profitability. Cash flow improved despite decline in profitability and reliance on long-term debt.

Large companies showed a decline in absolute working capital (working capital declined 92 per cent, and total assets increased 152 per cent), suggesting overcapitalization and excess capacity. A more constructive approach would have been improvement in working capital deficiencies before expansion into bigger operations. Small plants, perhaps became of restricted access to capital sources and larger markets, consolidated operations before expanding. Then working capital increased by 118 per cent, and declined by 42 per cent.

The subsequent period, 1981–4, showed gradual easing of the long-standing working capital deficiency in the large group (to 0.00 by 1984): management made adjustments to cope with low profitability, heavy debt load, and excess production capacity. The small and competitive groups improved performance early but by 1984 had evened out at levels that are reasonable for such a heavily capitalized industry (0.15 and 0.10 respectively).

Average collection period (days). Again distinct divisions appeared among plant types for 1974–84. Small companies tend to have the lowest collection period, and large companies the highest. Collection procedures and credit policies vary, as does the emphasis on cash as opposed to credit sales in the marketing mix. In general, as organizations grow larger, concern with day-to-day cash flow decreases in all of the above areas. Such a tendency is evident from these ratios.

Similar tendencies are visible during the crisis period, 1978–81. In the small group, average receivables increased by 89 per cent, compared to a 43 per cent decline in average sales per day, perhaps because of increasing reliance on credit sales or extension of terms of sale. The competitive and large groups remained relatively steady, with sales growth keeping pace with growth in receivables.

Recent trends indicate that only the competitive group has held its previous levels during 1981–4. Future financial improvements require internal changes. Lower levels are possible for all groups, and redoubled collection efforts and tightening of credit policies can facilitate recovery. Better management of working capital is essential for long-term improvements.

Days sales in inventory. Management of inventory levels was relatively efficient between 1974 and 1984. Small plants tended to be least efficient; competitive and large plants, relatively equal. Of course, more secure and established supply sources are available to larger operators, and so smaller plants require more stock relative to sales to maintain flexibility of operations and to meet demand. Even so, ratios for all three groups were far below target levels for the industry.

As discussed previously, variability of operating levels indicates inherent risk and sends out warning signals to investors and creditors alike. The small group varied greatly between 1978 and 1981, averaging significantly more time than the other groups (60.8 days v. 38.7 and 39.3 for competitive and large, respectively), mainly because of inability to reduce inventories carried. The large group became more efficient, while competitive companies adapted less quickly.

All three groups showed continued strong management from 1981 to 1984, all reducing levels from the end of the crisis. No long-term problems were evident. Performance was more efficient than target.

Equity position (Table 25)

Current / long-term / total debt ratio. From 1974 to 1984, the small group exhibited great variability, indicating the risk inherent for creditors and investors. The competitive group was relatively consistent – roughly comparable on average to the short/long-term debt-equity composition

TABLE 25
Analytic ratios in overall sample, 1978, 1981, 1974–84

	Small			Competitive			Large		
Equity position	1978	1981	1974–84	1978	1981	1974–84	1978	1981	1974–84
Current debt ratio	2.03	0.65	1.29	0.76	0.73	1.16	2.37	3.43	2.33
Long-term ratio	0.31	1.33	0.71	0.39	0.83	0.84	0.44	2.14	1.25
Total debt ratio	2.34	1.98	2.00	1.15	1.56	2.00	2.81	5.57	3.58
Equity (% net fixed assets)	0.05	0.47	0.58	0.86	0.88	0.77	0.69	0.31	0.45

of the small group. The large group was consistently higher in both short- and long-term composition. While variability was lower than that of the small group, greater debt relative to equity indicated more overall risk. In industries as inherently risky as fishing/processing the best capital structure would rely primarily on equity financing, because dividends are not fixed in amount or regular in payment. Long-term improvements are not possible until operations become consistently more profitable. Stable operations would ease internal cash flows, reduce reliance on debt, and improve perceptions about investment. This area poses the most significant problems and causes many of the other troubles in the industry.

Between 1978 and 1981, small plants relied more on long-term debt for financing assets and operations but less on short-term debt; total debt relative to equity declined. This trend indicates the strong earnings and greater access to equity investors. Greater reliance on long-term debt reflects increased investment in plant assets and has passed acceptable limits.

The competitive group used more long-term debt but stayed within target boundaries. Current debt levels give cause for concern, although current ratios indicate sufficient collateral.

For the large group, debt increased both absolutely and in relation to equity. Severe overexpansion of plant assets and poor earnings caused severe drains in cash flow and, given the shortage of equity funds, forced increased reliance on debt. In general, levels of long- and short-term debt are indicative of loan balances higher than desired and inadequate equity investment. The over-leveraged situation affects any assessment of risk of default.

A dramatic turnaround can be seen for 1981–4 for the large group.

Relative reliance on debt (versus equity) for financing decreased for current debt and for non-current debt, which began to approach target (0.89 by 1984). An infusion of equity funds in 1983, increased earnings retention, and easing of the trend to capital asset expansion all played a role.

Because of poorer earnings, both the small and particularly the competitive groups became steadily more reliant on debt. Debt levels substantially higher than targeted (1.11 and 2.26, respectively, by 1984), especially non-current debt, may present problems for years to come.

Equity as percentage of net fixed assets. Over the period 1974–84, the competitive group had the strongest equity base, and the large group the weakest. Results for all groups fell short of target, suggesting overexpansion beyond 'risk-free' sources. There is insufficient collateral for total debt and not enough security for shareholders' interest. Of course, such a situation harms risk perceptions for the industry and region.

Small and competitive groups consistently controlled expansion from 1978 to 1981, with little change in the relationships between equity and net fixed asset, which are quite close to the 1974–84 average. Large plants overexpanded, as shown by the marked decline in the ratio relationship, moving away from previous debt/equity proportions. Such a situation is self-perpetuating, as increased debt increases risk perceptions, which in turn make equity financing harder to obtain. This, in turn, increases reliance on debt. All three groups operated at dangerous levels, and all are deficient in this area, reflecting an industry with a poor equity base.

Small and large plants showed signs of recovery in the post-crisis period, 1981–4, with the large group getting back on par with the small group (0.64 and 0.63, respectively, in 1984). Stabilization of the equity base and slower expansion of net fixed assets contributed. The competitive group showed signed of decline (to 0.40 by 1984), probably because of lower profits and lack of equity funds; debt was used to finance a spurt in capital asset acquisitions.

Results of operations (Table 26)

Return on equity / return on total assets. The return provided on invested funds and assets employed tended to be highest in the small group

TABLE 26
Analytic ratios in overall sample, 1978, 1981, 1974–84

Results of operations	Small			Competitive			Large		
	1978	1981	1974–84	1978	1981	1974–84	1978	1981	1974–84
Return on equity	[0.22]	0.77	0.22	0.33	0.10	0.06	0.33	0.03	0.16
Return on total assets	[0.07]	0.22	0.08	0.12	0.03	0.03	0.06	0.00	0.04
Fixed asset turnover	2.98	2.45	3.63	5.35	5.48	7.06	7.27	4.75	5.10

over the years 1974–84, followed closely by large plants. While both groups yielded acceptable long-term returns (in comparison to target), annual variability (caused by seasonality, price fluctuations, reliance on export demand, and the like) would increase perceived risk for investors. The competitive group did not provide acceptable long-term returns.

The equity base for the sample and that for the industry were in poor condition. Returns on total assets for all three groups could not cover the funds used to finance their acquisition – equity (through dividends) and debt (through interest). The assets employed produced less profit than desired. The spread between return on equity and return on total assets is created by debt in the capital structure. The former return is lower than the latter, suggesting 'negative' leverage: the assets generated returns lower than the interest paid on debt used to finance them. The resulting adverse signals made it difficult to attract equity funds to the industry. Only several years of sustained profit could reverse this perception.

Competitive and large plants declined in profitability (by 61 per cent and 87 per cent, respectively) from 1978 to 1981. For the small group, returns improved consistently, indicating strong earnings. Comments made above regarding returns on total assets and negative leverage appear relevant for all three groups (observe that there is no increasing spread between the two ratios).

Trends set during the crisis continued for both competitive and large groups till 1984. The ratio reached negative levels by 1984, indicative of net losses in operations – a clear signal to the investment community. Small companies fared much better, although the upward trend levelled off and began to slide back down to more historic levels because of increased equity investment (perhaps reflecting relatively strong earnings during the crisis) and a gradual decline in earnings.

Turnover of fixed assets. In terms of ability to produce sales from net fixed assets during the years 1974–84, the competitive plants were most efficient, the small ones least so. The small group tended to be below desired levels, while competitive and large groups met or exceeded targets. A small organization would have greater difficulty in attracting off-season supplies of fish to fill lulls in production or shutdowns. The Task Force's study cites the recently high cost of financing asset acquisitions and the need to meet debt-service requirements regardless of whether or not processing is taking place.

Ratio results for 1978–81 were generally lower than for 1974–84, indicative of depressed selling prices and export demand. Significant decline among large plants reflected the overexpansion of plant assets beyond demand in a period of depressed markets. Results for the small and competitive groups were consistent with the overall 1974–84 period. Small plants were least efficient and below target, while competitive plants were more efficient and above target.

The small and competitive groups recovered well from the crisis to surpass targeted levels, especially the competitive group, which doubled its target in 1984. Recoveries in sales prices and demand for finished products helped greatly. The large group did not recover significantly despite apparent divestiture of a sizeable amount of plant assets; sales continued to slide, perhaps because of a shift in product emphasis or disinvestment in areas of more popular products.

Comparative income statement (Table 27). The cyclical average over the decade 1974–84 indicates the smaller organizations' efficiency in generating profits from operations, followed closely by the large and competitive groups. All three were within or close to target levels, despite noticeable variability year to year. Did 'good' years offset 'bad'? Over the long run, revenue tends to offset expenses so as to cancel out the negative information generated by year-to-year variability. The greatest differentiating features of the three groups over this period are gross margin on sales and general/administrative expenses.

Large plants tended to be more efficient producers, probably because of economics of scale created by large production volume and better access to competitive labour markets, greater control over supply sources

TABLE 27
Sample comparative income statement (percentage of sales), 1978, 1981, 1974–84

	Small			Competitive			Large		
	1978	1981	1974–84	1978	1981	1974–84	1978	1981	1974–84
Sales	100.0	100.0	100.0	100.0	100.0	100.0	100.0	100.0	100.0
Cost of sales	97.4	79.3	86.0	86.5	88.7	91.0	74.7	81.7	81.0
Gross margin	2.6	20.7	14.0	13.5	11.3	9.0	25.3	18.3	19.0
General/administrative	0.0	4.1	3.0	2.1	4.1	3.0	16.7	9.4	11.0
Interest	1.2	5.8	3.0	2.8	3.7	3.0	2.6	4.6	4.0
Depreciation	7.1	6.4	5.0	3.9	3.0	3.0	2.9	4.2	3.0
	8.3	16.3	11.0	8.8	10.8	9.0	22.2	18.2	18.0
Less: other income	2.0	10.2	1.0	1.0	1.0	1.0	0.1	0.1	1.0
Net operating costs	6.3	6.1	10.0	7.8	9.8	8.0	22.1	18.1	17.0
Net income	[3.7]	14.6	4.0	5.7	1.5	1.0	3.4	0.2	2.0

(vertical integration), and the ability to avoid seasonal product fluctuations because of a wider product base. However, higher general and administrative costs offset these benefits. The organizational effort, communication, and controls needed to tie together different plants are far greater than for operators with one or two plants near each other.

In comparison with the Task Force's target levels for the industry, small and competitive organizations were less efficient in terms of gross margin on sales (14.0 per cent and 9.0 per cent, respectively, against targets between 18.0 per cent and 20.8 per cent). However, they controlled general and administrative spending more effectively (3.0 per cent for both, against targets between 8.0 and 9.0 per cent). Depreciation expense was somewhat excessive for the small group (5.0 per cent, compared with targets between 1.7 and 2.5 per cent), suggesting excess capacity.

The small group was within acceptable profitability limits (4.0 per cent, against targets between 2.95 and 4.4 per cent); the competitive group, below (1.0 per cent). The large group was consistently within target levels, except in general administrative costs, which tended to push profitability slightly below target (2.0 per cent).

The crisis (1978–81) showed relatively strong earnings from the small group and declining earnings from the competitive and large groups. The small group greatly increased processing efficiency (gross margin) and 'other' income, which offset sizeable increases in interest costs and administrative costs. Depreciation, as percentage of sales, declined slightly, indicative of conservative expansion and more efficient use of existing facilities. The net result is a dramatic increase in profitability.

The crisis saw a drastic decline in profit levels for the competitive and large groups (from 5.7 per cent to 1.5 per cent and from 3.4 per cent to 0.2 per cent of sales, respectively) and decreases in gross margin (indicative of declines in labour productivity, depressed prices, and inflationary spirals of processing costs). Both groups were also hit by higher interest costs on increasing short- and long-term debt. The competitive group, similar to the small group, exhibited increased general/administrative costs and slight declines in depreciation charges. Large outfits showed greater depreciation charges, consistent with capital expansion and resultant excess capacity.

Operating results in the three-year post-crisis period indicate less profitability for all three groups, with the competitive and large groups suffering net losses in 1984. Small plants experienced substantial increases in general/administrative costs (to 14.2 per cent of sales in 1984), offset by slightly lower depreciation charges. The competitive group saw a drastic decline in gross margin (to 4.5 per cent in 1984) but decreased depreciation charges and easing of interest costs. The large group went through the same increases in general/administrative costs as the small group (to 18.8 per cent of sales in 1984), offset by lower depreciation charges and interest costs. These results for large operations reflect decreased acquisition of capital assets and reduced debt loads; sales do not seem to have rebounded as much as in the small and competitive groups.

Summary

While all three types of plants share broad similarities – high relative risk, poor capital structures, substantial excess capacity, relatively strong working capital management, and variable operating returns – each has its own distinct characteristics.

Small plants tended to vary most from year to year in most ratios, have the greatest excess capacity, and boast the best receivable collection records. These plants tended to be the worst at inventory management and at producing sales from investments in capital assets. Despite this, the group was the most efficient at turning profits on operations and achieved the highest returns on total assets employed and investments by owners.

Competitive plants tended to carry the best levels of working capital, often financed with long-term debt. They had the strongest equity base and were comparable to the small group in strength of capital structure, as well as being most efficient in generating sales from investments in capital assets. However, they managed production costs least well, which helps explain the lowest profits on sales and returns to shareholders.

In general, the large plants had the worst working capital position, the poorest cash flow, the worst equity base (and, in a connective fashion, the weakest capital structure), the greatest reliance on debt relative to equity funds, and the highest general and administrative expenses. On the positive side, again because of economies of scale, this group was the most efficient in production. This factor offset other higher costs to provide the second best level of profitability in the sample.

6 Supplying the US northeast

Richard Apostle
Gene Barrett
Leigh Mazany

Based on twenty-two in-depth interviews with fish brokers, processors, and distributors in the New York, Boston, and Maine areas in 1985, this chapter outlines some dimensions of the poorly understood market for fish products. This project's 1984 survey of fish plant managers (see Appendix to this volume) pointed out the significance of scale distinctions among processing operations in Nova Scotia for an understanding of Canadian marketing practices and, in particular, the fragmentation of the fish processing sector. The study of brokers revealed the diversity and complexity of the product market and the range of competitive and monopsonistic market segments.

This chapter explores relations between Nova Scotia processors and US distributors in various market segments: fresh fish, lobster, frozen and canned fish, and salt fish.

Sales and distribution

Nova Scotia fish is sold through a distribution network of brokers and commission agents, traders, value-added processors, and wholesalers/distributors.[1] Brokers and commission agents take fish on consignment and receive commissions on sales of the product (usually 3–7 per cent, depending on product and species). Traders take legal (though not necessarily physical) possession of the fish and make their money on the mark-up. This mark-up varies with terms agreed on with buyers regarding who pays for transportation, insurance, and so forth. Processors who fillet whole fish tend to buy through brokers, but processors who are taking frozen product to bread or to turn into frozen dinners usually buy directly from Canadian processors. Often Canadian processors will pack under a US processor's or even a distributor's label. Wholesalers/distributors can work on a commission basis or on mark-ups. They usually do not have their own brand name but market the processors' brands or fresh fish to chain stores, retailers, food service institutions, other processors, or other distributors.

Not having a sales force saves a processor costs in terms of salaries, physical overhead, and fringe benefits. Only the largest companies can afford sales personnel, and they rely on a mixture of the sales force and independent brokers/distributors. Using outside agencies for sales reduces control over sales, particularly when one is dealing with a broker as opposed to a trader or wholesaler. Brokers do not always have the incentive to push for that extra five cents a pound (7 per cent of an extra, say, $10,000, may not be worth the effort for the broker, although the extra $9,300 may be crucial for the processor). They may also be selling the products of the processor's competitors. A wholesaler or trader pays directly for the fish and takes possession of it (at least in a legal sense). Any problems in selling become the wholesaler's rather than the processor's.

The system used depends on species, product type, and historical accident. Fresh whole fish from Canada tend to be sold through brokers. Fresh fillets are usually sold directly to retail outlets or distributors. Frozen value-added product tends to be sold through a trader/distributor network or directly to chain stores or food service institutions. Salt fish is sold usually to traders, who then sell to distributors/wholesalers in the country of final sale.

Trade with the US northeast[2]

The Boston–New York axis absorbs a tremendous amount of North Atlantic fish from the United States, Canada, and Europe (in particular, Iceland, Norway, and some European Community countries). The Boston area channels fresh and frozen fish throughout New England and points south and west. New York handles salt fish going to the Caribbean and local New York markets and takes fresh and frozen fish for New York state consumption. Maine absorbs some fresh fish, which is then usually processed, and much of the live lobster. As we saw in chapter 3, many of these distribution systems have existed for thirty years or longer.

US and Canadian processors use the same distribution channels for some products, such as fresh fillets, but enter the market at different levels for other product types. For example, in the fresh whole fish market, US fishers sell to US processors. Most Canadian fishers sell to Canadian processors, who in turn sell primarily to US brokers (some may sell directly to US processors), who then sell to 'filleting houses.' In the frozen market, there is relatively little activity by US processors; Icelandic, Canadian, and Norwegian product dominates. US processors have tended to concentrate on the fresh fillet market, primarily because of the returns that can be earned, and so, depend on imports when domestic landings are low.

In 1985, fish product exports from Nova Scotia to the United States comprised 32 per cent (by value) of the province's total exports to that country.[3] Of this amount, slightly less than 17 per cent was exported whole (both fresh and frozen), 38 per cent as fillets and blocks (both fresh and frozen),[4] 6 per cent as 'preserved except canned,' less than 1 per cent in cans, and the rest (39 per cent) as 'other' (shellfish and salt fish), for a total value of approximately $381 million. This figure was up by 53 per cent over 1984, reflecting high prices in 1985 resulting from shortages in fish stocks (Canada, Statistics Canada, 1985).

The flow of fish products to New England often has a seasonal pattern.[5] Fresh whole cod is shipped to New England, primarily during winter (fourth and first quarters). In summer, fresh cod prices fall quite low, thus making it unprofitable to ship the cod fresh; instead, it is salted or frozen. In late fall and winter, prices for fresh cod rise as landings decrease and demand rises, and it becomes profitable to send cod to the fresh market. Exports of fresh whole haddock show more variability, because that species is usually not salted but is sold fresh or frozen. Variation over time probably reflects changes in supply, demand, and available processing alternatives.

Fresh cod fillets are exported primarily in the first quarter. In winter, there are few US cod fillets available: landings decrease because of weather. High demand for fillets – demand for fish is higher in winter, and Lent begins usually in February – is filled by fish caught by the large Canadian groundfish trawlers. Since 1977, exports of cod fillets have increased, suggesting more Canadian processing before export. US fillet production has also expanded. Thus rather than displacing domestic production, increased fillet exports are apparently a response to rising demand that cannot be filled by US supplies alone. Fresh fillets of haddock and flounder show less seasonality: neither species is salted, so supplies are not diverted during summer to salt fish. Also, in New England, haddock tends to be the preferred fresh fish, and so fluctuates less than for other species.

Frozen fillet exports have peak demand periods – Lent (first quarter) and when schools start (third quarter) – but excess demand can arise whenever bad weather keeps landings low. Fish dealers have reported that the Lenten and school seasons are starting to disappear, reflecting shifts to year-round consumption of fish products. Exports of blocks and slabs also do not show as much seasonality as for fresh: it can be stored; fresh fish cannot.

Live lobster is sold primarily through a system of dealers, each adding a mark-up. Comparatively speaking, most Canadian processors do not own pounds because of the costs and risks of ensuring healthy lobsters. It is primarily the distributors/wholesalers in Maine who own pounds. Exports of live lobster tend to fall into two categories: 'new-caught' and pound. 'New-caught' enters the market primarily in spring (up to May or June) and fall (from November on). Most Maine lobsters are caught from June through November. The high landings by both industries in May and June can cause substantial drops in lobster prices. The largest demand occurs in July because of tourists and the Canadian and US national holidays and December and January (the holiday season). Lobster supplies at this time tend to be from pounds, although, depending on weather, 'new-caught' may be available in December. The highest prices for lobster usually occur December through April, when supplies are scarce.

Salt fish refers to a wide range of products encompassing most species of groundfish.[6] We shall look at the four species – cod, cusk, hake, and pollock – that make up the largest share of Canadian production. There are two types of products: wet salt fish and dried salt fish. Wet salt or 'green' fish is produced by adding salt to a fish that has been 'dressed' and 'split.' The product will keep for up to one month without further processing. Sixteen per cent of Canada's export of salt fish in 1984 was green fish, 91 per cent of it produced in Newfoundland for export to Portugal. Dried salt fish is produced by mechanically or naturally drying the green salt fish. Salt-dried fish can be stored for up to eight months without refrigeration.

Three 'cures' dominate the dried Canadian product: heavy salted, light salted 'Newfoundland,' and light salted 'Gaspé.' The three differ in amount of salt used, drying techniques, and moisture content. Within the heavy salted cure (which has six subtypes according to moisture content), the 'ordinary,' from Nova Scotia, accounts for 56 per cent of all Canadian heavy salted exports by value, and the Newfoundland 'semi-dry' accounts for the rest. The Newfoundland light salted cure has three subtypes and accounts for approximately 50 per cent of light salted exports. The 'Gaspé' cure has two subtypes – hard dried and dried – and accounts for the remaining half.

The three Canadian cures are further graded. Each cure has an elaborate aesthetic classification: in the heavy salt cure, cod can be graded select, choice, standard, or commercial, and hake, cusk, or pollock is graded only choice, standard, or commercial; light 'Newfoundland' is graded choice, prime, madeira, thirds, West India, or Tomcod; and 'Gaspé' is graded select, choice, standard, or commercial. Five size categories (discussed below) are also used.

Heavy salted cod accounted for 37 per cent of total Canadian salt fish exports in 1984 (Canada, Statistics Canada, 1985). Twenty-four per cent of heavy salt cod is exported directly to Puerto Rico, and 24.5 per cent to the US mainland. The US International Trade Commission (USITC) estimates that the Puerto Rican market accounts for 90 per cent of all US consumption of heavy salted fish (USITC 1985). If this estimate is correct, then 44 per cent of Canadian heavy salt cod exports end up in Puerto Rico.[7] Salt cusk, hake and pollock represent a further 19 per cent of Canadian exports. While Jamaica is the prime export market for pollock, hake is consumed throughout the West Indies (excluding Puerto Rico). Ninety-two per cent of cusk exports are to the continental United States.

The salt fish products discussed so far – 'bone-in' salt fish – are semi-processed – split fish, scaled but with skin and bones intact. They are shipped either loosely in 50-pound boxes or wrapped individually in plastic bags and packed in 20-pound cartons. 'Boneless' salt fish is a highly specialized product produced largely in southwest Nova Scotia. All bones and skin are removed, and the split fish is filleted. Fillets are graded as fancy, choice, standard, or substandard and shipped in 30-pound boxes or packaged in one-pound cellopacks. Boneless salt fish represented 9.7 per cent of Canadian salt fish exports in 1984, and 88 per cent of the exports went to the United States.

Market segments

Fresh fish

The major fresh fish dealers in the Boston area are located near the fish pier, site of the Boston fish auction. Despite some stable relationships between Canadian processors and fresh fish buyers there, processors often move among buyers.[8] Boston dealers are in continuous telephone communication with Canadian suppliers.[9] Seven fresh fish and lobster dealers were interviewed. Six claimed at least one trip a year to Nova Scotia and, two of them make four or more. The exception is a broker who is major conduit for groundfish from Nova Scotia's major processor, National Sea Products, into New England. One dealer described what he did on his Nova Scotia trips as 'eating and bullshitting. I'm always looking for new suppliers, and I say "hi" to everyone. If I miss anyone, I hear about it.'

Approximately 16 per cent of fresh fish products (excluding lobster) exported to New England goes to Maine fish dealers.[10] Most of the fish goes to distributors, but some goes to processors for further processing. Fish distributors and processors are located primarily in the major ports (e.g., Rockland, Portland), although some, particularly lobster dealers, are in smaller ports. Maine dealers are more likely to have long-established contacts with Canadian suppliers, but most were open to new suppliers, and one said that he actively looked for new suppliers. All were in frequent telephone contact with suppliers. Maine dealers were less likely to travel to Nova Scotia; instead, usually the Nova Scotia processors would go to Maine. Boston and Maine dealers, as well as Nova Scotia processors, attend trade shows, such as the Seafood Exhibition in Boston; for some processors, these are an important source of new buyers.

Boston brokers have a variety of potential uses and outlets for fresh whole groundfish. They encourage as much business as possible and act as agents for processors, taking a commission on sales.[11] Fresh fillets have smaller markets, usually established prior to acquisition, and tend to be purchased at prices set by processors. The Boston auction is not regarded as much affecting prices received for Canadian groundfish.[12] Companies are aware of the sometimes substantial fluctuations in auction prices and will use them to set upper limits for Canadian fish, but not to determine specific prices. However, quality considerations frequently affect determination of prices for groundfish. Boston dealers say that they encounter inconsistent quality and have to renegotiate prices paid for material that is not of top quality. One dealer calls these problems a 'daily battle.' However, two Boston brokers who do not concentrate on fresh groundfish suggested that it is sometimes in the dealer's interest to overrate Canadian quality problems.

Maine dealers were more inclined to import whole fish from Nova Scotia to process into fillets. Their major complaints about quality concerned handling. Maine processors, however, tended to feel that they had more in common with their Nova Scotia counterparts than with Massachusetts processors, such as lack of perfect market information, especially when dealing with the Boston-based distribution network.

As for any defects in Canadian quality, dealers variously blame improper icing and handling by fishers, inadequate processing facilities, and mistreatment by American dealers. Two (of seven) dealers said that they no longer handle fish from Nova Scotia's largest processor because its sales department cannot describe the quality of material coming from its various plants.[13]

Differences in internal organization reflect broad variations in brokers' main product lines. Fresh fish companies tend to have low overhead, as reflected in relatively few salaried employees,[14] more casual record-keeping, and correspondingly little computerization.[15] Four of the eight brokers dealing predominantly or exclusively in fresh groundfish have just begun to use computers.[16] In sharp contrast, the six companies heavily or exclusively involved in frozen or canned fish use computers extensively, reflecting more intensive marketing and larger geographic markets. Dealers in fresh fish and lobster have more paternalistic relations with their employees and, with modest marketing efforts, tend to wear the more casual clothing typical of their suppliers.[17]

Lobster

Dealers in live lobster occupy an intermediate position in marketing operations, in both geographic range and organizational effort. While they are prepared to sell throughout the United States, demand in greater Boston and the northeast leads them to concentrate there. They do some checking on potential customers and attend regional foodshows around the country. However, potential customers do considerable comparison shopping, and dealers respond by trying to sell to more desirable types of customers. Dealers prefer wholesalers to retail outlets and restaurants, which are more difficult to check for credit and to collect from. By contrast with groundfish, lobster prices are relatively unaffected by quality differences. As one dealer put it, 'They're either live or dead.' There are, however, stiff penalties for relatively high proportions of dead lobster, or 'culls,' and for harm to lobster development caused by immersion in fresh water.[18]

Frozen and canned fish

Distribution channels are more clearly defined, and hence more formal, for frozen fish, which is not perishable and is always in supply. Product is more likely to be sold through a company's own sales force than through a broker/distributor network.

The frozen sector will likely continue to be dominated by large, vertically-integrated companies. The market – institutional, military, food service, and retail sectors – is much larger than that for fresh fish. Most buyers are large companies that insist on knowing supplies months in advance and hence require year-round supply. Smaller companies may enter the foreign market, but usually to supply specialty niches.

As suggested above, many dealers in frozen and canned fish sell directly to retail outlets or to chain stores and food distributors in a larger geographic area. Maine dealers also sell frozen products to restaurants and the military and devote more time and resources to researching and establishing markets. Despite focusing on the American northeast, because of ethnic markets[19] and geographic proximity, most regard the entire continental United States as their marketing area. They do little direct advertising but carefully evaluate potential new customers. One broker of canned fish explains that he finds customers only through the National Food Brokers Association. Non-members are not unethical, but 'they are not as strong in their capabilities or education.' One frozen fish dealer attempts to serve as an intermediary between supplier and customer to maximize research and development possibilities.

Salt fish

While the trade in salt fish represents only about 8 per cent of total fish exports by value from the Atlantic coast, salt fish has been the backbone of the independent competitive and small-scale fish processing sector.[20] Most of these companies produce salt fish as a complementary line along with fresh fish and lobster. Because wet and dried salt fish stores well, with slight overhead costs, it protects smaller processors against seasonal fluctuations, uncertainties, or downturns in the fresh fish market. The market is generally price inelastic because of its traditional ethnic composition, and so salt fish remains a steady, predictable component of annual sales.[21]

Unlike fresh fish and lobster, the US salt fish market is dominated by four distributing companies, which have an average of fifteen employees each but volumes of sales of $8 million (US) or more.[22] All but one broker leased office and warehouse space; only one owned any trucks. The two largest owned key fish processing plants in Nova Scotia (discussed below). All brokers interviewed had sales agents in major continental US cities and Puerto Rico.

The continental US market for salt fish is centred in the ethnic neighbourhoods of large cities, notably New York. For three of the six largest brokers, New York accounts for 60 per cent of sales and is the subject of intense competition. This market appears to be worth about $22 million annually and is clearly segmented. At the 'high' end are Italians, Greeks, and Portuguese, very conscious of quality and willing to pay a premium for select and choice grades of both 'bone-in' and 'boneless' cod. While Italians and Greeks traditionally eat salt fish only two or three times a year during the Christmas season, it is a staple in the Portuguese diet. Canadian fish has lost ground to Norwegian fish in this market because of inconsistent quality.

In the mid-market range are two West Indian groups: Puerto Ricans and Cubans. Puerto Ricans, who eat salt fish daily, particularly like 'Gaspé' cured cod, cusk, and hake. Cubans, who consume less, eat 'boneless' cod two or three times a year. At the 'low' end of the market are other West Indian nationals: Jamaicans, Dominicans and Haitians. Jamaicans would prefer to eat 'Gaspé' cod but eat the cheaper 'bone-in' heavy salted hake. Dominicans and Haitians eat 'bone-in' heavy salted pollock. In general terms, hake and pollock sold for about a dollar less per pound than cod in 1985. Dominicans have replaced Puerto Ricans as dominant customers in the New York market in the last decade. This change reflects a general trend, whereby the US market for dried salt fish has shifted from the higher to the lower end.

Canadian loss of the more valuable high-quality market was stressed by brokers. The crux of the problem is inconsistency and poor grading. One dealer's complaints ranged from poor handling of fish on board (e.g. washing, bleeding, icing) to poor processing (e.g. scaling, discoloration, use of 'dragger' fish, and the like). Problems in fish classification are chronic, particularly for 'bone-in' fish. Government regulations differ according to species and cure. For example, a heavy- or light-salted fish measuring 405 millimetres would be classified 'small' if it were cod, medium if it were hake or cusk, and 'large' if it were pollock. A cod measuring 430 millimetres would be classified as 'small' if it were 'Gaspé' cure, but 'medium' if it were heavy or light 'Newfoundland' cure. Shipments of fish are only spot checked, and this has led to substantial abuses. Fifty-pound boxes marked as fish of one grade and size will be filled with fish of various sizes and grades. The Gaspé Consortium EGC is particularly vulnerable, since eighteen plants may contribute production to a given load of fish for export.

Apart from conflicts between brokers and processors, wholesalers and retailers have ongoing disputes over price and quality in the New York area. A number of Nova Scotia processors decided that margins were getting too small in the 1980s and circumvented traditional brokers, dealing either directly with wholesalers and retailers or through their own commissioned agents. However, this only exacerbated general problems over price and quality. In one Jamaican market, Barrett was shown a range of boneless cod packages from Nova Scotia. The package marketed by a Nova Scotia processor directly through a new Korean wholesaler had deteriorated because of poor refrigeration and storage.[23] The wholesaler offered the processor a higher margin, thereby bringing overall prices down. According to a Cuban broker, even at the low end of the market relative quality is increasingly an issue with consumers. As a result, the lower end has become subject to more intense rivalry and competition.

Competition and restructuring

Government assistance

The New England fishing industry complains about government assistance perceived to be given to the Canadian industry.[24] Such assistance is felt to result in an unfair cost advantage and to mean that Canadian processing companies 'don't act like businessmen': they do not need to be efficient and can price-cut without worrying about going out of business, since the government will bail them out.

Feelings are strongest at the harvesting level, where US fishers think that low-priced imports keep US ex-vessel prices low. US processors tend to have mixed feelings. They need Canadian fresh whole fish to fill demand, particularly of late, given the decline in US landings. But to the extent that they compete with fresh fillets from Canada, they resent 'unfair competition.' Brokers and distributors, in contrast, are not greatly concerned. Their job is to move product at the best price possible, and so it is irrelevant to them whether the supplying company received government assistance or not.

The restructuring of the large Canadian companies in 1983–4 was particularly irksome to US processors who had to compete against them. Because of the substantial volumes involved, large companies' lowering of price to move product could seriously disrupt the market. Lobster dealers were incensed at establishment of a Canadian lobster company in Massachusetts, funded by government money. They found that the company was paying higher-than-average wages (by about 50 per cent), luring workers away from existing firms, and selling lobster at a lower price than they felt they could sell for and still stay in business. They argued that the company could survive on these reduced margins only because of financial assistance from government.

Dealers who do business primarily or exclusively with smaller, more independent Nova Scotia operations claim that some suppliers have been adversely affected in Canada by government support for large processors. Further, they believe that the large firms operate in American markets in ways that both disrupt normal market conditions and hurt their own business. One dealer in fresh groundfish said that the 'big heads' get all the money; another, that the big companies, buffered from market forces by government subsidies, 'show a strong ability to lose money.' One frozen groundfish dealer summarized these sentiments: 'One thing that upsets us about the larger processors who get more money out of the government is that they don't have to be so careful in marketing their product. We're in the area, so we see this all the time. If they have too much of one thing, or they need a little money from the bank, they can dump some stuff. So there's competition there. And they have so much money, they can drive your prices up on the wharf for your raw material. We're basically getting it at both ends.'

Competition and tariffs

Not all of the dealers interviewed saw unfair competition.[25] Several mentioned the exchange rate. Because the market is in the United States, fish prices are normally quoted in US dollars. Thus the Canadian processor will often decide on the Canadian dollar price it needs and convert that to a US-dollar equivalent. Because of the premium enjoyed by the US dollar vis-à-vis the Canadian, the quoted US dollar price may not be as high as what a US processor is demanding. It was reiterated many times that if the Canadian dollar were at par, Canadian processors would have a hard time competing against US processors. Of course, the Canadian processor would like to receive more for its product if possible, but there are so many prices in the market, depending on quality, shelf life, cut, packaging, and the like, that the cost of finding out the highest going price may not be worth the extra return. It may be less costly to accept a reasonable return and deal with a few buyers than to comb the market for the highest price, particularly for fresh fish, where perishability moves the product.

The US industry, especially the harvesting sector, initiated counter-vailing duty petitions against fresh fish exports from Canada. One was filed on 5 August 1985, with a final ruling given on 29 April 1986. Initially, an ad valorem tariff of 5.82 per cent (down from a preliminary 6.85 per cent) was imposed on fresh whole groundfish and fresh groundfish fillets. In its final determination, the United States International Trade Commission (USITC) ruled to keep the tariff on fresh whole fish but to remove it on fresh fillets. Because prices have been high as a result of other factors, it is difficult to tell whether prices to US fishers have increased because of the tariff. Most dealers, however, suggested that US fishers have not seen any increase in price caused by the tariff. A customs broker suggested that US dealers were requiring Canadian processors to absorb the duty, contrary to US law, which requires importers to pay the duty. If this were the case, one would not expect the US ex-vessel price to rise by the amount of the tariff.

Not all Canadian processors had to pay the duty. Some made arrangements to split it with buyers; others told buyers that the latter would have to pay the duty or else not receive the fish. In a situation of excess demand, processors are more likely to get buyers to pay the duty, particularly those who have alternative methods of processing the fish and thus do not have to rely exclusively on the fresh fish market.

Prior to imposition of the countervailing duty, the tariff rate on fresh whole and fresh fillets was, depending on species and product line, less than two cents per pound. This duty was intended primarily to raise revenue for fisheries development. Most processors and dealers claimed that this tariff had no effect on the flow of product. Clearly, its imposition is likely most to affect small and competitive processors, which are disproportionately likely to ship whole fish rather than fillets.

Of more consequence to trans-border flow of fish products are non-trade barriers: Canadian boats may not unload at US ports, military contracts have to be filled only with fish caught on US boats, and a USDC grade A designation can be given only to fillets produced in the United States. Restrictions like these caused the large Canadian companies to set up processing plants in the United States.

Since their peak in 1983, US landings have declined by about 25 per cent, while per capita US fish consumption has increased by approximately 8 per cent,[26] suggesting excess US demand for fish. This situation raised the price of fresh fish to historic levels in the mid-1980's – so high that wholesalers and retailers were reporting consumers shifting to high-quality frozen fish, freshwater fish, and other protein sources, such as poultry. Even a 6 per cent ad valorem tariff could shift demand away from fresh fish.

Restructuring the market

The newly adversarial relationship between the Canadian and American industries has led to discussions on restructuring the fresh fish market so as to increase its efficiency.[27] Because of the perishability of the product and the unpredictability of supply, co-ordinating supply and demand is difficult. The US market for Atlantic fresh fish is, if not the whole nation, certainly a sizeable proportion of it.[28] Distances, perishability, variability in quality, and the large numbers of buyers and sellers prevent acquisition of complete and accurate information on prices and demands for species and product types. Both formal and informal contracts between buyers and sellers reduce the risk involved, and so prices do not always reflect actual supply and demand but can be distorted by minor disturbances in the market.

Uncertainty about continuity of supply,[29] quality, and shelf life creates a hierarchy of prices based on perceptions about a supplier's ability to solve these problems. In an effort to reduce these uncertainties, the Maine Fishermen's Cooperative Association (MFCA) suggested that the two industries work together (1) to equalize access to each other's markets, (2) to discuss access to resources, (3) to equalize competitive advantages, and (4) to develop together North American and foreign markets.[30] The MFCA is concerned with improving the quality of fish put on the market in such a way that higher-quality fish receives a premium. In addition, it helped establish the Portland display auction, which opened at the end of April 1986, in hope of producing prices that would reflect quality differences and reduce some of the current uncertainties.

Intervention by the USITC on behalf of a Puerto Rican salt fish company led to imposition of a series of duties against Canadian companies accused of selling heavy salt dried fish at prices below the cost of production (USITC 1986). While the specific action was somewhat anomalous, the general context was one of extreme competition, a surplus of cheap Canadian fish, and deteriorating profit margins for US companies. In other words, the intervention of the US government is not surprising.

The regulations produced a major realignment in the Canadian processing industry. First, more fish entered the market in boneless or 'semiboneless' form to avoid the duty on 'bone-in' fish. In 1985 a US broker purchased a large plant specifically for this purpose. The plant produces the largest volume of 'boneless' product in Canada. Should the subsidiary operate as a captive of its parent, it will depress the US highend market. The move was clearly designed to secure for the broker steady supplies of consistently high-quality 'boneless' product, and two longliners were constructed by the plant. The duty necessarily raised the floor price of 'bone-in' fish. Brokers noted increased production of lower-priced, 'bone-in' cusk, hake, and pollock to offset these higher duty costs.

Second, the duty affected the corporate sector of the Nova Scotia industry. A number of larger companies chose to open their books to the USITC. In response, they were assessed with duties lower than the weighted average levied against other firms. They came to occupy preferential positions in the US 'bone-in' salt fish market, which situation substantially increased subcontracting. To compete, small processors marketed through processors with a lower duty. American brokers not tied to one of these Canadian processors had to post substantial bonds to maintain their sources of supply. That is, they paid the duty for the smaller processors in order to get their fish. The processing industry was effectively restructured.

Unanswered questions for observers of the industry remain. Is the deteriorating position of Canadian salt fish in the US market a consequence of poor Canadian quality or of the low prices paid by American brokers? If processors received greater price incentives, would they be more quality conscious? If processors paid fishers better prices, would they receive better-quality raw material? Price incentives are important; the Norwegian case proves the point. However, the problem in the salt fish market is not so easily solved in Canada. Consumer taste being what it is, salt fish will always be less desirable than fresh and frozen fish in the US market. As consumers' incomes rise and traditional tastes decline, resulting shifts in preferences will only lessen demand for salt fish.

The price of fresh fish and, more important, of other meats, such as beef, poultry, and pork, ultimately sets a ceiling on the retail price of salt dried fish. It is difficult to offer a sufficient price incentive to processors for them to put their best 'hook and line' groundfish into salt production when the fresh price is better. At best, fluctuations in the fresh price may lead them to do so. The attempt by some processors to eliminate the middleman can be seen as a way of increasing margins on salt fish, particularly in a depressed market. However, on a piecemeal basis, this has not led to improvements in quality at the processing or the retail end.

A number of options come to mind. Possibly Nova Scotia producers (or even US brokers handling Canadian salt fish) could break out of the cycle by entering other markets – Italy, Portugal, Brazil, and Nigeria being the most lucrative. Or Nova Scotia processors could organize a collective marketing strategy – privately, as Gaspé producers have done, or under the auspices of a crown agency, as in Newfoundland. Either outcome is unlikely in the current climate, given the virtual disintegration of intraindustry co-operation and goodwill because of the fractious effects of the duty on the South West Nova Salt Fish Packers Association. Most salt fish processors are competitive-scale producers engaged in fresh and lobster production as well: salt fish is a minor, but important, sideline when the fresh fish market is soft, and the market must be maintained.

Because there is so little US frozen production, there is much less competition between the two nation's industries in this area. US processors active in the field have complained that large vertically integrated firms had no restrictions on them: if they lost money, the government would simply bail them out. They saw such firms as predatory and unscrupulous, cutting prices to increase their market share.

Conclusion

This chapter has addressed the relationship between fish processors in Nova Scotia and dealers in the US market. In explaining the structural differentiation in the Nova Scotia industry, proximity and ties to the US market are clearly important. This study has provided a glimpse into the complex nature of these ties at an organizational level, particularly as they affect competitive-scale Nova Scotia processors.

Two central features are associated with the operations of these processors in the product market. First, these processors maximize profit margins chiefly by entering three production segments that involve minimal overhead in processing equipment or storage capacity: fresh fish, lobster, and salt fish. Each segment is largely discrete, and the market is characterized by heavy competition. Thus processors minimize dependence on any single broker or market. Second, flexible production schedules and the ability to respond quickly to seasonal and market changes in any single segment make up the key to maintaining a steady rate of profit.

7 Captains and buyers

Richard Apostle
Gene Barrett

Growing interest in the impact of formal and informal economic ties between fishers and fish buyers has characterized orthodox and radical perspectives alike. Economists are concerned with the process underlying price determination, while political economists have assessed class fractionalization and dependence.

This chapter explores the dynamics of port market ties between direct producers and fish buyers to test the proposition that economic ties are a primary mechanism used by capital to subsume fishers. On the basis of 404 interviews with boat captains in Nova Scotia, it develops a tripartite vessel typology based on ownership, scale, and technology. It then assesses the number and type of buyers and the service and input ties between each boat type and capital, controlling for geographical location. It concludes that formal ownership ties are a useful measure of structural dependence in the port market, while informal economic ties are not.

Modernization of the coastal-zone fishery since 1945 has transformed informal economic links. Where they exist, they indicate the power of direct producers to get capital to assume a share of the risks of production. Monopsony power rests in capital's control over outputs – grading and pricing arrangements – and through reduction of costly and risk-prone input services to fishers. Direct producers are able to extract such services where capital's position in the port market is weakest. The final section of the chapter explores some of the implications of these findings for the analysis of class differentiation.

The port market: a conceptualization

The port market refers to the point of contact between fishing captains and fish buyers. Sellers confront buyers with a variety of products under a range of market conditions. These relations influence the price of fish – an income for the seller and a cost of production for the buyer. All attempts to conceptualize port market dynamics consider the relative power each party brings to bear in the transaction. The logical range of possibilities is described in Figure 2.

A distinction has to be made between two types of port market dynamics: those based on arm's-length transactions, where no vertical ties bind sellers and buyers, and those that are not at arm's length. Two types of imperfect competition based on horizontal integration characterize arm's-length dealings: oligopoly (few sellers and many buyers) and monopsony (many sellers and few buyers). In each case, the decision to sell or buy is constrained by the limited number of alternative outlets. Geographical location, mobility, and ease of exit and entry into alternative species/product segments define the strength of these constraints. Perfect competition is the intermediate condition: 'In a competitive system with many buyers and sellers with simultaneous access to the goods, with freedom of entry and exit and with no organizations, collusion, or firms large enough to dominate a substantial portion of the market, price will settle at the opportunity costs of buyers and sellers' (Steinberg 1984: 35).

Non–arm's-length transactions reflect vertical ties that constrain decision-making. The most formal constraint stems from ownership of vessels by processors or of plants by fishing captains: minority interests secure resource or market access; majority ownership creates captive pricing arrangements. Informal economic ties primarily affect producers through their need for services – such as berthing, storage, haulout, refit, loans or loan guarantees, and credit for supplies – and inputs – such as bait, supplies, fuel, and ice.

Social constraints on port market transactions arise primarily from friendship and family networks and ethnic, religious, or linguistic ties that bond rural communities. Additional social links that tie buyers and producers may stem from commitments to certain ideals, such as co-operativism, free competition, a populist commitment to the 'small guy,' or community solidarity.

Another form of constraint is related to political factors, such as patronage and clientism or, at a more formal level, the state's regulation

FIGURE 2
Port market dynamics

Arm's-length transactions

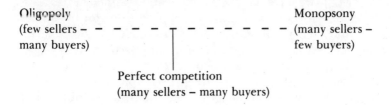

Oligopoly   Monopsony
(few sellers – – – – – – – – – – – – (many sellers –
many buyers)   few buyers)

Perfect competition
(many sellers – many buyers)

Non–arm's-length transactions

Economic ties
Formal (e.g. ownership and control)
Informal (e.g. services and inputs)

Social constraints (e.g. kith/kin relations, ideological commitment)

Political constraints (e.g. patronage, licenses/quotas)

of access and fishing effort. Licensing and quotas significantly affect the number of fishers and how much they catch.

This chapter explores the impact of formal and informal economic ties on port market relations in Nova Scotia. Chapters 8 and 9 present community-based studies of a number of other aspects of port market relations.

Theoretical perspectives

Two theoretical perspectives have emerged in the analysis of port market dynamics. For neoclassical economics, the central issue concerns the degree to which supply and demand influence fish prices. For political economy, the dynamics of fish prices and port markets reveal the fundamental class dynamic under capitalist production; price expresses the power dynamic between fishers and buyers.

Economists have been concerned primarily with the market imperfections that account for low fish prices, low producer incomes, and conflict between fishers and processors. Steinberg (1984) takes a fairly dualistic perspective, positing two distinct port markets in the Maritimes: offshore and inshore. Offshore vessels are company-owned, and so fish prices in this market are artificially 'administered' (1984: 34). Steinberg views the inshore sector as homogeneous in structure and largely without market power: 'The entire inshore fishery constitutes the port market, segmented by region, but not necessarily by species because, more often than not, buying cuts across fishery lines as fishermen themselves shift from one species to another to earn a living' (19).

Steinberg finds port market dynamics in the inshore fishery to be largely stable and based on contract sales. The buyer provides essential goods and services, inputs, and financing in return for exclusive sales by the fisher: 'buyers often finance vessel down-payments, provide the necessary equity in small fishing companies to aid fishermen in obtaining vessel loans, and cover the fishermen's overhead costs and the variable costs of operation, which the buyer may deduct from the value of the catch ... The stability of buyer-seller financial relations reinforces the stability of the price structure and rationalizes the market supply process' (19). Inshore fishers – undifferentiated price-takers – survive through mutually reinforcing, non–arm's-length ties to small buyers that are themselves price-takers.

According to Steinberg, the inshore market once had price stability and competition but now has monopsonistic imperfections.[1] The entire inshore fishery 'is a primary market that reflects the forces in the secondary market, but it is well insulated from their immediate effects by the dominance and resulting price-leadership of the major processors' (26). The central policy issue is to reintroduce balance into supply and demand so that prices again reflect market forces. While Steinberg explores a range of options, collective bargaining rights for fishers appear the most useful: 'Such a system would create the economic underpinnings for equalizing bargaining power in the port market by introducing monopoly conditions on the supply side to match the power of the major processors on the buying side' (41). 'Collective bargaining is a way to give each side the opportunity to be satisfied that they have got the best that could be obtained under the circumstances' (44).

The political economy model, in contrast, focuses on social differentiation in the fishery and its relationship to the larger socioeconomic structure of capitalism. Two recent contributions, by Sinclair (1985b) and Clement (1986), illustrate this perspective. Social scientists, in their concern with taxonomies of social class, have long been perplexed by fishers (McCay 1981; Sinclair 1985b: 14–30). Have fishers, like farmers, experienced complete or partial subsumption by capital, and, if so, what role do port market relations play in the process?

While taxonomies differ, most of the relevant variables used to classify fishers are threefold: (1) factors relating to scale, such as technology, cost, size of crew, and division of labour; (2) factors relating to the social organization of production, such as ownership and control of property and disposition of surplus; and (3) port market factors relating to input and output relations with capital. The logic of accumulation has driven capital toward centralization and concentration of economic power and gradual subsumption of independent producers,[2] leading to emergence of four class types: (1) a *fishing proletariat*: crew members working largely for wages on large, company-owned boats (Clement 1986: 63–4); (2) a *dependent producer* class: fishers on small-scale vessels who nominally own their own means of production but depend on capital for inputs and outputs (Clement 1986: 64, 67; Sinclair 1985b: 44–56, 118–19); (3) an *independent producer* class: small-boat fishers largely 'free of contractual obligations to capital' and who sell their catch on a competitive market (Clement 1986: 64, 71); and (4) a *petty capitalist* class: intermediate-scale boat captains whose social position is defined by an exploitative relationship to their crew, the scale and ownership of their vessel, and a dependent market relationship with capital (Clement 1986: 63–4, 78, 79–81; Sinclair 1985b: 95–9).

Paradoxically, the neoclassical and political economy perspectives share a couple of views. Each sees the port market as monopsonistic and stresses the informal economic ties that make fishers dependent on capital. Clement, for example, argues:

Most fishers face a port market that is a large processing company with intricate controls over the conditions of sale. The most obvious example is the crews of trawlers directly owned by the processors and whose 'prices' have no market other than labour market rate negotiations carried out with unions. Often when fishers are the nominal owners of their own boats, they are tied to particular processors through a variety of bonds such as loans and advances, servicing arrangements, or, quite simply, a buyer's monopoly ... there are, to be sure, some competitive port markets where prices are set in the interface between many producers and many buyers such as in the sale of ... some lobster on the East Coast. Overall, however, these are fairly minor markets. (1986: 49–50)

Sinclair (1985b: 99, 115–20) points to the multiplicity of capitals – boat builders, banks, insurance companies, marine supply companies – on which fishers come increasingly to depend as their technology becomes more complex: 'There is some pressure to sell to local processors rather than to companies which truck out fish, because the latter practice leads to reduced employment for relatives and friends in local fish plants. Other than that, the only grounds for choice would lie in the grading practices, unloading facilities and services provided by the various plants' (Sinclair 1985b: 123–4). Fishers are 'price-takers within the commodity production chain in which they operate (Sinclair 1985b: 99): 'Fishermen face a limited number of buyers at times when they have an abundance of fish to sell. Each fishing enterprise is too small in relation to the total catching capacity of the fleet to exert any influence on the price offered by the buyer' (124–6).

Like Steinberg, both Clement and Sinclair feel that the existence of small or intermediate fish processors not owned by the large companies has little significance. The logic of capital accumulation pits processors of all sizes against fishers in a structural contradiction. The price-taking, dependent position of smaller companies in a 'weak staple commodity' makes them inconsequential (Sinclair 1985b: 124–6; Clement 1986: 53). And, like Steinberg, both authors place unionization of fishers at the centre of any attempts to redress this power imbalance.

Research problems

As we saw in chapter 4, the research program informing this book is directed toward explaining the differentiated structure of capital in the fishery of Nova Scotia. This chapter explores some factors that may underlie the independence or dependence experienced by types of fishers and relates these back to the plant scale typology. The key appears to be adaptation to resource/product-specific niches. The terms 'price-leaders' and 'price-followers' deflect attention from the heterogeneous structure of the industry.[3] This project's plant survey revealed a more fruitful line of inquiry: the degree of participation by fishers and processors alike in species/product segments and the importance of each of their overall income/revenue position. Research carried out for this project has shown that the large corporate sector does not dominate many such segments. In some cases, such as herring roe, a smaller number of independent capitals might be characterized as 'price-leaders.' However, other segments, such as fresh fish, salt fish, and lobster, experience heavy competition. Both small and competitive processors and a majority of coastal-zone fishers participate in these segments.

The importance of formal and informal economic ties posed a significant question that this project set out to assess through a survey of boat captains selling to a number of the plants in the 1984 survey (see Appendix). The two prevailing perspectives tend to operationalize dependence in terms of these ties. What seems clear is an image of dependent relations between capital and producers rooted fundamentally in the struggle over the price of fish. This view has been extrapolated from examinations of the truck system in the mercantile and early industrial ages. As we saw in chapter 2, credit and debt dependence tied fishers quite profoundly to capital, allowing capital to manipulate prices to the point that fishers were kept impoverished (Wadel 1969; Brox 1972; Ommer 1981; Sinclair 1985b: 43–51). The perspective that such ties were disproportionately in capital's interest and were the primary mechanism used by capital to subsume fishers has carried over into contemporary views of informal economic ties. One proposition is clear from this work: greater monopsony in the port market increases the likelihood that producers are subsumed through formal and informal ties to capital.

Based on 404 interviews with boat captains, this chapter develops a boat typology differentiated along ownership, scale, and technological lines. It then assesses buying, service, and input ties between each boat type and capital.

Vessel typology

The class position of fishers in this typology depends ultimately on two factors: ownership of the vessel and the enterprise-level scale of production. Other factors, relating to the labour process, crew size, and the like, appear to be secondary, differentiating factors.[4] Whether or not fishing capital is independently owned by individuals who calculate profit and loss at the enterprise level affects fishing effort, marketing, and reinvestment. When fishing capital is owned and controlled by external actors, operational and management decisions are incorporated into the hierarchical structure of land-based capital, different accounting principles are used, and enterprise-level costs and revenues are internalized and reflect the consolidated position of capital as a whole.

Scale of production among independent boat owners forms a second distinguishing feature in this typology. Size of vessel and level of technology, gear, and so on indicate fixed capital investment and the scale of productive property. The important distinction here is based on scale, not on relative income or profitability. At issue is disposition of the surplus by boat owners: is it accumulated in productive property or dissipated through personal consumption? The resulting taxonomy is presented in Table 28 and distinguishes three types of fishing enterprises: a branch-plant sector, a petty-capital sector, and a direct-producer sector.

Branch-plant sector

Given the position of large capital in the industry, vertical ties to processors through ownership and control make offshore vessels branch-plant facilities, run by non-owning managers (captains) with a proletarianized work-force. While the labour process is defined largely by scale, unit-cost accounting and the accumulation and disposition of surplus are all exogenously controlled. Port market ties are therefore a function of ownership and are quite rigid. Fish prices bear no relationship to supply and demand; crew shares are negotiated 'piece' wages; and captains are highly paid functionaries with substantial productivity incentives but little latitude as to where and when to fish, for what species, where to dispose of it, or how much to sell it for. The largest offshore vessels are wholly owned subsidiaries of onshore capital. Vessel construction and operation costs and, more important, the largely reciprocal insecurities of large-scale demand and supply have created this situation. The same logic behind vertical integration of large offshore vessels affects smaller vessels. However, as we saw in chapter 4, such ties are much less necessary where processors have adequate access to fish and fishers have access to markets. The economic factors that press fishers and processors toward or away from such integration relate to the abundance of fish, particularly of certain species, quality demands, seasonal availability, and fish prices.

This branch-plant category included fifteen groundfish trawlers, three scallop draggers, two lobster boats, and one longliner, all over 95 feet in length. There were also five groundfish draggers, one scallop drag-

TABLE 28
Descriptive profile of boat types*

Boat type	Length (feet) Median	SD	Weight (tonnes) Median	SD	Value of fixed capital Median	SD	Number of crew Median	SD	(N)
Direct producers									
Lobster	31.6	7.3	7.8	5.1	$24,975	$30,666	1.62	0.81	(97)
Longline	36.3	6.4	11.6	6.8	$33,000	$46,051	1.71	1.00	(27)
Lobster/hand-line	36.5	5.9	12.4	6.1	$50,400	$48,595	2.19	1.00	(79)
Lobster/long-line/net	32.2	6.9	8.1	5.7	$29,800	$44,085	1.67	0.83	(121)
Total	32.5	7.0	9.6	5.9	$30,044	$43,229	1.79	0.91	(324)
Petty capital									
Inshore draggers	44.6	10.3	25.5	45.9	$235,000	$216,622	2.91	0.87	(27)
Inshore scallo-pers	60.5	4.0	64.0	48.4	$255,000	$136,784	3.06	1.07	(12)
Longliners/sei-ners	45.5	6.2	32.5	16.1	$225,000	$138,677	4.17	1.14	(12)
Total	47.7	9.8	48.5	43.6	$248,200	$179,368	3.29	1.12	(51)
Branch-plant vessels									
Intermediate vessels	54.5	10.7	57.5	30.1	$300,000	$155,930	2.90	1.13	(8)
Offshore vessels	150.0	37.7	633.5	307.6	$7,050,000	$4,145,570	15.60	3.85	(21)
Total	120.0	52.3	363.5	367.3	$2,050,000	$4,355,713	14.25	5.88	(29)

* Gear type classifications are based on gear type actually used in 1985 and not simply gear licences held. Multiple gear combinations are based on gear types that produced landings greater than 20 per cent of total landed value in 1985.

ger, one longliner, and one multiple-gear, small boat, all between 38 and 65 feet in length.

Petty-capital sector

The petty-capital sector is defined by independent, non-processor ownership of the vessel; intermediate-scale technology; and petty accumulation of capital by the owner on a unit basis. While crew size, division of labour, and authority structure may vary,[5] what is important is the level of surplus appropriation that occurs vis-à-vis the crew and how this profit is used. Scale takes the form primarily of intermediate-level industrial technology, often single-gear in nature, so that fishing is highly species-specific. Geographical mobility is this fleet's primary advantage. Differentiation may occur through varying levels of pro-ducer self-exploitation and reinvestment of profits through family-based enterprises, such as other boats (horizontal integration) on onshore handling and processing facilities (see chapter 9).[6] Such vessel owners may 'retire' from active fishing with a plurality of boats fishing for them (see chapter 8). Success depends on availability of fish and fish prices. Vessel construction and operating costs are relatively high and must always be covered on a unit basis. Prospects for transfer pricing and certain economies of scale may exist for the most successful family enterprises. However, the 'artisanate' origins of these owners make them preoccupied with whether the 'boat is making money.' This cat-egory included ten longliners over 42 feet in length, twenty-seven in-shore groundfish draggers, twelve inshore scallopers, and two herring seiners.

Direct-producer sector

Multiple-gear. The direct-producer sector is defined by independent ownership of the vessel and a small-scale fishing unit. Fishing vessels are labour-intensive, with small crews, high self-exploitation, and little differentiation between captain and crew. While mobility is constrained by the size of vessels, flexibility and adaptability allow pursuit of a variety of fisheries during different seasons and circumstances (see chapter 9). Combinations of line, trap, and net technologies are most typical. However, as we saw in chapter 3, the backbone of this sector in Nova Scotia is seasonal lobstering. Crew shares tend to be much higher than on larger vessels, and captains' incomes much closer to those of the crew (Barrett 1983). Some recent ethnographic accounts posit an 'image of the limited good' of extended effort and capital accumulation among these boat owners: disposition of surplus and di-version of effort for community- or socially defined purposes may pre-sent a community-based moral economy (A. Davis 1985; J.C. Scott 1976). This category included 79 fishers who used some combination of trap and line technology and 121 who used trap, line, and net gear.

Single-gear. In this sector, otherwise like the first, a particular fishery is pursued exclusively. The lobster fishery and the ground fishery offer this opportunity to direct producers in Nova Scotia. The survey covered ninety-seven lobster fishers and twenty-seven small-boat line fishers.

Fishers and buyers

Number and type of buyers

Table 29 distinguishes captains selling to one buyer and to more than one buyer. In overall terms, 52.4 per cent of the sample were in the first category, and 47.6 per cent in the second. Comparative percentages indicated a strong relation among scale, single gear/species segments, and number of buyers. With the exception of single-gear lobster boats, all vessels using just one gear type were much likely to sell to just one buyer. Among such vessels, 76.9 per cent of small boat longliners, 59.3 per cent of inshore groundfish draggers, 75.0 per cent of inshore scallop draggers, and 66.7 per cent of petty-capital longliners and herring seiners sold to just one buyer. While gear combinations varied, lobster seemed to allow most small-boat fishers to sell to more than one buyer.

For branch plant vessels, 76.9 per cent sold to just one large-scale plant; 65 per cent of these were located in Lunenburg/Queens (see Map 5). Nearly 53 per cent of single-gear small-boat longliners sold to just one competitive plant, and 42.1 per cent sold to just one large plant. Fifty per cent of inshore draggers and 37.5 per cent of petty-capital longliners and seiners sold to just one small plant. Half of these inshore draggers were located in Annapolis, Digby, and Yarmouth, while all of the longliners and seiners were in Shelburne.

Lobster seemed to allow most small-boat fishers to sell to more than one buyer. For the majority of multi-gear boats, lobster was sold to one buyer only, groundfish or pelagic species to others. Single-gear boats having more than one buyer had, by definition, a competitive relationship to the port market – particularly lobster fishers and, to a lesser degree, small-boat longliners in the southwest. Nearly 45 per cent of multiple-gear boats and 6.7 per cent of single-gear boats in Shelburne County, having multiple buyers, sold to competitive capital as their first most important buyer. One-third of multiple-gear boats

TABLE 29
Boat typology by number of buyers* (percentages)

Boat type	Number of buyers		
	One only	Multiple	(N)
Direct producers			
Lobster	49.5	50.5	(97)
Longline/handline	76.9	23.1	(26)
Lobster/longline/handline	38.5	61.5	(78)
Lobster/longline/net	44.5	55.5	(119)
Petty capital			
Inshore draggers	59.3	40.7	(27)
Inshore scallopers	75.0	25.0	(12)
Longliners/seiners	66.7	33.3	(12)
Branch plant			
All vessels	89.3	10.7	(28)
Total vessels	52.4	47.6	(399)

* The chi-square value of 35.12 is significant at the 0.001 level.

and 54.5 per cent of single-gear boats in this area sold next to competitive capital.

These patterns confirm those developed in chapters 3 and 4 concerning locational and scale variations in the relation between harvesting and processing sectors. In the data provided by boat captains, two anomalies went against the general pattern based on scale but confirmed the locational trends. Most multiple-gear boats had multiple buyers. However, in Lunenburg/Queens, 66 per cent of fishers sold in single-buyer port markets; 77.7 per cent of these vessels sold to large capital. The reasons for this are explored below.

The second anomaly characterized the eastern shore. As we have seen, most lobster fishers in the province had multiple buyers. However, 72.7 per cent of lobster fishers in Halifax and Guysborough counties sold to just one buyer; in 95.8 per cent of the cases this single buyer was large. This pattern also characterized multiple-gear, line, trap, and net fishers on the eastern shore, where 38.2 per cent sold to single buyers. In 80.8 per cent of these latter cases, the buyer was large capital. Again, this locational specification is explored below.

MAP 5 Boat typology zones

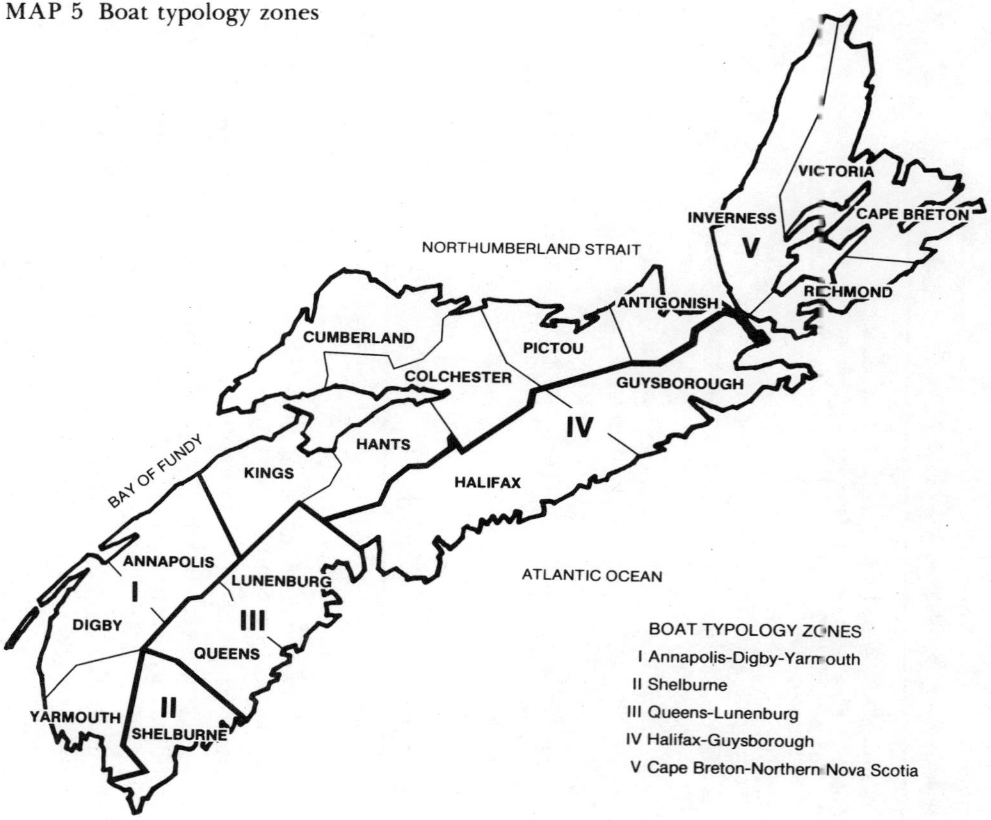
VICTORIA
INVERNESS
CAPE BRETON
V
NORTHUMBERLAND STRAIT
RICHMOND
ANTIGONISH
CUMBERLAND
PICTOU
COLCHESTER
GUYSBOROUGH
IV
HANTS
BAY OF FUNDY
KINGS
HALIFAX
ANNAPOLIS
LUNENBURG
I
DIGBY
III
QUEENS
ATLANTIC OCEAN
II
YARMOUTH
SHELBURNE
BOAT TYPOLOGY ZONES
I Annapolis-Digby-Yarmouth
II Shelburne
III Queens-Lunenburg
IV Halifax-Guysborough
V Cape Breton-Northern Nova Scotia

Informal economic ties

As mentioned above, informal economic ties between fishers and buyers affect port market dynamics. The survey asked respondents about such ties on eleven different measures, including berthing, refit and maintenance; haulout; purchases of fuel, bait, or ice; loans or loan guarantees, bookkeeping, and bait shed space. Two forms of linkages seemed important: service ties and input ties. In what follows, each of these two dimensions is explored in a number of port market settings.

For service ties, the sample split roughly in half: captains sold to one buyer or more than one buyer. Table 30 illustrates a strength-of-link continuum from none to strong.

Three patterns are salient among the single-buyer-only subset. First, vessels using longline gear, irrespective of scale or use of other gear, were moderately to highly service-linked to their only buyer. Second, the boats showing the lowest service ties were inshore scallopers, small boats, multiple-gear boats, and lobster boats. Third, branch-plant vessels had the highest service index scores, reflecting the scale of servicing required and the significance of vertical integration.

Captains able to sell to another buyer were distinctly less tied through service to their primary buyer.[7] Reduced ties appeared most clearly in service links between boats with multiple buyers and their second biggest buyer: nearly 30 per cent were completely independent.

The factor service index (Table 31) measures four items that showed a high degree of association:[8] berthing, haulout, refit and maintenance, and co-signing of loans. The scale of production affected the degree to which boats might become tied to particular processors for these services. Eighty-nine per cent of the company-owned fleet and 56.6 per cent of the inshore dragger fleet had such ties. The company-owned fleet was concentrated in Lunenburg/Queens and Annapolis/Digby/Yarmouth, while some service-linked inshore draggers were concentrated in Annapolis/Digby/Yarmouth.[9] The latter region also displayed high linkage for the small-boat, multiple-gear fleet; 50 per cent scored high, considerably higher than the provincial average for this group.

Two separate trends seem to underlie these patterns. First, capital internalized the high costs of maintenance, refit, haulout, and berthing associated with the large-scale operations of the offshore fleet.[10] Second, the extreme west (Annapolis/Digby/Yarmouth) had extensive

TABLE 30 Boat typology by summed service index* (percentages, with numbers in parentheses)

| | Ties to first important buyer | | | | | | | | | | Ties to second most important buyer | | | | |
| | One buyer only† | | | | Multiple buyers‡ | | | | | Multiple buyers§ | | | | |
Boat type	None	Weak	Mod.	Strong (N)	None	Weak	Mod.	Strong (N)		None	Weak	Mod.	Strong (N)
Direct producers													
Lobster	14.6	37.5	37.5	10.4 (48)	18.4	24.5	42.8	14.3 (49)		10.2	46.9	36.7	6.3 (49)
Longline/hand-line	5.0	20.0	50.0	25.0 (20)	16.7	0.0	66.7	16.7 (6)		66.7	0.0	33.3	0.0 (6)
Lobster/long-line/handline	6.7	16.7	43.4	33.4 (30)	12.5	27.1	43.7	16.7 (48)		26.2	42.9	21.4	9.5 (42)
Lobster/long-line/net	15.1	37.7	35.8	22.3 (53)	16.7	37.9	45.4	0.0 (66)		42.9	47.6	9.5	0.0 (63)
Petty capital													
Inshore drag-gers	0.0	12.5	56.4	31.3 (16)	9.1	18.2	54.6	18.2 (11)		22.2	33.3	44.5	0.0 (9)
Inshore scallo-pers	11.1	44.4	33.3	11.1 (9)	0.0	66.7	33.3	0.0 (3)		0.0	50.0	50.0	0.0 (2)
Longliners/seiners	0.0	12.5	50.0	37.5 (8)	0.0	50.0	25.0	25.0 (4)		50.0	25.0	25.0	0.0 (4)
Branch plant													
All	8.0	4.0	16.0	72.0 (25)	0.0	33.3	0.0	66.7 (3)		0.0	0.0	50.0	50.0 (2)
Total vessels	10.0	26.3	38.2	25.3 (209)	14.7	30.0	44.1	11.0 (190)		28.8	42.9	23.7	4.5 (177)

* Constructed on the basis of affirmative responses to eleven service-related measures. The four categories of responses were none (negative responses on 11 items), weak (positive responses on 1 or 2 items), moderate (positive responses on 3 to 5 items), and strong (positive responses on 6 to 11 items). The questions asked were: '61. Did the buyer provide any of the following services to you last year?

a. Birthing priority (no/yes)   d. Refit/maintenance (no/yes)   g. Ice (no/yes)   j. Co-sign loans (no/yes)
b. Unloading boat (no/yes)   e. Bait shed space (no/yes)   h. Trucking (no/yes)   k. Other (no/yes)'
c. Haul out (no/yes)   f. Refrigeration (no/yes)   i. Bookkeeping (no/yes)

† The chi-square value of 51.95 is significant at the 0.001 level when one collapses the longliner categories, on the one hand, and the petty capital categories, on the other, for cell number considerations.

‡ The chi-square value of 17.37 is significant at the 0.043 level, when one collapses the longliner categories, on the one hand, and the petty capital and branch plant categories, on the other, for cell number considerations.

TABLE 31

Boat typology by factor service index* by area (percentages, with numbers in parentheses)

Boat type	Annapolis/ Digby/ Yarmouth	Shelburne	Lunen- burg/ Queens	Halifax/ Guysbor- ough	Cape Bre- ton, eastern and north- ern Nova Scotia	Province	Total†
Direct producers							
Lobster	38.5 (13)	33.3 (21)	66.7 (3)	3.0 (33)	9.1 (11)	46.7 (15)	24.7 (96)
Longline/handline	0.0 (1)	53.8 (13)	66.7 (6)	0.0 (6)	–	–	42.3 (26)
Lobster/longline/hand-line	50.0 (4)	50.0 (52)	50.0 (4)	0.0 (10)	40.0 (5)	0.0 (4)	40.5 (79)
Lobster/longline/net	50.0 (18)	30.0 (10)	30.8 (13)	12.1 (58)	6.3 (16)	0.0 (3)	21.0 (118)
Petty capital							
Inshore draggers	72.7 (11)	42.9 (7)	0.0 (2)	0.0 (1)	66.7 (6)	–	56.6 (27)
Inshore scallopers	20.0 (10)	–	0.0 (1)	–	–	–	25.0 (11)
Longliners/seiners	100.0 (1)	55.6 (9)	0.0 (1)	0.0 (1)	–	–	50.0 (12)
Branch plant							
All	87.5 (8)	100.0 (3)	85.7 (14)	100.0 (3)	–	–	89.3 (28)
Total vessels	51.5 (66)	47.0 (115)	54.5 (44)	9.8 (112)	21.1 (38)	31.8 (22)	35.2 (397)

* This measure was constructed through a combination of factor and reliability analysis. The factor service index consists of the summed scores on the following questions: '61. Did the buyer provide any of the following services to you last year?

a. Berthing priority (No/Yes) d. Refit/maintenance (No/Yes)
b. Haul out (No/Yes) j. Co-sign loans (No/Yes)'

The index is scored from 0 to 4, with a score of 4 representing four yes answers. The percentages represent high scores on the index (1 to 4). The Cronbach's alpha value for this index is 0.87.

† The chi-square value of 59.04 for the 'Total' subtotal is significant at the 0.001 level. Although some of the other subtotals represent insignificant differences, each of the first and third area differences is statistically significant at the 0.05 level.

service ties to capital among the independently owned inshore dragger and small-boat, multiple-gear fleet. This trend is discussed further below.

For input ties, the index (Table 32) measures three items that showed a high degree of association:[11] purchases of bait, of fuel, and of supplies. In sharp contrast to the service index, this index revealed how the small-boat fleet was linked primarily to buyers. Boats engaged in the lobster fishery, either on a single- or multiple-gear basis, scored highest. Longliners, both small and intermediate, scored high as well. These trends reflected the importance of bait and supplies generally and of credit to cover outfitting costs at the beginning of each season.[12] A number of regional specifications are worthy of note. First, lobster boats most likely to have a competitive relationship to the market – that is, selling to multiple buyers in the Annapolis/Digby/Yarmouth and the Cape Breton east/north zones – showed the highest input linkage. Multiple-gear small boats with multiple buyers in Annapolis/Digby/Yarmouth also show very high input links to their first buyer. Second, the vessels with the lowest input index scores were lobster boats in Halifax/Guysborough and small, multiple-gear boats in Lunenburg/Queens. Third, longliner input ties were strongly associated with single-buyer connections, regionally concentrated in Shelburne County.[13] Port market evidence mentioned earlier indicated above-average single-buyer dependence.

Obtaining input and services

Vessel scale and catch technology had the most obvious impact on service and input ties to capital. Generally speaking, the larger the vessel, the greater its need for costly, professional servicing. Most vessels over 95 feet were vertically integrated into the land-based operations of fish processors and dependent on them for a number of services. Longline technology involved substantial land-based supports, such as bait freezers, shed space, and bait. Inshore scallop boat captains enjoyed considerable independence. While for scallopers, vessel scale and catch technology were clearly industrialized and highly specialized, geographical mobility and a luxury market sustained extremely high incomes – and distance from buyers in terms of servicing and inputs.

The remaining boat types were differentiated by geographical location. Small lobster and multiple-gear boats in Lunenburg/Queens

TABLE 32
Boat typology by factor input index* by area (percentages, with numbers in parentheses)

Boat type	Annapolis/ Digby/ Yarmouth	Shelburne	Lunen- burg/ Queens	Halifax/ Guysbor- ough	Cape Bre- ton, eastern and north- ern Nova Scotia	Province	Total†
Direct producers							
Lobster	84.6 (13)	61.9 (21)	0.0 (3)	28.1 (32)	90.9 (11)	100.0 (15)	61.5 (95)
Longline/handline	0.0 (1)	69.2 (13)	50.0 (6)	33.3 (6)	–	–	53.8 (26)
Lobster/longline/hand-line	75.0 (4)	82.7 (52)	25.0 (4)	20.0 (10)	100.0 (5)	25.0 (4)	69.6 (79)
Lobster/longline/net	83.3 (18)	40.0 (10)	46.2 (13)	44.8 (58)	93.8 (16)	66.7 (3)	57.1 (118)
Petty capital							
Inshore draggers	27.3 (11)	33.3 (6)	0.0 (2)	0.0 (1)	66.7 (6)	–	34.6 (26)
Inshore scallopers	0.0 (10)	–	0.0 (1)	–	–	–	0.0 (11)
Longliners/seiners	100.0 (1)	55.6 (9)	0.0 (1)	100.0 (1)	–	–	58.3 (12)
Branch plant							
All	42.9 (7)	0.0 (1)	16.7 (12)	0.0 (2)	–	–	22.7 (22)
Total vessels	55.4 (65)	67.9 (112)	28.6 (42)	36.4 (110)	89.5 (38)	81.8 (22)	55.4 (389)

* This measure was constructed through a combination of factor and reliability analysis. The factor service index consists of the summed scores on the following questions: '62. Did you purchase bait from this buyer last year? (No/Yes) / 63. Did you purchase fuel from this buyer last year? (No/Yes) / 64a. Did you have an account with this buyer last year? (No/Yes)' The index is scored from 0 to 3, with a score of 3 representing three yes answers. The percentages represent high scores on the index (2 and 3). The Cronbach's alpha value for this index is 0.75.
† The chi-square value of 37.05 for the 'Total' subtotal is significant at the 0.001 level. Although some of the other subtotals represent insignificant differences, each of the three area differences discussed on page 289 are statistically significant at the 0.001 level.

and Halifax/Guysborough were least linked to buyers, in spite of high dependence on single-buyer port markets. By contrast, small lobster and inshore multiple-gear boats in the southwest, especially Annapolis/ Digby/Yarmouth, were relatively strongly linked, despite a predominance of multiple-buyer port markets.

The analysis of buyer plant scale sheds light on service and input linkage. The connection between large capital and the branch-plant-vessel sector has been well documented. The strong service linkage was a function of economic necessity, not of a power relation (as we saw with the independent-fishing-vessel sector). Otherwise, service and input ties between large capital and the independent fleet fall into two types: those where capital has sufficient monopsony power to ensure adequate supplies of fish and those where it has to compete for fish. In each case location was the key variable. On the eastern shore, as already mentioned, large capital has a monopsony relationship to small-boat and single- and multiple-gear fishers. In each case, these fishers have the fewest service and input ties to capital in the province. In the southwest, however, large capital had high service and input ties to small-boat lobster/longline boats, to intermediate longline and herring seine vessels, and, to a lesser degree, to inshore draggers. Competitive capital was largely responsible for the high input links with the small single- and multiple-gear boat fleet. Lobster boats had the highest input ties.

Small capital's ties to the independent fleet were unusual. While one-quarter of direct producers sold fish to small capital as primary buyers, 43 per cent of petty capital sold to small capital – especially among inshore draggers. Fifty per cent of inshore draggers had high service ties to small capital, and 46 per cent had high input ties to small capital. Both figures were comparatively high and indicated 'desperation' on the part of small capital. This trend contrasts sharply with the weak relation between inshore draggers and competitive capital in Shelburne County. The surprising thrust of these patterns, particularly in light of the political economy perspective, lies in the underlying rationale for post-war informal economic ties. Services and inputs represent cost and risk for processors. Processors get involved in these activities primarily to get the fishers fish – especially lobster. The price of fish is largely independent of these factors. Credit for inputs or free services helps fishers offset operating costs or gearing-up costs. What then is happening in Nova Scotia?

Fishers in Halifax/Guysborough experienced substantial overall contraction during the 1970s. Stocks and landings, numbers of buyers, and prices and incomes dropped. Remaining large buyers became less interested in the overhead costs and risks associated with credit for supplies and services. In some locations buyers disappeared altogether. Such marginalization led many fishers to become quite independent by purchasing bait freezers and provisioning their own trucking, bait shed space, and the like.[14] In Lunenburg and Queens, the number of buyers fell dramatically. Unlike on the eastern shore, however, this has been a consequence of industrialization and of shrinkage in the reserve labour pool on which competitive- and small-fish plants depend (see chapter 10). Remaining are buyers relatively large and, with one or two exceptions, enjoy monoposony positions.

Southwest Nova Scotia, by contrast, has witnessed overall expansion of the fishery since the 1950s. By the 1980s, markets and prices for fresh fish, salt fish, and lobster remained strong, even though landings had started to decline. The numbers of independent buyers increased, and competition for fish became acute. The intermediate mobile fleet is moving further afield in search of groundfish. In some cases, fishers are landing their fish 500 miles away, in Cape Breton, and buyers are trucking iced fish back to the southwest for processing. Thus the need to secure dependable supplies of fish from the independent fleet is greater than ever. Buyers therefore are ready to provide whatever services and inputs are needed to attract fishers, except for the inshore dragger fleet landing in Shelburne County. Small and competitive processors in that area are involved primarily in the fresh and salt fish sectors, where demand for high-quality groundfish is great. 'Dragger' fish is of lower quality and is purchased mainly by large plants (for the frozen fish market) or by small plants in the Digby Neck region. Low demand and competition for dragger fish in Shelburne County make it difficult for dragger captains to get credit and services from fish processors. These captains are in a structural position similar to that of the small-boat fishers to the east.

Conclusion

This study has addressed a number of conceptual and theoretical issues. In support of the neoclassical position, it has found that the offshore fleet, of ships over 95 feet in length, is a distinctive entity, defined by formal and informal integration into onshore processing. Despite Steinberg's analysis, however, the 'inshore' sector is highly differentiated –

by scale, technology, location, enterprise-level social relations, and the nature of vertical ties to the processing sector. While Steinberg argues that these ties are widespread and underwrite a contractual pricing arrangement, we have seen that the ties are the opposite of what would have been expected: they are directly correlated with the demand for fish and the degree of competition among processors. The price of fish is not necessarily associated with such links.

In support of the political economy perspective, analysis of social differentiation in the fishery seems fruitful. The tripartite distinction among small-, intermediate-, and large-boat sectors, and the corresponding classifications based on domestic, petty capitalist, and proletarian relations of production, are also useful. It appears that the subsumption of direct producers by capital – the proletarianization process – has to be reconsidered. One extreme is largely understood: when capital owns the fishing boat, its crew are proletarianized. Offshore, this is done largely to control costs and to gain access to large volumes of fish. The other positions in the process are less well understood.

A number of points are worth considering in this regard. First, provision of inputs and services to fishers is an overhead cost to capital and has come to be avoided except where it is necessary to get the fishers catch. Where buyers are in a monopsony position, why assume this cost when fishers have no alternative outlets? Under the truck system, informal economic ties between fisher and merchant were a fact of life. Geographical isolation, monopsony control, and poverty made the merchant the only source of supply and left the fisher no choice. For the merchant, the risks and costs of provisioning services or inputs were relatively low and a consequence of poverty-level prices in the industry. Oligopolistic control of supplies ensured merchants a margin of security against fluctuations in resource or markets.

As we saw in chapters 2 and 3, post-1945 modernization changed these relations. The emergence of alternative outlets for fish, alternative service and input suppliers, state-based income support, and loan institutions improved the fisher's position. Fish became more important than imposing low prices. Fishers still needed services and inputs, but the terms had improved in their favour. Their dependence on particular buyers may have continued to enhance their short-run pricing power, but not necessarily so. More important, informal ties became a measure of the market power exercised by fishers under conditions of scarce supply and buoyant demand. As an unqualified index of proletarianization/dependence, therefore, they are an inadequate measure.

Second, capital strives to subsume direct producers completely only under very specific circumstances associated with securing access to fish. As we saw in chapter 1, direct ownership of vessels represents a substantial investment risk and is largely avoided by all but the largest companies. It is preferable to let direct producers assume risks. The weaker the competition in the port market, the greater will be capital's ability to control costs and increase its margin of profit. Monopsony power is therefore enhanced primarily through tight control over grading and pricing arrangements and through reduction in costly and risk-prone services to fishers.

Third, the position of independent fishers needs to be clarified. Inshore scallop fishers seem to embody everything in the term 'petty capitalist.' They are also independent in port market relations, though selling primarily to one buyer. This group is everything that other groups in the petty capitalist sector want to be, and more. However, prices for groundfish – both longline- and dragger-caught – and herring do not reach the same levels that scallops have. The difference, in monetary terms, represents the level of insurance that longline and dragger fishers need in inputs and services from capital to offset their weaker market position. So, in a sense, the structural-class proposition – that the fraction of petty capital most independent of capital is also the most independent in class terms – is correct. The mistake is to assume the converse: that the producer that is the most linked to capital for services and inputs is the most proletarianized.

A second twist to the notion of independence is represented by marginalized fishers, who are left to their own devices by capital. In the absence of buyers, fishers are also independent and thrown back on their own resources. This case is not unique historically, nor entirely exceptional. Independent producers with marginal input and production costs can remain viable, given access to the fishery and some outlets for their fish. While the independence of scallop boat captains was a function of access to the high incomes of a luxury market, that of marginal direct producers is a function of the subsumption of costs by the domestic unit of production.

8 Modernization in Digby Neck and the Islands

Anthony Davis
assisted by
Leonard Kasdan

As a man from Centreville exclaimed: 'Around here ya either fish, work with fish or hang around and throw rocks at gulls. That's all there is.' This sentiment reflects the dependence of the people and communities of Digby Neck and the Islands on fishing. The organization and characteristics of the fishing industry pulse through the annual cycle of social and economic activities in community and individual life, and changes in the industry have dramatically affected livelihoods. This chapter examines modernization of the fishing industry in the area since the 1940s and the impact of dragger technology and government policy on fishing communities there (see Map 6).

Government policy and industrialization

While small-boat, drag-net fishing in Digby Neck and the Islands began quite independent of government, its full-fledged development was clearly encouraged by federal and provincial policies. With the onset of the Second World War, the push to increase wartime food production spilled over into Atlantic Canada's fishing industry, which government saw as inefficient and archaic:

With the appointment of Ernest Bertrand (Laurier) as Minister of Fisheries in 1942, the modernization policy mushroomed with subsidies for dragger construction and schooner conversion to trawling gear. Such offshore development programs were supplemented by government sponsorship of an experimental inshore boat design – the longliner. The intent of these programs was to upgrade the efficiency of both the independent producer and the corporate fishery ...

MAP 6 Digby Neck and the Islands

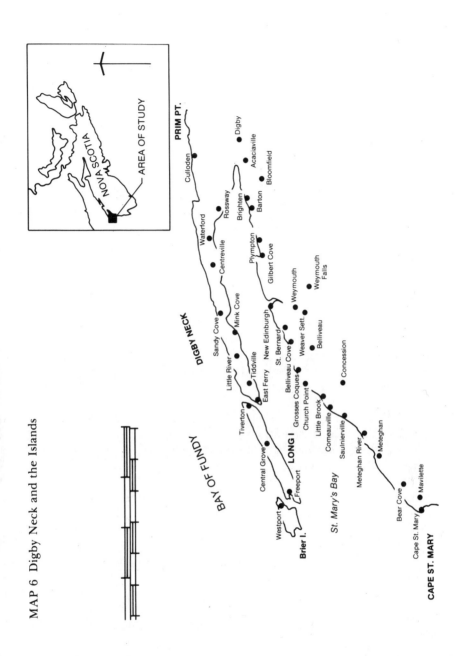

Stewart Bates' Report on the Atlantic Sea-Fishery for the Nova Scotia Royal Commission on Provincial Development and Rehabilitation marked the call-to-arms for the province. His analysis was essentially that underdevelopment in the fishing industry was the legacy of regulations such as the trawler restrictions of the 1930s, the chronically under capitalized primary and secondary production. With the growing prospects of substantial federal transfer funds to underwrite their activities, the province created the Fisheries Division of the Department of Trade and Industry to deal with the dual problem of encouraging large-scale development, and re-equipping smaller-scale fishermen. (Barrett 1984: 79)

Government financial assistance and policy picked up momentum during the next two decades:

The [Nova Scotia] Fisheries Loan Board, with federal assistance, launched a subsidy program for independently owned, medium-sized longliners and draggers. Between 1947 and 1960, 125 longliners and 34 draggers were built in Nova Scotia with such assistance. Federal subsidies for larger vessel construction averaged $7,732 per vessel between 1947 and 1965, a period in which approximately 32 vessels were added to the Nova Scotian fleet. A notable regulation passed in 1953 made federal subsidies contingent upon the affiliation of large trawlers with processing companies. (Barrett 1984: 79)

In 1970, a federal study (Mitchell and Frick 1970) of programs for vessel construction and equipment purchase loans and subsidies pointed to the impact of financial aid the intermediate-sized fleet, which contained small boat draggers:

Although it is difficult to assess the economic performance of vessels ... it has been found that intermediate-sized vessels are capable of realizing adequate returns on investment ... From recent studies on returns to large trawlers of 100 gross tons and over it seems that intermediate-sized vessels rank as the most efficient vessels in the off-shore fleet on the Atlantic Coast ... The subsidy program at the outset, by covering vessels from 55 to 65 feet, therefore promoted a desirable size of vessel. (50)

Between 1956 and 1964, federal and provincial governments provided an average of 84.0 per cent of the financing (17.0 per cent through federal boat-construction assistance and 67.0 per cent through provincial loan boards) for Atlantic Canadians purchasing 50- to 60-foot fish draggers, each costing around $50,000.[1] Fishers provided, as cash down payments, only about 16.0 per cent ($8,372 on average) of the purchase price – a substantial sum, however, given producers' questionable ability either to generate surplus income or to develop equity in a dismal fish-price environment.

These programs, as Mitchell and Frick showed, greatly accelerated the transformation of fishing efforts. By 1968–9, federal assistance for vessel construction had aided in the construction and acquisition of 285 vessels falling between 25 and 99.9 gross tons in Nova Scotia alone (36). Many of these were small boat fish draggers. In addition, alterations in government loan and subsidy programs during 1964–5 extended assistance to fishers purchasing boats down to 35 feet in length and under 25 gross tons, giving coastal-zone fishers ready access to financing. Between 1965 and 1968, 192 Nova Scotia boat purchases were assisted (39).

Similar initiatives were directed at the processing sector: 'Plant expansion and fleet expansion were mutually reinforcing; new plant capacity needed the assurance of a large and continual supply of fish; loan and subsidy assistance for the construction of fishing vessels made these easier to obtain, while new processing capacity ensured a port market for the catch of the additional vessels' (Mitchell and Frick 1970: 52). Paradoxically, economists criticized changes that now gave coastal-zone fishers access to the programs:

Only small numbers of ... inshore craft were subsidized under the Department of Fisheries program when the minimum length of vessels eligible was 45 feet. The inclusion in the program in 1964 of 'experimental' vessels down to 35 feet minimum length opened the flood gates, however; the term 'experimental' was liberally interpreted in practice and the provincial boards responsible for administering the program lost control. However, the main criticism would be levelled against the federal government authorities for introducing the amendment without due regard to its probable consequences and its incompatibility with the basic objectives of the Department's subsidy program. (35)

As long as assistance was restricted to intermediate- and large-sized vessels, the programs were considered consistent with rationalization. Coastal-zone fishers and their communities were considered to represent an archaic and economically inefficient from of production: 'By encouraging an increase in the number of inshore vessels the subsidy program contributed to the perpetuation and exacerbation of problems in the inshore fisheries. This is considered to be a major short-coming of the program ... The subsidy program, insofar as it facilitated the movement of inshore fishermen to offshore fishing by the use of larger vessels, has been beneficial' (Mitchell and Frick 1970: 50–1).

However, while these government programs allowed some fishers to modernize and industrialize, they often deepened producers' dependence on and obligation to fish buyers/processors. To cover their share of a vessel's purchase price, fishers often had to obtain from buyers co-signed bank loans, outright cash contributions, and letters of reference for lending institutions. Moreover, the large loan component of government assistance saddled individual fishers with a tremendous debt burden, motivating many to increase their fishing efforts dramatically. This could only deplete fish stocks and harm relations between fishers employing different technologies. Indeed, the frequently discussed conflict between fishers using different technologies resulted from this government intervention.[2] The following examination of the fisheries of Digby Neck and the Islands analyses this approach to fisheries development.

Modernization

Compared with the multi-purpose Cape Islander, small boat draggers are the 'new kids on the block.' The first draggers, late in the 1940s, were converted Cape Island–type fishing boats. By the late 1960s they had become the most cost-efficient and productive fishing boat in Atlantic Canada.[3] Moreover, they have helped to transform the Digby Neck and the Islands fisheries and, as a consequence, to the socioeconomic dilemmas currently confronting the industry.[4]

Until the late 1940s, the area's small-boat fisheries appeared little different from those in other parts of southwestern Nova Scotia. People fished hook and line, lobster trap, and net technologies out of owner-operator Cape Island–type boats. Crews were commonly kin-related and from the same communities. They mainly exploited seasonally available stocks of herring, various groundfish (especially haddock, cod, pollock, and hake), and lobster. The lobster fishery was the only major activity regulated by Ottawa. Regulations specified the season (i.e. late fall through spring) and saleable catches (i.e. minimum carapace length and no 'berried' females), while licences governed entry.

Most fishers sold to one buyer – to whom they were frequently connected by credit-debt relations. The buyer provided credit so that fishers could outfit. In return, the fishers, able to repay only once fishing, had to sell to debt-holding buyers. Moreover, prices during the mid-1940s were very low. For instance, several fishers reported prices per hundred pounds of $1.50 for cod, $2.00 for haddock, $0.50 for pollock, and $0.30 for hake, and $0.25–0.30 per pound for lobster. Similar prices were paid by buyers throughout Digby Neck and the Islands. Even fishers with some economic independence would be hard-pressed to realize any advantage from selling to buyers of their choice. Only consistently large catches would create meaningful income sur-pluses, over and above the costs of operating a fishing enterprise and supporting a household. Such conditions did provide substantial mo-tivation to improve productivity. Several men from one East Ferry family tried a foreign technology when the opportunity arose.

The German invasion of Norway in 1940 led many Norwegian fishers to emigrate to Canada. One such man who had worked on draggers for many years married the daughter of an East Ferry fishing family and settled in East Ferry. At first, he tried unsuccessfully to establish a herring processing facility at Mink Cove. Next he and a local fisher/buyer obtained a small boat and rigged it to go fish dragging. These tentative efforts demonstrated the potential of fish dragging to local fishers – especially his wife's cousins. As one of them observed: 'My brother and I got started about a month or so after that. Rigged a boat up and tried it out.'
  Unlike many local fishers, this family had few debt obligations to any one buyer. These men started fishing at an early age. By age fourteen they were longline and lobster fishing full-time in their fa-ther's gasoline-powered boat: 'He was basic to our start. He taught us all about the ground for lobstering and longline fishing. He showed us how to rig gear, how to use gear and how to repair and care for a boat.' Like other fathers, he had his sons as crew-apprentices. One son observed that, while his father had an account with a fish buyer, 'He never was one to borrow much. He believed that it's best to pay as you go.' Their father 'always sold around,' seeking the best prices available. They followed his practice after purchasing their own boats and gear. One son recalled:

We'd start in the spring fishing in St. Mary's Bay. Usually from about mid-

March to the last week of April. Then we'd begin fishing trawl in the Bay of
Fundy. We'd fish two tides every day. We'd go from before daylight to after
dark for at least six weeks. Usually by mid-June we'd be offshore in the Bay
of Fundy after hake. We used to stick at this until sometime around the end
of October. Then we'd go back to haddocking until the lobster season. If
there wasn't anything going in lobstering, we'd get right back to haddock
fishing. It was drive, drive, drive!

He had difficulty keeping new members: 'They never agreed with what
I did. They never wanted to go as hard as I wanted to go. Seemed
like I had a different man every year.'
   This man financed his first boat through an interest-free loan from
an East Ferry fish buyer. By the time he sold the vessel two years later,
the loan was paid and he had enough savings to reduce the buyer's
contribution to the next boat – a three-year-old, 38-foot–1-inch Cape
Island, powered by a new eight-cylinder Buick gasoline engine. He
rigged this boat for fish dragging. The Norwegian helped the brothers
rig the boat and gear, teaching them how to assemble the numerous
components of trawl gear. He taught them to handle and to fish a drag
net, to select appropriate ocean bottom, to keep the trawl doors properly placed, and to regulate the speed of tows and hauling back. In
return he got half the catch. These men learned quickly; soon the
catches were filling their holds. 'We'd drag all night and sell over to
Belliveau Cove. Then we'd drag all day and sell in East Ferry.' Most
crew members did not agree with the demands placed on them. Nonetheless, the brothers quickly established small boat fish dragging as a
permanent feature of the local fishery. They even helped other men
rig their boats and gear.
   At first, fish dragging was a summer activity, restricted to St Mary's
Bay. The rigged-over Cape Island boats were small and underpowered.
Only protected, flat-bottomed St Mary's Bay would permit such boats
to work drag nets: 'Then there was plenty of fish in St Mary's Bay –
flounder, haddock, cod, catfish. You could make a go of it there then.'
During the rest of the year the brothers still went longlining and lobster
fishing.
   In 1952 one of the brothers bought a 'state of the art' vessel for
$8,000. It was a 42-foot-long, 14-inch-wide Cape Island–type hull, out-
fitted with a six-cylinder GMC diesel on a reduction gear and equipped
with radar and sounding machines. The brother paid one-third down
and financed the rest with a government-subsidized loan. This boat

was designed to enable continued participation in the longline and lobster fisheries: 'We still weren't certain drag fishing was going to last. We wanted to be able to do other things in case it failed.'

But by the mid-1950s, fish dragging had emerged as an established form of small boat groundfishing. After 1954 this brother was fish dragging year around. He fished from St Mary's Bay south to Trinity Ledge from March until about mid-June. Thereafter, his effort shifted to the Bay of Fundy until the following March. He fished the Bay of Fundy shore from Digby to Brier Island.

The small boat dragger fleet developed gradually, unregulated until the late 1950s, when the first federal licences and regulations designed specifically for this activity began to appear. Early in 1956, a fishery officer reported only eight draggers fishing out of the area.[5] Seven of these were 40-foot boats, operating mainly in St Mary's Bay; but the eighth was one of the new, technically equipped, and powerfully driven 60-foot vessels. Its appearance signalled development of a specialized and mobile fishing machine able to exploit distant as well as nearshore grounds twelve months of the year. By late 1962 there were twenty-five draggers, many in the 40-foot class, but the shift to the larger, more technically proficient fish dragger was well under way. In January 1963 a fishery officer reported: 'The only difference [in the small boat dragger fleet] being three 40 ft. class draggers were replaced by 65 ft. vessels and one 54 ft. converted. There is a strong tendency toward 55 ft. and 65 ft. class draggers at this time but fishermen report that it is very difficult to get builders to accept their orders.'

By now the small boat dragger was a specialized technology with a distinctive hull and superstructure design – the Cape St Mary's type of vessel.[6] It contained the latest electronic aids – echo sounders (fish finders), radar, and Loran A (a navigation/location identification device). A new, completely outfitted, wooden-hull fish dragger cost between fifty and sixty thousand dollars during the early 1960s – a princely sum for men paid six cents a pound for haddock, four cents a pound for cod, and two cents a pound for pollock. The production potentials of such costly vessels, however, more than compensated for low fish prices.

The explosive growth in the dragger fleet since the late 1950s and its dramatic technical development were closely associated with changes in fish buying and processing and fishers' access to government financing and subsidies. Buyers were cultivating northeastern US markets for fresh fish, as well as moving processing away from salt fish and toward fresh fish fillets. Consequently, they helped develop fish catching techniques that would provide them with high-volume supplies of groundfish, especially haddock (the fish of preference for the fresh markets). In fact, the local fish processing industry was also undergoing structural and technical change. New federal and provincial programs offered fishers a combination of subsidies and low-interest loans to finance purchase of new boats and equipment. However, most fishers were able to make the transition only after processors became interested in high-volume catches.

Without question, government intervention facilitated development of the small boat dragger fleet and reorientation of fish processing, which consolidated and reorganized throughout the 1960s. A few new, highly capitalized plants employing fish-processing machinery and refrigeration technology replaced many operations producing salt fish. Salt fish gave way to fresh and frozen products, in large part because of small boat draggers.

During the 1940s and most of the 1950s the processing sector produced salt fish from groundfish species such as hake, cod, and pollock. Haddock was either turned over fresh for a commission to Maritime Fish (later National Sea Products) in Digby or trucked to the Boston fresh fish market. Buyers purchasing lobster sold mostly to northeastern US brokers and pound operators. Herring was sold to fishmeal and reduction plants.

In 1961, twenty-three fish firms operated in Digby Neck and the Islands, with all but one open ten to twelve months of the year. A senior federal fisheries officer in district 37 noted in 1962 that many processors were moving from salt fish to fresh and fresh frozen products (Narrative Reports). However, by the start of 1962, 'seventeen salt fish firms with thirty-four dryers still remain in operation; but, these are not producing the same quantities as a few years back' (Narrative Reports). Production of salt fish was declining by the early 1960s. Late in 1966 the same officer reported: 'In this District now there is a trend to larger and more modern vessels and ashore the trend is to more modern processing methods. A large percentage of all fish landings now are processed fresh and shipped out frozen. A very few years back salt fish was the main part of the industry' (Narrative Reports).

While retaining salt fish production, many firms moved into fresh and fresh frozen product lines. Late in 1969 this same officer noted seventeen fish firms, three of them specialized enterprises in business only during the lobster season. Between 1961 and 1969, eight year-round plants had either gone out of business or amalgamated with other firms (twenty-two full-time firms in 1961 were reduced to fourteen in 1969). By 1979, several more had either shut down early or gone out of business (Narrative Reports). Collapsing groundfish stocks forced marginal enterprises out of business and closed down viable operations until conditions improved.

With the productivity of small boat fish draggers already demonstrated, buyers/processors began to encourage acquisition of this technology. 'Independent' fishers' control of large-volume supply provided fishers with a significant lever in price negotiations. Recognizing this, several buyers were directly (through outright ownership) and indirectly (through financing agreements with fishers) acquiring fish draggers by the late 1950s. They thus secured their position and tapped into increased supplies of groundfish, especially haddock. Many fishers who moved into draggers now worked under much larger capital debt, but buyers were much more directly involved in boat acquisitions, as minority and sometimes majority owners. Buyers, especially those moving into fresh fish product lines that are dependent on large supplies, could remain as debt-obligation holders and, consequently, as price-givers.

These practices have remained common in Digby Neck and the Islands. Indeed, direct participation by buyers/processors has, if anything, increased since the 1960s. By 1984, every major processor owned at least one small boat dragger. Many owned two or more, while also holding minority positions in other vessels. Since processors are not given formal access to government programs for vessel acquisition, they have established independent boat companies. When asked about the advantages of boat ownership, most of the eight buyers interviewed replied that guaranteed supply was the primary attraction. There are also other advantages:

Owning boats gives us both continuity and assurance in our fish supplies. To begin with, this way we're guaranteed that our plant will have supply. But, this also lets us get involved in planning supply. We can direct captains to fish for certain fish. We can space fishing effort so that we don't find ourselves in a situation of too much oversupply or undersupply. By the way, owning boats also lets us have some control over the quality of fish landed. We insist that our captains take plenty of ice and that they ice the fish properly.

The people who work these boats receive shares of each trip's catch value after the processor has taken at least 45 per cent as a boat share. The processor's portion will reach as high as 60 per cent if it pays operating expenses such as fuel.

The relation between the growth of the small boat dragger fleet and fish processors' increasing equity position in vessels mirrors several fundamental shifts in product lines and markets. Markets for Atlantic Canadian salt fish slumped, if not collapsed, during the late 1950s and early 1960s, just as many processors in southwest Nova Scotia were diversifying into fresh and frozen products. Proximity and ease of access to the northeastern US market enabled them, before most others, to participate quickly and fully in the provision of fresh frozen fish, which, however, has different production requirements. Such participation usually involves contracts with and obligations toward Canadian and US buyers which must be satisfied regularly and on time and places greater emphasis on increasing capitalization, in the form of processing machinery (e.g. filleting and skinning machines) and freezers/coolers. Volume output is needed to pay for, and to use, such expanded capacity. Small boat dragger technology was ideally suited to the supply needs of the modernized fish plants in Digby Neck and the Islands. In short, the interests of the processing sector had become wedded to maintenance and future development of the fleet. Corresponding with this has been the dramatic decline, except in the lobster fishery, of the coastal-zone fisheries.

The coastal-zone fishery in decline

Coastal-zone fishers of Digby Neck and the Islands employ mainly Cape Island–type boats in pursuit of resources adjacent to their home ports (see Map 7). To these fishers, the marine environment is a complex system. Often the most pertinent constituent parts are named by local fishers. 'The Pinnacle,' 'the Rip,' 'the Deep Hole,' and 'Head and Horns' simply describe physical features. Other names, such as 'Sandy Cove Ground' and 'Whale Cove Ground,' are derived from their proximity to specific harbours. A few spots are named after the individual who 'discovered' the place, while others reflect the onshore 'mark' used as a location reference point.[7] Every name conveys to the fisher topographical features of the ocean bottom, water depth, location from port, availability of specific types of resources, accessibility at certain

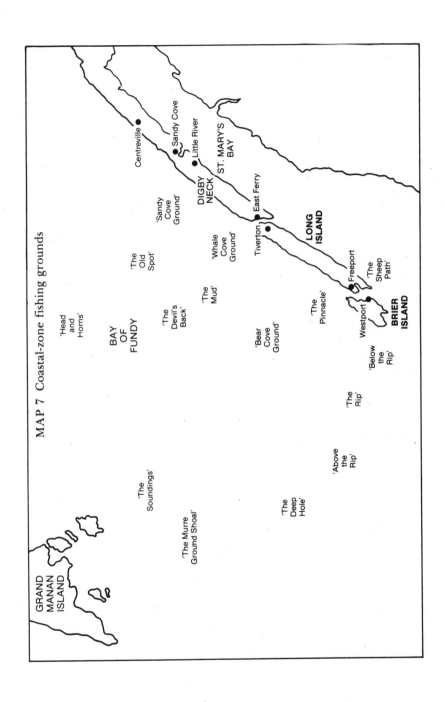

MAP 7 Coastal-zone fishing grounds

times of the year in specific weather conditions, and the biological profile of the ocean floor.

Fishers frequently exploit a portion of the coastal resource zone – 'our' fishing ground. Those fishing out of each harbour claim rights of first access in areas that have been continuously exploited by generations of fishers. Every harbour's fishers know and generally respect their own and others' boundaries, particularly in the high-value lobster fishery. An older Tiverton fisher recalled: 'We always fished lobster on ground to the sou'west in St. Mary's Bay and the Bay of Fundy. All along these shores of Long Island. Men from East Ferry go lobstering along the nor'east side of the Neck. Westporters fish down below and all around Briar Island while Freeporters lobster up the nor'est sides of Long Island. We all know where we lobster and look out if you edge into the other fellow's grounds.'

Harbour-focused property claims are also expressed for groundfish and herring, especially for fixed-gear types of fishing such as longline, herring net, and herring weir activity, which have more flexible boundaries. Often limits are defined simply in terms of the geographical area that can be exploited effectively in a day fishery, though producers from neighbouring harbours often exploit the same grounds. So, property claims in the groundfish and herring fisheries are expressed more generally and reflect interests, activities, and limitations shared by neighbouring harbours.[8] These property relations were disrupted by the introduction of mobile, high-production technologies such as drag net fishing and herring seining.

Conflict between hook-and-line and drag-net fishers reflects, in part, competition between different technical approaches to exploitation of a common property resource. While the early drag net vessels could fish only ocean bottom that was relatively smooth and sheltered (e.g. in St Mary's Bay), technological advances soon permitted exploitation of many grounds fished by longline and handline fishers. Longline is a set gear; drag nets are mobile gear. They can peacefully exploit the same ground at the same time only if dragger captains monitor the course of their tows and divert fishing activity away from set gear. The onus is on dragger captains, because their technology is mobile.

Hook-and-line fishers on Digby Neck and the Islands have complained about drag net fishing since its outset, often directly to federal fisheries officers. For instance, one officer in 1955 reported numerous complaints that line fishers were being interfered with by draggers.

There were similar reports in many succeeding years.[9] Opposition was widespread. Line fishers complained that dragging killed small fish, destroyed ocean bottom habitats, broke up fish shoals, and overexploited stocks. Several hook-and-line fishers claimed that the destruction of enormous quantities of small fish was such an integral aspect of fish dragging that early boats had special holding bins into which catches were dumped. Moreover, it was argued that such practices, coupled with the mobile character of drag net operations, forced line fishers to fish only those marine environments inaccessible to draggers. Many also argued that draggers landed such large catches in the port market that they pushed down fish prices. In short, line fishers strongly felt that dragger fishing eroded the basis of their livelihoods.

Without question, small boat draggers changed the fishery of Digby Neck and the Islands. Landed weights for species such as cod and haddock increased substantially by the mid-1960s (Table 33). For instance, cod landings in 1965 were over one-fifth higher, and haddock landings over two and one-half times higher, than those reported for 1952. Conversely, by 1965, hake landings had declined from 1952 by almost 82 per cent. The growth in cod and haddock corresponds with the rise of the small boat dragger, while the collapse in hake is associated with either a shift away from the species of preference in the hook-and-line fishery or a collapse in stocks.[10]

In December 1966, the senior fisheries officer in district 37 reported: 'Hake, one of the chief species that supported this line fishery through the years have practically disappeared. Since 1963 the overall landings of hake have dropped approximately 80 per cent' (Narrative Reports). Some line fishers agree, asserting that the damage caused by drag fishing destroyed the hake fishery. Earlier in 1966, the same officer described the gradual replacement of hook and line by draggers (Narrative Reports). Increased drag net fishing forced hook-and-line fishers to exploit less-productive grounds that were inaccessible to draggers.

These trends continued unabated until the early 1970s. The general collapse of fish stocks (particularly haddock) from overexploitation in Atlantic Canada dramatically reduced the number of small boat draggers fishing out of Digby Neck and the Islands; by 1972, only sixteen were operating. However, the local fleet began to build again following the recovery of fish stocks and, especially, Canada's 1977 declaration of the 200-mile exclusive economic zone. By 1983 there were thirty-five small boat draggers, most of them over 50 feet in length.[11]

TABLE 33

Landed weights (000 lb) and landed values ($000 – 1983) for selected groundfish species, Digby Neck and the Islands (district 37), 1952–83

Year	Cod		Haddock		Pollock		Hake	
	Weight	Value	Weight	Value	Weight	Value	Weight	Value
1952	2,831	91	2,179	119	3,543	79	5,805	135
1953	2,467	65	3,275	140	4,258	75	5,874	105
1954	2,678	78	2,367	121	4,712	77	6,518	121
1955	3,152	95	3,494	149	4,673	93	5,790	116
1956	3,756	113	4,501	212	4,351	87	5,555	127
1957	3,017	94	4,159	198	7,184	169	6,054	162
1958	2,414	88	3,815	232	6,127	128	3,686	95
1959	2,901	108	3,506	217	4,141	108	4,730	133
1960	2,953	102	3,291	173	9,414	250	3,264	93
1961	2,950	115	3,774	231	4,823	126	3,561	113
1962	3,792	153	6,980	385	8,844	310	2,889	97
1963	3,330	151	8,677	492	9,729	357	3,519	134
1964	4,950	235	7,590	476	5,363	213	2,786	113
1965	6,249	315	7,858	586	4,312	180	1,049	43
1966	6,779	384	16,146	1,206	4,760	204	651	30
1967	8,002	415	12,005	704	3,241	131	579	28
1968	6,826	357	9,905	750	5,599	200	894	45
1969	5,383	325	6,185	639	5,017	195	980	33
1970	4,033	331	4,395	601	3,337	157	859	36
1971	3,202	298	3,088	421	2,972	172	1,074	55
1972	3,903	383	2,502	463	4,024	240	1,985	119
1973	2,819	322	1,760	360	10,629	651	2,517	161
1974	3,155	398	3,217	650	6,835	483	2,677	202
1975	3,821	397	5,068	869	6,717	390	2,597	157
1976	3,111	381	3,627	1,734	7,385	466	1,839	128
1977	7,033	932	4,621	1,021	6,409	513	1,351	105
1978	5,496	835	8,501	1,806	4,341	411	1,462	154
1979	5,769	1,124	7,948	1,858	8,067	1,156	1,759	277
1980	6,404	1,378	8,724	2,381	8,353	1,288	1,945	286
1981	7,540	1,657	8,536	1,851	6,517	1,060	1,488	248
1982	9,365	2,316	8,093	2,189	7,685	1,397	1,964	336
1983	11,023	2,419	9,632	3,121	8,406	1,116	1,561	157

Source: Selected fisheries statistics, district 37, unpublished data 1984

Stock collapse had created an industry crisis which compelled government to de-emphasize industrial development and focus strategies for management of stock and access. Supposedly, too much capacity was uncontrollably pursuing too few fish, and so new policies were intended to match fishing effort with the biological capacity of stocks,[12] through controlled access to marine resources and mechanisms such as annual licences and quotas. Greater emphasis was placed on 'policing' fish catching and processing (Barrett 1984; R.D.S. MacDonald 1984). Many people without a history of participation in specific fisheries obtained licences. For example, groundfish entry permits for otter trawl were introduced during 1977–8. In 1979, sixty-nine such permits were issued to fishers from Digby Neck and the Islands – far more than the actual number of participating vessels. In 1979, only twenty-four small boat draggers fished out of ports in Digby Neck and the Islands. Realizing that this situation contravened limited-entry management, the Department of Fisheries and Oceans (DFO) attached conditions for licence renewal: demonstrable preparation to use the licence or actual participation. Several people responded by entering the fishery. The number of draggers working out of local ports increased by six between 1979 and 1980 – a 25 per cent jump. As one dragger captain explained. 'It was either go into it or lose my license. I'd been thinking about doing it anyway. Things looked good. You could make some money at it then. So, I made the jump. You could say I was sort of pushed I suppose.' Others responded by purchasing some of the equipment associated with drag net fishing, creating the illusion of pending participation.

Many remaining coastal-zone fishers as well as dragger captains must service the large debts resulting from their purchase of boats and equipment at dealer-inflated prices. Ironically, federal and provincial subsidy programs for vessel acquisition stimulated increases in the prices of boats. Boat builders and equipment suppliers have consistently increased prices in relation to the level of subsidy available, leaving the fishers who receive assistance without any real savings. Moreover, fishers who fail to obtain assistance pay a boat price that is inflated by the level of subsidization available (Baker 1979).

By 1985, cod landings in district 37 totalled over eleven million pounds, almost a fourfold increase over the 1973 low (see Table 33). By 1983, haddock landings were almost four times higher than in 1973.[13] Indeed, fish landings are fast approaching the level of productivity associated with the late 1960s' collapse of fish stocks.[14]

Line fishing, because of the increase in dragger fishing, has experienced another, perhaps fatal, blow (Table 34).[15] Except for increased shares in total landings for 1980 and 1981, the coastal zone has experienced gradual and persistent erosion in landings since the mid-1970s and seems to be completely collapsing. At the end of 1983, it

TABLE 34
Inshore landings as percentage of total landed weights for
selected groundfish species, Digby Neck, 1967–83

Year	Cod	Haddock	Pollock	Hake
1967	31.0	8.2	22.9	96.9
1968	26.4	10.0	23.4	80.0
1969	31.3	23.4	24.6	70.9
1970	36.9	21.1	18.8	84.5
1971	6.3	11.8	18.8	83.0
1972	29.8	17.0	5.8	93.7
1973	37.3	17.8	7.5	94.8
1974	36.4	10.6	16.2	94.7
1975	38.8	8.5	12.0	97.5
1976	37.3	6.1	12.1	98.0
1977	21.9	7.9	13.9	96.6
1978	19.6	3.0	9.4	96.7
1979	27.8	3.6	15.5	84.7
1980	38.8	12.0	17.7	88.0
1981	34.4	12.5	7.8	94.0
1982	18.9	6.6	9.0	62.7
1983	9.0	4.9	4.7	37.9

Source: As Table 33

accounted for only 9.0 per cent of cod landings, 4.9 per cent of haddock, 4.7 per cent of pollock, and, remarkably, only 37.9 per cent of hake (once the major line fishery). Moreover, line fishers are not simply shifting from hake to other groundfish.

During the summer of 1984, many line fishers confirmed the prevalence and persistence of this condition. Most attribute it to the growth and fishing effort of small boat draggers. While the small boat dragger fleet may not be the only cause, many line fishers blame it. Compounding this situation is the dearth of alternative fisheries. For instance, herring fishing with weirs and gill nets at one time provided an alternative or supplement to groundfishing. Yet while the inshore sector currently (Table 35) accounts for the majority of herring landings, the collapse of stocks as well as the recent fall in prices prevents most coastal-zone fishers from turning to the herring fishery.

Only through lobster fishing can coastal-zone fishers compensate for income lost because of the collapse of line fishing (see Table 36). Increased landings during 1982 and 1983 may indicate greater effort or

TABLE 35
Inshore herring landings as percentage of total
herring landings, Digby Neck, 1967–83

Year	Inshore landings (%)	Total landings (000 lb)
1967	19.5	46,340
1968	10.1	40,790
1969	11.9	30,698
1970	31.8	16,426
1971	38.9	15,697
1972	24.5	12,841
1973	28.3	29,733
1974	16.3	23,310
1975	12.9	26,230
1976	56.8	6,625
1977	100.0	2,266
1978	88.3	5,844
1979	55.0	3,598
1980	62.3	2,714
1981	91.8	4,687
1982	85.1	2,610
1983	87.2	2,264

Source: As Table 33

simply reflect a point on the upward side of the cyclical swings in lobster populations, as is suggested by data for earlier years. However, coastal-zone fishers now must place much greater emphasis on lobster fishing in order to satisfy income needs. Many approach total dependence on this fishery.

Thus a form of fish production that once could pursue various fisheries is now moving quickly toward specialized focus on the lobster fishery. Because lobster fishing is a federally regulated activity restricted to a specific time of year, the local coastal-zone fishery has become largely a seasonal, part-time pursuit.[16] As one older fisher noted: 'Many of the boats here make up what we call the mosquito fleet. Part-timers that run around here and there hand-lining the odd fish, some of them just putting in time until lobstering begins.'

Many small boat dragger fishers are aware of the situation facing coastal-zone fishers. After all, not so long ago either they or their fathers were part of that fishery. Indeed, they acknowledge that their own fishing activity has helped destroy the line fishery. Some claim that what is currently happening to coastal-zone fishers would have

TABLE 36
Landed weights (000 lb) and landed values ($000 – 1983) for
lobster, Digby Neck, 1952–83

Year	Weight (000 lb)	Value ($000)	Average price per lb ($)
1952	579	265	0.46
1953	649	284	0.44
1954	561	261	0.47
1955	509	272	0.53
1956	627	361	0.58
1957	582	278	0.48
1958	513	276	0.54
1959	602	321	0.53
1960	559	256	0.53
1961	613	326	0.53
1962	642	384	0.60
1963	751	491	0.65
1964	710	538	0.76
1965	726	662	0.91
1966	689	517	0.75
1967	696	555	0.80
1968	774	584	0.75
1969	1,142	955	0.84
1970	818	820	1.00
1971	844	967	1.15
1972	550	809	1.47
1973	676	1,050	1.55
1974	568	847	1.49
1975	620	1,120	1.81
1976	463	1,003	2.17
1977	653	1,514	2.32
1978	487	1,500	3.08
1979	602	1,586	2.64
1980	461	1,497	3.35
1981	549	1,705	3.11
1982	609	1,949	3.20
1983	800	2,794	3.49

Source: As Table 33

befallen them had they not begun drag fishing. A number insist that they did not have much of an option. One captain stated: 'It didn't seem I had much of a choice. There wasn't much going in the line and herring fisheries while things were looking pretty good at dragging. So, it was either do this or get out of it altogether.' Ironically, several small boat dragger captains are convinced that their fishery will be the next to fail and are trying to sell their boats in order to buy a Cape Island–type vessel for line and/or net fishing. Many dragger captains and crew members already own licensed lobstering boats which they use every fall – tying up their draggers to pursue the potentially quick and bountiful dollars of the fall lobster fishery.

During the mid-1970s, processing, like harvesting, experienced recovery, but it emerged with a new orientation. Many of the remaining firms adopted automated processing equipment, such as filleting, deboning, skinning, and heading machines. Also, the 1970s saw widespread introduction of plant freezers and coolers. By the summer of 1984, ten year-round fish plants remained in operation, including a reduction plant that obtains its raw material from the other enterprises. Several plants – in effect, branches – are specialized filleting operations which turn their entire product over to either the Digby branch of National Sea Products (NSP) or Comeau Seafoods, a large company based in Saulnierville. However, the majority of the nine fish-buying enterprises remain independent, family-owned businesses, diversified in their product mix: salt fish, fresh fish, fresh and frozen fish fillets, and, in several instances, smoked fish as well as food herring products. A number also buy and sell lobster.

Fish processing has undergone structural change of magnitude similar to that experienced by the fish catching sector. Today's fish plants in Digby Neck and the Islands are the outcome of a process of consolidation and concentration that started in the mid- to late 1950s. Most employ over forty people during their peak periods; several occasionally employ more than one hundred. Their product lines are diversified, and some use automated production and sophisticated infrastructural technologies. The few specialized filleting plants feeding products to the large, economically dominant fish companies represent significant recent consolidation – symbolizing perhaps a major future development. The remaining independent processors report increasing pressure from the large companies, especially NSP. As one processor asserted,

If I want to stay independent I've got to compete with National Sea Products who are competing with me with taxpayers' money. It makes me mad! If I don't sell them all my product they won't deal with me. I called them [National Sea, Digby] one day, wanted him to take a load of codfish we had coming in.

The son of a bitch wouldn't take them. He had the ability to take them. Almost wanted them. But he wouldn't take them because I wouldn't make a commitment to sell them all my fish in the future. They got the markets, they got the money and they're making it harder every day for us small companies to stay in it.

Processors' dependence is so great that they feel compelled to increase their dependence in order to survive the crisis.

The impact of modernization

Regardless of the future fate of the small boat dragger, its development has helped transform fisheries of the Digby Neck and the Islands. Fish dragging has been concentrated in East Ferry, Little River, and Tiverton (Table 37). In recent years, draggers have also been working out of Freeport and, particularly, Westport. Centreville, an early locus, has experienced rapid decline, from eight vessels in 1967 to one by the early 1980s. Overall, there were over 40 per cent fewer small boats fishing out of the major ports in 1983 than in 1957. Even more dramatic declines are apparent for Little River and Freeport between the early 1970s and 1983; the former port shows a 59.3 per cent decrease, and the latter a 47.5 per cent decline. Between 1957 and 1983, the fleet shrunk in Centreville by 50.0 per cent, in Freeport by 32.3 per cent, in Little River by 26.7 per cent, in Sandy Cove by 22.2 per cent, in Tiverton by 55.6 per cent, and in Westport by 53.7 per cent. The decline of the inshore fishery mirrors the growth of the dragger fleet and the demise of the line fishery. In many places, the boat fleet today is a mere shadow of its former self. Indeed, Centreville is on the verge of extinction as a fishing port. Many boats are fished either part-time or seasonally, as in the lobster fishery, and so it appears that the inshore fishery is in an irreversible decline.

It could be argued that the change in fleet structure signifies no more than the switch by line fishers to another technology. During the early years, dragger fishers did come from a line and coastal-zone fishing background, but the decline of the boat fishery was part of a larger, more systematic process. Data in Table 38 reveal how change in the fisheries affected participation. By 1983, 35.7 per cent fewer fishers worked out of these ports than in 1957. Compared with 1957, the number of fishers had by 1983 declined in Centreville by over 70 per cent, in Sandy Cove by over 40 per cent, in Tiverton by over 50

TABLE 37
Fleet structure by selected communities, Digby Neck, various years 1957–83*

Communities	1957	1967	1971	1972	1977	1979	1980	1981	1982	1983	Percentage change 1957–83
Centreville											
Draggers	1	8	4	3	N/A	3	1	1	1	1	0
Boats	18	13	16	14	N/A	9	8	8	8	9	−50.0
Other	–	–	–	–	N/A	–	–	–	–	–	–
East Ferry											
Draggers	4	5	4	2	N/A	4	6	7	7	5	+25.0
Boats	8	12	11	11	N/A	10	15	10	10	10	+25.0
Other	–	–	–	–	N/A	1	1	–	–	–	–
Freeport											
Draggers	–	–	–	1	–	–	2	2	2	2	+200.0
Boats	31	34	37	40	37	25	26	24	25	21	−32.3
Other	4	3	–	1	1	–	–	2	2	2	−50.0
Little River											
Draggers	3	5	7	5	N/A	12	14	14	17	17	+467.0
Boats	15	17	20	27	N/A	15	14	13	11	11	−26.7
Other	1	–	–	–	N/A	–	–	–	–	–	–
Sandy Cove											
Draggers	–	1	–	–	N/A	–	–	1	–	–	–
Boats	18	19	16	16	N/A	17	14	15	15	14	−22.2
Other	–	–	–	–	N/A	–	–	–	–	–	–
Tiverton											
Draggers	2	3	3	4	3	4	5	3	4	5	+150.0
Boats	36	23	24	26	24	16	8	19	7	16	−55.6
Other	–	–	–	–	N/A	–	–	–	–	–	–
Westport											
Draggers	–	–	1	1	1	1	2	1	3	5	+500.0
Boats	41	34	32	26	20	25	27	23	28	19	−53.7
Other	2	6	–	1	1	1	4	3	2	–	−100.0
Totals											
Draggers	10	22	19	16	N/A	24	30	29	34	35	+250.0
Boats	167	152	156	162	N/A	117	122	112	114	100	−40.1
Other	1	13	2	3	N/A	2	5	5	4	2	−81.8

Source: As Table 33
* The years represented are the only years for which these data were available.

TABLE 38
Number of fishers by status* in selected communities, Digby Neck, various years 1957–83†

Communities	1957‡ ft	pt	t	1972§ ft	pt	t	1980 ft	pt	t	1983 ft	pt	t	% change 1957–83 (total)	% change 1972–83 ft	pt	t
Centreville	–	–	67	28	14	42	14	9	23	12	7	19	–71.6	–57.1	–50.0	–54.8
Sandy Cove	–	–	54	29	9	38	14	17	31	24	7	31	–42.6	–17.2	–22.2	–18.4
Little River	–	–	38	27	21	48	31	19	50	54	14	68	+79.0	+100.0	–33.3	+41.7
East Ferry	–	–	27	15	10	25	18	2	20	18	6	24	–11.1	+60.0	–40.0	–4.0
Tiverton	–	–	114	39	13	52	26	22	48	33	19	52	–54.4	–15.4	+31.6	0.0
Freeport	–	–	103	50	20	70	21	31	52	23	18	41	–60.2	–54.0	–10.0	–41.4
Westport	–	–	116	50	8	58	51	24	75	75	24	99	–14.7	+50.0	+200.0	+70.7
Total	–	–	519	238	95	333	175	124	299	239	95	334	–35.7	+0.4	0.0	+0.3

Source: Canada, Department of Fisheries and Oceans, Statistics Branch

* Status is a category developed by the Department of Fisheries and Oceans (DFO) to measure extent of participation. (See DFO, *Annual Statistical Review of Canadian Fisheries*, various years, for a detailed explanation.) While defined in various ways through the years, it most frequently reflects either length of participation through the years or income dependence. For example, those deriving less than 75 per cent have been classified as part time. The category was, no doubt, initially developed to assist DFO policy makers assess, design, and apply licensing programs when management interventions accelerated during the late 1960s. Full time = ft; part-time = pt; total = t.

† 1957 and 1972 are the only two years available prior to 1980.

‡ The data do not distinguish between full-time and part-time for this year.

§ Fishers categorized as 'occasional' (e.g. less than 50 per cent of income or less than five months of the year) are collapsed into the part-time category. DFO altered its basis for including men in this category through various years. Moreover, status classification often affects access to resources such as new licences, licence renewals, and boat and equipment loans/subsidies. That is, priority access is given to those classified as full-time. Consequently, the link between access and status classification frequently compelled producers to strive toward or create the illusion of full-time participation.

per cent, and in Freeport by over 60 per cent. East Ferry and Westport had lost 11.1 per cent and 14.7 per cent respectively. Only Little River, the centre of small boat dragging, had more fishers: 79 per cent more in 1983 than in 1957. The increases in numbers there do not approach the decreases in fishing populations of the other ports. In 1983 there were only thirty more fishing out of Little River than reported in 1957; by 1983, the major ports had 185 fewer persons participating in the fisheries. Thus, the changes in the boat fleet described earlier, as well as the decline of the line and inshore groundfisheries, cannot be explained simply by producers shifting from one fishing technology to another.

These developments challenge many communities, particularly since the region depends on the fishery for income and employment. On a per-fisher basis, over 225 per cent more groundfish (by weight) was landed in 1983 than in 1952.[17] Yet this increased productivity is neither reflected in increased total employment in the fisheries nor distributed evenly among local communities.

The dramatic increases in total landed weights have not expanded employment in fish processing, which has fallen over the last ten years. For instance, a 1975 study reported that local fish buyers and processors employed a maximum of 343 workers (Canada, Environment Canada, 1976b). Three years later, maximum employment was reported to be 298, down by 15 per cent (Nova Scotia, Department of Fisheries, 1978). In the 1981 census, 245 persons indicated manufacturing work (Canada, Statistics Canada, 1983a) – and fish processing is the only major manufacturing activity in the area. Overall, between 1975 and 1981 employment in the industry fell by as much as 29 per cent.

Fish processing too was adopting technologies and labour processes which vastly increased productivity per worker but reduced the total numbers of employees required. Those communities affected most keenly will participate less in the fishing industry. Currently, they do not contain many job opportunities; many people, especially the young, must leave in pursuit of work elsewhere.

The relation between small boat fish draggers and fresh or frozen fish product suggests interdependence involving increased capital commitments and specialization. Both sectors have become increasingly vulnerable to vagaries in the market, shortfalls in preferred fish species, and imperatives of the large corporate sector. For instance, recent declines in groundfish landings have compelled processors to obtain supply from other sources, especially buyers in Cape Breton and Newfoundland. In 1984, processors in Digby Neck and the Islands were trucking in substantial supplies through the summer, a season normally of optimum supply and peak production. Unlike during the stock collapse of the late 1960s and early 1970s, the inshore line fleet had been so thoroughly reduced by the early 1980s that it could no longer compensate for supply shortfalls from draggers.

The only abundant species during the summer of 1984 was pollock. Because it is limited in its fresh market appeal, processors lowered prices to compel dragger captains to reduce their effort on this stock and catch other groundfish species, particularly cod and haddock. But stocks of locally available cod, and especially haddock, have been decimated – indeed, many draggers have been fishing out of ports in Shelburne and Yarmouth counties, in order to gain access to exploitable stocks on Browns Bank, as well as eastern grounds, such as LaHave and Sable Island Banks. Fishers spend more time and money commuting; processors have increased transportation and handling costs, as well as more concern over quality of supply, because of the greater time between landing and processing. Several dragger captains insist that the condition of cod and haddock stocks has become so bad that boats are increasingly exploiting underaged, small fish.

Several independent owners of small boat draggers report such poor catches recently that it has been a struggle to pay operating costs, let alone generate sufficient income to satisfy basic needs. A few stand to lose their boats to the provincial Fishermen's Loan Board for failing to make loan repayments. Some insisted that a number of processors are buying up reclaimed boats for very low prices, taking advantage of the fishers' plight. While fishers argue that processors should not be permitted to buy boats, processors are now so dependent on dragger supply that they have few options remaining, particularly since the line fishery has ceased to be significant.

Conclusion

'All that draggers done around here is make a few millionaires while the rest of us have been left paupers' (Digby Neck and Islands fisher). The local fishing industry currently faces several difficulties rooted in the relation among increasing capitalization, overexploitation of resources, and specialization. By dramatically increasing production of certain groundfish species, the small boat fish dragger has contributed to overexploitation of stocks. But to make loan payments, dragger captains have to increase exploitation of fish stocks, thereby worsening the situation.

Many processors confront similar difficulties. Their enterprises are committed to high-volume production of fresh and/or frozen fillets. Such operations require the large and consistent supply produced by fish draggers. Persistent shortfalls delay debt repayment, threaten markets, and, eventually, close plants. The future fate of processors is wedded to that of fish draggers. Each requires the other; yet continuation and consequent worsening of existing conditions can only reduce the ability of any to survive. Further consolidation is the most optimistic outcome of this process.

The material presented here shows that the coastal-zone fisheries, once the industry's backbone, have been reduced to a part-time, seasonal, and specialized activity. Both the development of the dragger-processor nexus and the thrust of government management and development policies have exacerbated this process. By 1984, marginalized coastal-zone fishers were more dependent than ever on government income supplements. Ironically, in the fresh fish boom of the late 1980s, these circumstances effectively eliminated coastal-zone, small-boat fisheries as a back-up form of production when dragger effort fails. This change only deepens processors' dependence on dragger supply, thereby accentuating the push for draggers to fish more. Sustained reduction in lobster stocks would probably push the coastal-zone fisheries over the edge, essentially removing them as a significant feature of the local industry.

9 Captains and buyers in 'Gangen Harbour'

Lawrence Willett

Statistical surveys of the port market too often assume that actors make economic decisions as individuals operating in a social and political vacuum. This chapter looks at sellers and buyers, at technological and social factors affecting the local port market, at methods of selling elsewhere, and at a sample decision-making process. The study is based on participant observation fieldwork conducted in an area of southwest Nova Scotia during August and September 1984. Research was confined almost totally to the 'work place' in an area commonly considered a 'competitive' port market. Rapport with fishing captains was established through intensive daily interaction on fishing trips as a member of the crew and ashore unloading the catch, working on gear, and so on.

The harbours and settlements along the south coast of Nova Scotia are quite similar both geographically and historically. They extend like gangens[1] from the southern coastline. The peninsulas extend on average for about eight miles and indent the coast from Liverpool to Yarmouth. Except for the Argyle district of Yarmouth County, the communities are populated by English-speaking descendants of pre-Revolutionary 'Nantucketeers' and United Empire Loyalists. Gangen Harbour (a pseudonym) has had approximately 200 years of continuous habitation.

For this study, Gangen Harbour is delineated by an imaginary line across the entrance to the harbour. It comprises a small town and five unincorporated settlements. The total population was nearly 4,000 in 1981. The mother tongue of 97 per cent of the residents is English. In a labour force of about 2,000 people, approximately 200, or just over 10 per cent, are full-time fishers. More than half of these fishers are crew members who do not own vessels. The people are predominantly Protestant and Baptist. This is a legacy of the Puritan background of the New England settlers. The New Light revivalist movement also generated many converts to the Baptist persuasion in the eighteenth century. The resultant ethos involves hard work and dedication to duty and is rooted in Calvinism (Apostle, Kasdan, and Hanson 1985: 263, 266).

Sellers and buyers

Captains and boats

According to a federal Department of Fisheries and Oceans registry of fishing vessels in the district, 103 boats were recorded for Gangen Harbour as of 2 January 1985. Since then, the total number of registered fishing vessels has remained close to 100. Small boats, between 12 and 26 feet, make up one-third of these vessels; they are largely open, 20-foot boats used in the inshore lobster and groundfish handline fishery. Forty-nine per cent are small groundfish draggers and longliners between 29 and 44 feet; these craft are highly versatile, employing a variety of fishing gear, such as lobster and herring traps, groundfish handline and longline, gillnets, and groundfish trawl. They fish inshore, nearshore, and offshore grounds. 'Midrange' groundfish draggers and longliners between 47 and 64 feet often fish in the eastern waters of the province for halibut and make up 10 per cent of the fleet. Finally, 8 per cent of the fleet consists of offshore trawlers between 86 and 112 feet.[2]

The handliner must purchase fresh or frozen bait, gasoline or diesel fuel, rope, twine, and other gear. Because it is not worth the time, trouble, or expense driving to some distant wharf (perhaps one from which neither friends nor family fish), a fisher usually sets out from the wharf nearest home. Also, handliners fish inshore and usually return by midday. Usually two relatives fish together, seldom more than two hours from home port. However, the bulk of annual income is derived from the winter lobster fishery (see chapter 3).

One focus of this research was the subset of groundfish dragger captains with vessels between 40 and 55 feet (rather than the larger, company-owned 'off-shore' trawlers). These captains need no bait, use government wharfs for off-loading, and make up only a small portion of the overall fishing fleet. They drag otter trawls over the smooth bottom of the banks to which they have access and concentrate on haddock and cod. The owner/skipper usually gets one share, and the remaining shares for crew are taken out of the profits only after the vessel's share (40 per cent) has been deducted.

The largest group of fishers in the harbour – and another focus of this study – have longliners. They fish both inshore and offshore. A trip may be as short as overnight and less than a dozen miles, requiring a haul-up from dawn until noon or slightly later; voyages 'to the east'ard' may last up to ten days. Much depends on the size of the boat. Onshore these boats require fuel, bait, storage space, freezer space for tubs of baited trawl (the longline), room to bait trawl, and room to repair and work on other gear (see chapter 7). Services and space can be provided on government wharfs, even by plants that lack wharfs themselves. These facilities must be close and accessible, since the key to success is flexibility. For example, if dogfish come inshore and are ruining the cod and haddock longlining, then even if baited, that type of trawl can be left in cold storage and heavier or halibut gear[3] can be used until later in the season, when the dogfish have gone. Of course, the latter gear requires a change of fishing grounds and all that such a shift entails – for example, increased fuel bills and longer periods at sea.

Groundfish gillnetters require no bait and much less storage space. Vessels are independently owned. Gillnetters from Gangen Harbour avoid 'gear conflict' with trawlers while on the banks by setting their nets over bottom too rough ('hard') for an otter trawl. They avoid 'prey conflict' with both longliners and trawlers by concentrating on pollock. Gillnetters typically unload at private wharfs. They use government wharfs when repairing gear and when visiting or helping kin – usually brothers who fish longliners or draggers. The three most active gillnetters share common problems. Some time ago, the government tried to take back their licences to upgrade the quality of fish. They banded together, sought legal counsel, and retained their permits.

Eight other local fishers own and operate a trap for herring, while two others use herring gillnets. The former have been operating the trap for eight years in a location where the net is made fast to the shore, less than an hour by boat from home port. One of the oldest herring fishers is the net boss, and he makes the decisions about when to clean the net. Crew members (generally related by blood or marriage) all fish from the same cove in four boats. Herring trappers and gillnetters would like to catch mackerel stocks as well, but that species is not sufficiently dependable to make catching them profitable. Therefore, for both groups, the herring fishery is augmented by lobster fishing during winter. The only herring fisher who does not have a lobster licence prefers to hunt game and seafowl on the islands along the coast during winter.[4]

After the last Tuesday in November, which marks the onset of the season, lobster fishing is pursued from small Cape Island boats, with or without 'houses,' or from skiffs near the shore. Many small lobster vessels are pulled ashore before the summer fieldwork, and the owners fish as crew members of longliners and draggers. Yet when late November arrives they leave that employment to fish lobster. Even captains of draggers leave for a month or so for this lucrative endeavour. Part-time fishers, in contrast, fish only during lobster season.

The central adaptative strategy involves changing boats or gear to meet seasonal requirements or availability of fish.[5] For example, fishers may stop longlining and fish for lobster or herring. They remove the longline chute or 'shooter' – the device used to control the laying out of the trawl at the stern – to make room for herring or lobster trap work. Alternatively, they may switch longline gear, exchanging the light gear used for cod and haddock for heavier line used for halibut. Draggers, however, are not 'geared' to other fishing techniques by simple changes of equipment. If the captain/owner decides to fish lobsters for a month, the most qualified crew member will become master for the pre-Christmas part of the lobster season and the captain will go lobstering on another boat.

Gangen Harbour is an occasional port market for fishers from other harbours – some even winter there because of its proximity to the grounds. Usually these visiting draggers (see chapter 8) unload groundfish or scallops. The groundfish draggers represent little direct competition to local handliners, gillnetters, or inshore longliners. However, they do compete for winter grounds with local offshore longliners and draggers. The outside groundfish draggers also supply local buyers. While this depresses local prices somewhat, many captains take a philosophical attitude, seeing the outsiders as a stimulant to the fish business. As one buyer exclaimed: 'You want to be here in February if you want to see fish fly!'[6] The scallop 'draggers' have no competitors in Gangen Harbour.

Buyers

Gangen Harbour is rimmed with six fish plants to which fishers sell their catches. As well, four[7] small-scale independent buyers vie for business. The independents buy both for themselves – to ship or process in their own plants – and for other, usually larger, buyers elsewhere. None of them owns wharfs, but they do rent and control space on government wharfs. Two of the six larger fish plants own wharfs and buy from independent fishers; two others neither own nor are situated on wharfs. The latter group buys fish at government wharfs and transports it by truck to plants. The two remaining firms, with plants at their own wharfs, are private family-owned companies which usually buy fish only from their own vessels. They either process their catch or ship all or part of it to the Boston market.

The fish plants range from large, with their own offshore draggers, to small cold-storage facilities, where catches are prepared for shipment. The smallest 'plant' is supplied by one groundfish dragger. The plants are open 'all year'; but only the larger ones handle both fresh and salt fish. The smallest plants do not handle salt fish or buy lobsters but deal exclusively in fresh fish. The plants and buyers are separately located around the harbour, except that the smallest buyers and the largest plant are situated at the government wharf near the town. The sheds of the small buyers are dwarfed by the massive processing plant. According to one small buyer and various fishers who were interviewed, this has been the situation for at least five years.

The port market

The impact of gear

Fishers' decisions to sell to a particular buyer are related directly to the type of gear used to catch fish, because, as a rule, certain gear catches certain species. Because of working space, storage, and bait and ice requirements, most inshore handliner and longliner captains have limited marketing choices. The handliner sells fish to the plant nearest home, because it might be further to another buyer up the harbour. Regardless of size or distance travelled to fishing banks or grounds, all longliner fishers have arrangements with fish plant managers in order to ensure sales, in exchange for services and inputs. Fishers are reluctant to ship or sell elsewhere 'for a few cents more [per pound]' when prices are high, because they may have to move all their gear to other shelter, lose privileges to a berth and have to strike new arrangements with a different company – 'avoidable trouble.' Also, the longline fishers tend to live near the plants with which they have made arrangements. Thus personal convenience affects who sells to whom.[8] (See chapter 7.)

Such is not the case for dragger fishers. The captain of a dragger can steam across the harbour, or to another harbour, to gain a few extra cents per pound – multiplied by perhaps 30,000 pounds of fish. The fisher's product must fit the buyer's requirements. For example, some buyers specialize in salt fish. Other processors want only high-quality herring for the food market (as opposed to fishmeal). Considerations of catch quality usually limit the dragger captain to 'bulk sales' – that is, to a plant with a major frozen processing capability.

Local groundfish gillnetters will probably not see a trawler or a longliner during a full day of fishing, because they fish different kinds of gear over different kinds of bottom. However, their selling arrangements resemble those made by the longliners. They too have pacts with buyers or fish according to company, depending on whether the catch is to be shipped fresh or processed or sold locally. Though gillnetters do not need bait, the location and species landed, and the limited number of buyers, restrict their options. A percentage of the catch usually will have to be salted.

With gillnetting, fish quality varies because slime eels and other organisms feed on the fish that is left in the net overnight. This effect is lessened by hauling the nets as frequently as possible. If the number of nets fished is kept near the forty currently permitted, then captain and crew are more apt to haul them frequently in order to produce high-quality fish. Unfortunately, the gillnetters suspect each other of setting as many as sixty. Where examples of hauling nets only once a week are reported, the whole catch must be sold for salt fish. Often captains and crews are simply lax. Wastage or poor quality is disregarded because the product market is predetermined. Fishers who catch pollock for salting see little advantage in working harder to ensure higher-quality fish, which will not earn a better price.[9] Gillnetting limits the numbers of buyers to which catch can be sold. The company-owned gillnetter mentioned above fishes only for its family firm, but the other two gillnetters have few plants with which to bargain and make pre-arrangements.[10] No gillnetter leaves port without knowing to whom the catch will be sold, and for how much.

Those who trap herring sell either to a small local company, which processes small amounts as food, or to a large company, which sells to ships from the Soviet Union, Germany, and Japan. These sales are 'over-the-wharf' and include both barrelled and filleted herring. The trap fishers have large sales only when a plant has contracts with foreign buyers. Therefore herring are left to swim around in the trap for up to three days before either being sold or released. Some of the trapped herring are sold as bait to halibut longliners,[11] but most are processed. Gillnetted herring go almost entirely for bait. The large local plant that processes most of the herring will not buy gillnetted herring: herring left overnight in gillnets can spoil easily and quickly. Therefore, the two herring gillnetters have to sell their catch – either to longliners as bait or to other fish plants, which buy bait and then sell it to 'their' longliners.

Lobster fishers are less pressed to sell their catch because of perishability. However, only two of the eight buyers that purchase groundfish also buy lobster, thus limiting 'in-harbour' competition. The catch can be kept in floating crates near the wharf and sold alive when the price goes up.[12] More important, regardless of to whom the groundfish is usually sold, the fisher is free to sell lobster to the highest 'bidder.' As the season advances, and as the string of floating crates extends along the cove, much time is spent discussing prices with other fishers and dickering with lobster buyers.

Thus particular gear types shape and channel many decisions in the port market, just gear and vessel type affect the duration, frequency, and distance of trips.

Social factors

Government statistics listing owners and captains in Gangen Harbour suggest that there are 100 independent fisher–decision-makers. Closer examination, however, reveals just forty-three family-based information units (see Figure 3). Of this group, eight units own and operate forty vessels: four consist of 'individuals' and four are families of fathers, sons, or brothers. As Table 39 indicates, over 10 per cent (eleven vessels) of the local fishing fleet is owned by one family – all brothers. This family has a registered company as well. It neither sells fish to Gangen Harbour buyers nor buys from other local fishers; it sells in bulk to other Nova Scotia plants or ships to Boston. Another 6 per cent of the fleet is owned by two brothers, and another 4 per cent by

FIGURE 3
Kin-based vessel ownership

A Longliners and gillnetters (non-shippers who sell to one plant)

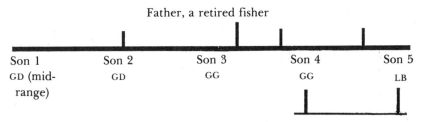

B Groundfish draggers and groundfish gillnetters (fisher/processors who supply their own plant)

a retired fisher. These eight 'individuals' own one-fifth of the fleet but constitute only three information units.

Local fishing families are traditional in their outlook, rural in orientation, and, if not part of an actual family firm, then part of a network communicating fish prices and evaluating buyers. Family members tend to sell to the same buyer. As one of the small buyers put it: 'Older family members, older men, influence the decision of the young. What do you suppose happens to a buyer if an old, respected fisherman tells the young fellas – "Don't sell to him; he's a crook!"''

Many fishing families own vessels that fish from other harbours as well. One family has a vessel in the next harbour west; one owns sixteen elsewhere. As Table 40 indicates, 18 families control 35 local and 109 outside vessels.

Successful families and individuals greatly influence local people and, some feel, the price of fish.[13] Fishers felt that the existence of a 'Fish Club' was proved, as buyers paid the same price (when allowances were made for off-loading costs, a service some buyers did not provide). One

TABLE 39
Families fishing vessels only from Gangen Harbour

Number code for family	Number of individuals with vessels registered under that family name	Number of vessels registered for that family
1	5	11
2	2	6
3	1	2
4	1	2
5	4	7
6	1	2
7	1	2
8	5	8
8 families	20 individuals	40 vessels

TABLE 40
Vessel ownership by families with vessels registered at
Gangen Harbour and elsewhere

Number code of family	Vessels at Gangen Harbour	Vessels at other harbours
1	1	6
2	3	5
3	1	2
4	1	16
5	4	5
6	2	19
7	1	2
8	1	2
9	2	4
10	1	2
11	4	8
12	4	10
13	1	2
14	1	5
15	4	9
16	2	4
17	1	3
18	1	5
18 families	35 vessels	109 vessels

buyer claimed that only the 'big boys' belong to the Fish Club. He saw himself outside the club but controlled by its price-setting: 'If it's not collusion, why do the prices not differ at times, like it does among the small buyers when one of them has a pressing need for supply?' One captain did not see club members as following the largest company there, but as mutual partners in price-setting, establishing prices not based on demand for fish at Boston. Under such circumstances, many feel that it makes little difference to whom they sell their fish.

By selling fish to a particular buyer in Gangen Harbour, a fisher becomes welcome at a beach party held every Friday and Saturday night during the summer. 'The Party' consists of fishers and their wives or girlfriends. Fishers are expected to bring drinks and food, and instruments, should they be musicians. The same area of marram grass is used each week. The buyer's trailer is set up, and lights are strung on poles for the summer. Fishing craft entering or leaving Gangen Harbour pass the scene. The party centres around one plant manager/ buyer and his wife, who own the business. The party is regrouped every weekend and features the type of 'blue grass' music loved by many folk of southwest Nova Scotia. (This combination of 'gospel' and 'country' music attracts devotees to various festivals[14] throughout the summer.) The buyer's trailer becomes a backdrop, and, if the stories are true, the playing never stops – day or night.

There is only one tavern (and one liquor store) in the vicinity. There are dances and other functions at the Royal Canadian Legion hall, but taverns, lounges, bars, and Legion halls cater to people of mixed economic and professional or trade backgrounds and contemporary ('rock') or urban-oriented music predominates. Not so for 'The Party.' It features people from the same occupational and ethnic background (with many being relatives) and with the same preference in music.

'The Party' reinforces solidarity among fishers and between them and the buyer. The fishers belong tacitly to a fellowship of the sea. They are dependent on each other to a far greater degree than people in other occupations. They appreciate each other for their industry and courage. Yet they know and dislike the way many locals fail to appreciate their efforts and contributions. They know that they are often ridiculed and even scorned.[15] Fishers are well aware that, along with hog farmers, woods people, and other primary workers who smell at times of sweat, manure, or offal, they are looked down upon by others. This might help to explain why fishers prefer the company of their own kind.

Selling elsewhere

Shipping fish

Fishers often ship their own fish and so know the importance of price spreads: a margin must be left to ensure profits between the price paid to fishers and the price received on the market. A skipper and his wife explained that after shipping for a few years they now see both the fishers and the processor's positions. They understood both the fisher's mistrust (based partly on lack of information about prices and markets) and the difficulties of operating a business. They have experienced market fluctuations between truckloads of product sent to Boston, even in the time between loading in Gangen Harbour and arrival in Boston. These rapid changes must be either borne by the buyer or charged to the fisher, through lower prices. At least one fish buyer/small processor had to sustain losses in order to remain in business. He could not go back to the fishers to cover his losses, yet had had to keep his prices up in order to compete with other buyers.

Fishers' suspicion of buyers is exacerbated when fishers themselves will not share price information. For example, the shipper's wife will not tell the other fishers, who do not ship, the current prices in Boston. Shippers, by contrast, have a telephone number which they can use 'twenty-four hours a day' and such excellent rapport with their brokers that should the price for a species tumble the broker will call Gangen Harbour (see chapter 6). Other shippers have emphasized the integrity of their brokers: 'We missed a box or two of fish, and he paid for the real count. He didn't have to – we'd have never known!'

Fishers no longer have a common view of selling fish because the industry is changing (see chapter 14). When prices paid in Boston were favourable for trucking, at least five captains shipped there directly. This has led to a slight rift between those who ship and those who do not. One family of shippers who fish small draggers 'jokingly' claimed that longliners were not welcome at their wharf. Fishers wonder about the assistance that some shippers receive from government – for instance, money to dredge coves, build wharfs, construct fish sheds, and buy trucks and forklifts. Some feel that contracts for government-sponsored research, or for educational programs for fishers in less 'progressive' regions, unfairly help the owner to become independent, to pay off his loans, or to get more boats. They suspect patronage-style politics of having too great a role in such endeavours.

Gangen Harbour fishers in general feel that everyone should have the right to make a living, but they resent profiteering. One of the first local fishers to begin shipping to Boston started to buy fish from his fellow fishers. One species he bought was grey sole. For nearly five years he claimed that it was not in demand in Boston – supposedly brokers did not want to handle it at all. Yet, as a favour to his fellows, he would 'take them off their hands' and pay them twenty cents per pound. Because he was a local, he was trusted by his 'parties,' and he paid more than the established prices for all species. Then other fishers began shipping to Boston. They discovered that grey sole sold for around sixty cents per pound – and 'had never dropped below fifty cents per pound for years.' In fact, the price had never fallen since the fishers had been shipping. Many fishers in Gangen Harbour were hurt by the display of bad faith. This example increased their suspicions about business. They no longer sell to that buyer, and remember that even a fellow fisher might lose his integrity when 'good' business becomes sharp practice.

A third option

Historically Gangen Harbour fishers have had more than two options of selling locally or shipping. Changes in fisheries regulations sometimes resulted in unusual arrangements. According to one longliner captain, when Canadian authorities condemned swordfish because of high mercury levels, American captains regularly bought swordfish, over the side, from Gangen Harbour crews outside the international limit. The fish were tagged inside the napes, so that the captains who sold them would receive compensation. They were illegal fish taken to Boston in American vessels. No money exchanged hands at sea.

A couple of weeks later, two airplane tickets would arrive by post from Boston. Two from the group of fishing captains involved were selected to make the trip. On arrival they were taken by automobile to a broker's office and were sometimes searched for firearms. A deal would be struck, and they would be given a tour and a night or two 'on the town.' Just before their return flight, they would be handed a briefcase containing as much as $85,000 (US) by their 'chauffeur.' On one occasion, payment for one whole shipment was lost when the transshipper's boat was reported 'lost at sea.' No compensation was offered or received. The informant, like others, was convinced that some brokerages in Boston are controlled by 'the Mafia.' But defiance of government ordinances and 'control' and sheer excitement were as important in this episode as the monetary incentive. It was a 'victimless' crime: nobody was hurt, nor was it against their religious beliefs.

Fish will be sold if at all possible. For example, one crew had caught more haddock than allowed. Even after giving all it could to its 'parties' (i.e. when it had filled these quotas as well), it still had 3,000 pounds left over. It could have 'dumped them' and avoided further trouble, but the captain had an idea. He put one of the crew ashore at a small jetty on an island outside Gangen Harbour and surrounded him with the surplus haddock. Leaving him some food and a handline, the captain and remaining crew members went to unload the main part of their catch. Federal fisheries personnel had either overheard the fishers on the radio or received a complaint and went to investigate. They found the lone fisher, who was having such terrific 'luck.' The fishers went so far as to ask the officer to radio his captain and ask him to come and collect his catch.

A decision to sell

An independent fishing captain entering Gangen Harbour with no obligations to sell to anyone might sell part of his catch – say, cod or other groundfish, like cusk or hake – and ship (export) the rest to Boston via truck and ferry. This might be possible if the captain has made prior arrangements with a broker and with a trucking firm, has ice, plastic shipping bins or boxes, and a forklift to load the fish on the truck, and has a wharf or the use of a boom and space on the government wharf; and if the price paid in Boston can cover costs and generate profit. Haddock and halibut are the preferred species shipped.[16]

What, however, if conditions are not conducive to shipping? Should the captain have a longliner (and work space and cold-storage facilities) or a dragger, then there might be a choice between shipping or selling locally. If there is not catch enough of one species or enough high-value species to load a truck, or if brothers or friends do not produce a sufficient amount when their catch is added in, the captain might sell locally. Or, if there is enough fish for shipping, perhaps higher-priced species will be shipped and lower-priced fish will be sold locally.

Alternatively, the captain might not ship at all, particularly when a buyer, formerly a fishing captain, owns more than one boat. Then the captain might agree with one plant manager to have all the fish handled for a minimum price for the whole season.[17] The decision appears to be made for reasons of expediency in dealing in bulk fish. So much time is spent in the operation of a fleet that the owner has little wish to worry about shipping. As well, a fleet of longliners (should one or more be going eastward for any length of time) can produce a lot of fish which become too stale for shipping in the summer heat.[18]

The following case illustrates the manner in which fishers seek as much price information as possible in order to make an informed decision while still at sea. A small, 42-foot dragger returns to port after three nights of fishing on Georges Bank. In its hold are 27,000 pounds of groundfish (25,000 pounds of cod and 2,000 pounds of haddock). The weather is fine and hot. Crew members are worried about the effects of the heat and the melting of the ice in the hold. The water on Georges Bank is warm. The latest news on the marine radio is about a dragger-load of groundfish that was condemned that morning. The skipper and crew lost the whole catch. Receiving this information had caused the captain of the dragger, now steaming toward Gangen Harbour, to stop fishing and head for port. Otherwise he would have stayed longer,[19] and a few more thousands of pounds could have been stowed aboard.[20]

The captain has a choice among four large buyers and four small ones in the harbour. To whom does he sell? He calls his wife and, without actually saying how many or what kind of fish he has, asks her to check on the prices being paid around the harbour. She calls local buyers and their personal broker in Boston.[21] After she compares prices received for the previous shipment to Boston, she relays this information to her husband at sea. However, the Boston price is kept secret. The information is transferred to the captain in code.[22] During the twelve to seventeen hours it will take to get to home port, the captain has to make a decision about where to sell.

With so many buyers in Gangen Harbour and two more close by in neighbouring ports, how does the captain decide where to sell, and to whom? The captain replies that his decision is based '100 per cent' on economics. Whoever pays the best price will get the fish. He is independent and can sell to whomever he chooses. The researcher goes below and finishes a game of cribbage interrupted by the last 'tow' of fish. A newly arisen crew member, rejoining the game, is told what the captain said. This crew member (the youngest aboard, and non-kin of the captain), whose relatives manage a large fish plant, smiles and says that the captain is wrong. 'It's more like 90 and 10,' he claims, explaining that the captain sells to his relatives whenever all the buyers are paying the same price.

On retiring, the captain tells a crew member to take the vessel to the government wharf in a harbour just west of Gangen Harbour.[23] The following morning, about eight o'clock, captain and crew reassemble at the wharf. En route to the wharf the captain had learned from his wife that every buyer in the home port was paying the same price for cod. By checking the Boston price over the telephone, she learned that the difference between the local and Boston prices for cod was too small for them to bother shipping. The captain headed for a particular buyer's wharf rather than go to his own wharf and unload, pack in ice, and ship.[24]

The cod were off-loaded along with a small amount of 'shack' and weighed by the bucket-load on the wharf in full view of fishers and other onlookers. Two thousand pounds of haddock and two small halibut were laid aside. Then the captain had a crew member back up the captain's three-ton truck near the boat, and a fish plant employee loaded the two one-thousand-pound boxes of haddock onto the truck using the company-owned forklift. This service was done 'for free,' just as the ice used to pack the haddock was 'free.' The captain stated: 'They'd do it for anybody.' The haddock was taken to the captain's cold storage shed at his private wharf in Gangen Harbour until either his group of shippers had accumulated a truckload or the day, usually Thursday, arrived when trucks departed for Boston.

Conclusion

In this examination of Gangen Harbour, we have seen the importance of technology, the role of the family (especially wives on shore) as an information and decision-making unit, the relative autonomy of independent dragger fishers, and the significance to longliners of pre-arrangements with kith and kin for selling fish. Within the constraints of gear used, and species sought, economic factors are counterbalanced by socio-cultural factors.

The discussion substantiates the view of the Task Force on Atlantic Fisheries (Canada 1982: 8), which found the 'economic fishery' and the 'social fishery' interwoven so tightly that such a distinction made no sense to the fishers themselves. Steinberg (1984: 26) would agree that Gangen Harbour is not a port market in the neoclassical sense.

For him, the 'entire inshore fishery constitutes the port market.' Further, the findings here on the Fish Club indicate, in keeping with his analysis, that price-leadership by the major processors is felt throughout the fishery.

But why does the small-scale fishery persist? It is suggested here that a more structural view, one that incorporates socio-cultural dimensions, would help explain the endurance of small-scale fisheries in southwest Nova Scotia. Merely creating employment will not necessarily satisfy the social needs of people or create 'maximum economic efficiency.' Nor should many independent fishers be pushed out of the fishery in order to create higher salaries for wage-earners on factory-trawlers. This would probably result in unemployment for the former small-scale fishers and force them to leave the province for work elsewhere. This would also disrupt the very kin-based infrastructure[25] that was overlooked by Steinberg and the task force. And as for the alleged need for high-quality fish products, the fishers of southwest Nova Scotia are already sending excellent fish to Boston.

10 Surplus labour

Richard Apostle
Gene Barrett

In chapter 4 surplus labour was identified as a central factor affecting the location and industrial structure of capital in fish processing. Regional political economy, like dependency theory generally (see chapter 1), rests much of its analysis on surplus labour in the reproduction of underdevelopment (Mandel 1969; Veltmeyer 1978, 1979).[1] It often, however, treats surplus labour simultaneously as cause and as consequence. This chapter seeks to explore regional surplus labour in one concrete case – that of fish processing – in order to flesh out variations and descriptive dimensions and the reciprocal causation among industry, location, and labour.

By all standard economic and social indicators, the Atlantic region of Canada is underdeveloped (Canada, Economic Council of Canada, 1977; Phillips 1982; R. Matthews 1983).[2] A central theme in the literature on regional problems has been the role of surplus labour.[3] Many political economists see distorted capitalist development that both reproduces and exploits a rurally based labour surplus. Average wage levels are lower than in other regions, and the work-force faces underemployment and prolonged unemployment. The entire structure is propped up by social security and the rural family and household economy (Veltmeyer 1979; Sacouman 1980, 1981; Clow 1984; Connelly and MacDonald 1985; Porter 1987).

Nowhere is the connection among capital, surplus labour, superexploitation, and underdevelopment said to be more clear than in fish processing, particularly among women, who are exploited as an industrial reserve army in large-scale mechanized fish plants and subordinated at home under patriarchical family relations (E. Antler 1977b; McFarland 1980; Connelly and MacDonald 1983, 1985; Porter 1985a,

1985b; Ilcan 1986; Neis 1988). This literature analyses the long-standing problem in the fishing industry: its reliance on surplus labour.

However, most of the research has concentrated on women in large-scale plants. This project's 1984 survey of fish plant managers found fish workers highly differentiated by wage and employment levels, ethnicity, gender, and labour process. Occupational distribution in the industry seemed to vary by localized labour markets. To assess reciprocal effects between occupational structures and labour markets, and the nature of these new patterns of differentiation, we interviewed workers in a core of the plants surveyed in 1984 (see Appendix). The findings are presented below in terms of the impact of scale and of location, using data from both surveys.

Scale of operations

Four clear patterns (in large plants, competitive plants, small plants, and piecework) characterize fish plant work in Nova Scotia, based on segmentation of capital. First, large plants, while highly diversified, are characterized by mechanized production lines where work is routinized and the labour force is closely monitored in terms of productivity. Large plants employ more machine-related workers in trades, technical work, and light labour than do small and competitive plants.

Second, competitive plants have flexible operations with up to three major product lines. Workers are more autonomous, handle tasks in all production lines, and shift (often at a moment's notice) from one line to the other. Competitive plants employ skilled cutters[4] and splitters in their fresh and salt fish production lines,[5] and most work is classified as 'multiple job task.'

Third, small plants tend to be highly seasonal, with one basic product line and a small work-force responsible for tasks such as salting, splitting, packing fish, and driving a forklift. These plants tend to be more specialized, with more light labour.

Fourth, in piecework, competitive plants in the Barrington-Argyle district of southwest Nova Scotia often contract out production of boneless salt fish to the most skilled 'splitters' in the area.[6]

Small plants employed 6 per cent, competitive plants 19 per cent, and large plants 74 per cent of the total work-force (Table 41). While women comprised 47.7 per cent of all fish plant workers, 80 per cent of them were employed in large plants, compared with 69 per cent of males. Competitive and small plants relied disproportionately on male

TABLE 41

Occupational categories by gender and plant type, 1984 plant survey (percentages, with numbers in parentheses)

Plant type	Supervisors			Trades/technical*			Cutters/splitters			Light labour†			Other labour‡			Multiple jobs			Total		
	M	F	(N)	M	F	(N)	M	F	(N)	M	F	(N)	M	F	(N)	M	F	(N)	M	F	(N)
Piecework (N = 9)	0.0	0.0	(0)	0.0	0.0	(0)	80.0	20.0	(15)	0.0	100.0	(3)	0.0	0.0	(0)	0.0	0.0	(0)	66.7	33.3	(18)
Small (N = 28)	83.3	16.7	(12)	100.0	0.0	(9)	85.7	14.3	(35)	20.0	80.0	(65)	12.0	88.0	(25)	74.3	25.7	(191)	61.4	38.6	(337)
Competitive (N = 35)	90.2	9.8	(41)	100.0	0.0	(18)	70.2	29.8	(215)	31.3	68.7	(128)	75.0	25.0	(36)	64.5	35.5	(668)	63.7	36.3	(1106)
Large (N = 37)	86.1	13.9	(137)	100.0	0.0	(250)	85.8	14.2	(359)	27.7	72.3	(2432)	83.1	17.9	(236)	62.4	37.6	(834)	48.6	51.4	(4248)
Total (N = 99)	86.8	13.2	(190)	100.0	0.0	(277)	80.3	19.7	(624)	27.7	72.3	(2628)	76.1	23.9	(297)	64.6	35.4	(1693)	52.3	47.7	(5709)

* Includes plant and electrical maintenance and refrigeration and boiler-room operators
† Includes machine feeders, packing, trimming, and weighing
‡ Includes clean-up, forklift operators, and machine operators

workers. Eighty-one per cent of women in large plants were classified as light labour – machine feeders, packers, trimmers, weighers, and so on – compared with only 33 per cent of men. Men tended to be much more evenly distributed, though overrepresented in higher-paid, higher-skilled jobs, such as supervisors, skilled trades, and multiple jobs. Males represented only 48.6 per cent of the total work-force in large plants but 86 per cent of all cutters and splitters and 86 per cent of all supervisors. While women in competitive and small plants tended to be segregated in light labour, men dominated multiple-job tasks.

Women constituted 68 per cent of all persons earning $5.99 or less per hour, while men comprised 80 per cent of all persons earning $7.00 or more per hour. Moreover, 83 per cent of all persons earning less than $5.00 per hour were women, while 95 per cent of those earning $8.00 or more per hour were men (see Table 42). In large plants, women constituted more of those earning $5.99 or less per hour (72 per cent) than was the case in competitive firms (53 per cent). Conversely, men comprised a somewhat larger share of those earning $7.00 or more per hour in large plants (84 per cent) than in competitive ones (77 per cent). Because of the more developed task specialization in larger plants, pay rates there were linked to tasks ranging from unskilled to skilled categories.

Large plants

Large-scale capital faces technological imperatives to maximize throughput and minimize costs. Large volumes of fish are needed to supply highly mechanized production structures on a regular, predictable basis, but costs of raw material and labour have to be kept as low as possible. Overfishing and enterprise allocations in the 1980s made excess processing capacity common. Corporate capital's tolerance of unions which characterized the late 1960s and the 1970s has changed. Its drive for greater efficiency has led to production quotas and more part-time work.[7]

Less than one-third of small- and competitive-plant workers experience incentive systems, compared to over half of workers in large plants.[8] Of this latter group, 42.7 per cent are sometimes unable to meet production standards. Fully 62.0 per cent of workers facing incentive systems were not satisfied with their jobs, while 48.3 per cent of other workers were dissatisfied. Ilcan (1985b) reports that production demands particularly affected skilled cutters, putting considerable

TABLE 42
Worker pay scales by gender and plant type, 1984 plant survey (percentages, with numbers in parentheses)

Plant type	Less than $5.00			$5.00–$5.99			$6.00–$6.99			$7.00–7.99			$8.00 or more			Total		
	M	F	(N)	M	F	(N)	M	F	(N)	M	F	(N)	M	F	(N)	M	F	(N)
Small (N = 28)	9.1	90.9	(77)	37.1	62.9	(132)	90.9	9.1	(186)	81.5	18.5	(27)	0.0	100.0	(5)	57.8	42.2	(427)
Competitive (N = 35)	34.6	65.4	(110)	50.6	49.4	(399)	76.6	23.4	(243)	74.3	25.7	(222)	87.9	12.1	(58)	62.2	37.8	(1032)
Large (N = 37)	14.3	85.7	(488)	35.0	65.0	(938)	47.2	52.8	(2087)	76.2	23.8	(399)	100.0	0.0	(183)	45.7	54.3	(4095)
Total (N = 90)	17.0	83.0	(675)	39.4	60.6	(1469)	53.2	46.8	(2516)	75.8	24.2	(648)	95.1	4.9	(246)	49.7	50.3	(5554)

pressure on workers further down the line. Management interviews showed some smaller plants employing older workers who could no longer tolerate the pressure under these systems.

Although large plants are somewhat more likely to provide full-time employment (see Table 43), nearly half of workers are classified as seasonal and women comprise the majority of this group (64.7 per cent). In large plants, seasonal work makes up well over half of total employment (Table 44). The winter period, from November to March, is particularly acute. Even during summer, however, full employment is enjoyed by only 50 to 60 per cent of the work-force. During the peak months of August and September over one-quarter still worked on a temporary basis.

The structural basis of occupational sex-typing becomes clear in irregular work (Table 45). From January to March 1986, the proportions of women and men in large plants who did not work were roughly equal. Between April and July, and again in November and December, women experienced substantially higher levels of unemployment. Male workers consistently had the most continuous plant employment; this differential averages 13.1 percentage points over the course of 1986. However, women, in every month but three, had more intermittent work.

Women also had, on average, fewer hours. Fifty-five per cent of workers working over forty hours per week on average were male, while 63 per cent of workers working less than forty hours were female. Men got much more of available overtime work. For example, 37 per cent of the men report having more than 166 hours of overtime during 1986, compared to only 18 per cent of the women. Further, while just 16 per cent of the men had hours of work changed on a seasonal basis, 27 per cent of women did.[9]

In 1984, 85.8 per cent of workers in large plants earned less than $6.99 per hour (Table 46).[10] However, women working in large, non-union plants made considerably less than those in large, unionized plants. Thirty-nine per cent of women in large, non-union plants made less than $5.00, and a further 47 per cent of them earned from $5.00 to $5.99. By contrast, 92 per cent of women working in large, unionized plants made over $6.00 an hour. These differences were not related to differences in types of work. The spatial dimension of this specification is explored below.

While hourly wages were comparatively high in the unionized large plants, intermittent employment and occupational sex-typing kept

TABLE 43

Seasonal employment in fish plants, 1984 plant survey (percentages, with numbers in parentheses)

Plant type	Regular workers			Seasonal workers		
	M	F	(N)	M	F	(N)
Small	60.6	39.4	(142)	50.7	49.3	(217)
Competitive	66.1	33.9	(440)	59.6	40.4	(597)
Large	57.5	42.5	(2234)	35.3	64.7	(2014)
Total	59.0	41.0	(2816)	41.6	58.4	(2828)

workers' incomes low and dependent on unemployment insurance (UI).[11] Forty-six per cent of male workers earned $15,000 or over from all sources in 1986; 92 per cent of female workers earned less than $15,000. While 46.3 per cent of men replied that UI was important to their incomes, 71.1 per cent of females did so.

All managers of large plants complained about turnover.[12] One stated:

We find that we are getting a lot of young workers. There are hardly any older people .... I don't think that 10 per cent would be over 40 years old ... So what we find is in their twenties or someone who had just got out of school ... We're finding them very restless .... If we wanted to maintain a regular yearly staff of about 30 to 35 workers we would have to go through ... 100 or more employees ... They're restless, they're undecided. Last week we had three who just up and quit. They wanted their separation papers ... Their attitudes are all wrong. They're on a road to nowhere.

Paradoxically, some managers blamed workers for being too lazy and going on UI: 'in [o]ne year we ended up with about 25 per cent of our workforce on unemployment. We had to ... get 'in bed' with the U.I.C. people and Manpower who chased them out of the woodwork ... I can remember 5 years ago they wouldn't work the summer. They just work in the winter, get their stamps and away they go for the summer. There were a number of people around here who were "picked off" on U.I.C.'

Managers preferred women workers for several reasons: 'The male work-force is not anything like the female work-force. They're just so restless they can't stay in one spot ... [Females] are more settled. If you hire a female you're more apt to keep that work-force. We could prob-

TABLE 44
Seasonal employment (weeks worked) in fish plants, 1986 worker survey (percentages, with numbers in parentheses)

Month (1986)	Small				Competitive				Large				Total			
	0	1-3	4	(N)	0	1-3	4	(N)	0	1-3	4	(N)	0	1-3	4	(N)
January	100.0	0.0	0.0	(11)	92.3	7.7	0.0	(26)	57.2	8.8	34.0	(159)	64.3	8.1	26.6	(196)
February	100.0	0.0	0.0	(10)	88.5	11.5	0.0	(26)	51.9	9.6	38.6	(158)	59.3	9.3	31.4	(194)
March	90.0	10.0	0.0	(10)	84.0	12.0	4.0	(25)	48.4	7.7	43.9	(155)	55.3	8.4	36.3	(190)
April	90.0	10.0	0.0	(10)	72.0	16.0	12.0	(25)	37.0	14.9	48.1	(154)	44.4	14.8	40.8	(189)
May	90.0	10.0	0.0	(10)	41.7	50.0	8.3	(24)	36.0	10.0	54.0	(150)	39.9	15.2	45.1	(184)
June	90.0	10.0	0.0	(10)	39.1	34.8	26.1	(23)	32.5	16.5	51.0	(151)	36.4	18.5	45.1	(184)
July	60.0	30.0	10.0	(10)	21.7	52.2	26.1	(23)	25.2	17.6	57.1	(147)	26.7	22.7	50.6	(180)
August	10.0	40.0	50.0	(10)	4.3	43.5	52.2	(23)	11.0	29.0	60.0	(145)	10.1	31.5	58.4	(178)
September	20.0	10.0	70.0	(10)	13.7	36.3	50.0	(22)	18.1	25.5	56.4	(149)	17.7	26.0	56.3	(181)
October	60.0	20.0	20.0	(10)	26.1	39.1	34.8	(23)	32.2	18.8	49.0	(149)	33.0	21.4	45.6	(182)
November	80.0	0.0	20.0	(10)	57.7	26.9	15.4	(26)	42.7	21.6	35.7	(157)	46.6	21.3	32.1	(193)
December	81.8	9.1	9.1	(11)	73.1	19.2	7.7	(26)	49.4	14.1	36.5	(156)	54.4	14.5	31.1	(193)

TABLE 45
Gender-based seasonal employment (weeks worked) in large fish plants, 1986 worker survey (percentages, with numbers in parentheses)

Month	Male				Female			
(1986)	0	1–3	4	(N)	0	1–3	4	(N)
January	54.6	4.8	40.6	(64)	58.9	11.6	29.5	(95)
February	52.4	3.2	44.4	(63)	51.6	13.7	34.7	(95)
March	50.0	1.7	48.3	(60)	47.4	11.6	41.0	(95)
April	32.8	11.5	55.7	(61)	39.8	17.2	43.0	(93)
May	28.8	10.2	61.0	(59)	40.7	9.9	49.4	(91)
June	23.7	13.6	62.7	(59)	38.1	18.4	43.5	(92)
July	21.1	19.3	59.6	(57)	27.8	16.7	55.5	(90)
August	10.5	22.8	66.7	(57)	11.3	33.0	55.7	(88)
September	17.2	17.2	65.6	(58)	18.7	30.8	50.5	(91)
October	27.6	13.8	58.6	(58)	35.2	22.0	42.8	(91)
November	34.9	22.2	42.9	(63)	47.9	21.3	30.8	(94)
December	39.7	7.9	52.4	(63)	55.9	18.3	25.8	(93)

TABLE 46
Worker pay scales by plant type, 1984 plant survey (percentages, with numbers in parentheses)

Plant type	Less than $6.99	$7.00–$7.99	$8.00 and over	(N)
Small	92.5	6.3	1.2	(427)
Competitive	72.9	21.5	5.6	(1032)
Large	85.8	9.7	4.5	(4095)
Total	83.9	11.7	4.4	(5554)

ably operate this plant on a total female work-force ... which is probably something to look into ... [Turnover] has been an extreme problem since 1980. Before that it was a different situation because our labour force ... at that stage ... was primarily older employees.' Another manager was more explicit:[13] [Women] are certainly more dextrous and they are better workers. I would use women everywhere if they were physically strong enough to do the work because they are easier to work with. They just seem to be easier to deal with. They are not rowdy ... They are easier to talk to. They don't give you a lot of lip. They're much better people to deal with.'

Competitive plants

The labour problem in competitive plants continues to be a function of production flexibility and market adaptation. Labour has to be versatile and compliant; management's ability to make radical daily and hourly decisions has to be unchallenged; and, most important, labour has to be able to absorb intermittent fluctuations in production schedules caused by irregular supply and an erratic competitive market.

In contrast to large plants, fixed costs here are low and the margin of profit from high-value market segments is potentially great. The cost of labour can therefore be allowed to rise to relatively higher levels. When closed, of course, such plants are not faced with absorbing labour costs or with significant costs for idle machinery. Workers are expected to be self-motivated and versatile. A number of managers echoed one respondent, who observed:

We try to train all the men who come here to be masters of all trades ... and when a man comes in here and he can't operate a forklift we give him every opportunity to learn. We give him a six-month trial period and then we boost him up close to the top (wages).

When you are operating a small plant you can't have a man for every job. So we have had to double up, triple up, and quadruple a lot of them.

I don't like organizing my workforce such that people specialize in just one task. I try to train enough people so that if one or two leave we can still put another person on the machine and ... keep those machines going ... You've got to have good machine operators. Your whole production rides on that ... But there's not one person over there that couldn't go on a machine ... or couldn't cull fish or couldn't pack fish. They are quite capable of doing more than one function.

Unions are anathema to these managers. One operator described 'union-free' flexibility: 'The foreman is also the guy who works in the freezer and does some maintenance, and the rest (thirty-seven men) are jacks-of-all-trades. We switch them around from place to place, and they do most anything we ask them to.' The worker survey in 1986 revealed that 70.2 per cent of workers in such plants had two or more job tasks, while 45.9 per cent of workers in large plants had just one job task.

As we have seen, managers of competitive plants often rely on the independent fishery (independent longliners or intermediate draggers) as a flexible source that is limited in financial liability. This type of source generates irregular work hours because fish supplies are unpredictable. Thus, the 'versatile worker' is not only competent and independent but constantly available for work – day or night: 'We have a good working force. I would say the best on the South Shore ... Because I can call them anytime of the day or night, they'll come in and work, they'll do the job efficiently and fast. If I need a truck to go out at a certain time, I've never had one ... that didn't get out on time with the amount of fish on it that you usually wanted. I find them an awful [good] crew of men to work with and to get work out of.'

While 47.1 per cent of competitive plant workers indicated that they worked over forty hours per week on average, less than 10.0 per cent in either small or large plants did so. As we saw in Table 44, this work time was clustered intermittently between May and November. At no time was more than 52.2 per cent of the work-force employed on a permanent basis. Throughout the peak summer months the proportion of competitive workers employed intermittently ran, on average, 23.2 percentage points higher than that in large plants.

Given these patterns, how to explain the greater job satisfaction among workers in competitive plants? The combination of high average hourly wages and 'high stamps' in the qualifying period for UI went hand-in-hand with congenial working conditions. Workers replied the most strongly of any group when asked how important UI was to their income. Seventy-nine per cent indicated somewhat or very important, by comparison with 64 per cent of workers in small plants and 56 per cent in large. Many managers use the qualifying period for UI as a consideration in giving people work. As one stated:

When the season wound down and we had some workers who lacked two or three weeks and we had some workers who, say, had five weeks to spare, we would say to them: 'So and so needs three more weeks and things are slowing down.' ... Most people would say, 'Sure, I don't mind, I'll go on unemployment and let him work for two or three weeks to get his stamps.' They usually work back and forth and workers do that amongst themselves a lot. A lot of times a guy would come to you and say, 'I think I'll be all through in another week's time because so and so really needs the stamps and I don't and if you don't use me or you don't think you'll be that busy he can fill my place.' Basically we would go along with it.

Use of the UI system seems to be another community dimension in competitive-plants not evident in the large mechanized plants, with their seniority-based call-in systems. Managerial discretion and flexibility are vital components of patronage-clientism in the family-business segment of the industry. In 1984, 97 per cent of managers indicated that the UI qualifying period influenced the call-in procedure for non-union work-forces.

Small plants

The labour problem in small plants appears to be a variant of that facing competitive plants. Similarities in operation include low fixed machinery costs, direct managerial control, and seasonal labour. The key differences arise because small operations subcontract for large-scale capital and so must keep costs at a minimum. Therefore, labour has to be cheap and available for intermittent work. Further, because small plants tend to be specialized and extremely seasonal, versatile labour is less important and seasonal labour more so.

Small plants had the highest proportion of seasonal workers (Table 47). Over 60 per cent of workers were seasonal, as opposed to 50 per cent for the entire industry. Employment was concentrated in August and September – particularly during the herring roe season (see chapter 11). Hours of work were also limited: 79 per cent of workers in small plants indicated no overtime, as compared with 21 per cent in competitive plants and 16 per cent in large plants.

The 1984 plant survey showed that 92.5 per cent of small-plant workers earned less than $6.99 per hour; the 1986 worker survey found 92.9 per cent making less than $6.99 per hour. In both cases, lower wage levels make small capital an attractive subcontractor. In 1984 there were differences of six points (in the under-$5.00-per-hour category) and eight points (in the $5.00–5.99-per-hour category) between pay scales in small and large plants. In 1986, there was a thirty-six point difference in the under-$6.99-per-hour range.[14]

Workers in small plants are the poorest in the industry. Hourly wages, weeks worked, and total hours of work are the lowest, and, consequently, UI benefits are the smallest. Fifty per cent of these workers earned less than $4,000 per year from plant employment, and 69 per cent earned less than $7,000 total income from all sources.

Location

The significance of geographical differentiation to the fishing industry has been raised in preceding chapters. The 1984 plant survey revealed

TABLE 47
Seasonal employment in fish plants by plant type, 1984 plant
survey (percentages, with numbers in parentheses)

	Regular	Seasonal	(N)
Small	39.6	60.4	(359)
Competitive	42.4	57.6	(1037)
Large	52.6	47.4	(4248)
Total	49.9	50.1	(5644)

a tripartite division of the 'labour market' in Nova Scotia. Based on
variations in both industrial and wage structures, three distinct labour
markets were examined: two 'competitive' zones and one 'non-competitive' zone (see Map 8). A 'rural' competitive zone comprises the
Shelburne, Barrington, and Argyle municipal districts,[15] and an 'industrial' competitive zone comprises Yarmouth town and municipal
district, Shelburne town, Lockeport town, Queens County, Lunenburg
County, Halifax census subdistricts A through E,[16] and Cape Breton
County. A rural non-competitive zone comprises Halifax census subdistricts F and G and the counties of Guysborough, Antigonish, Richmond, Victoria, Annapolis, and Digby.

The fishing industry located in the industrial competitive zone has
been discussed in chapters 3 and 4. First, large-scale capital industrialized its onshore and offshore operations and expanded plant facilities.
Except in parts of Cape Breton County, this has gone hand in hand
with a contraction in small and competitive capital. Second, since the
1970s, large-scale capital has also vacated Halifax, and the area within
easy commuting distance of it, to avoid competition in wages and employment from the burgeoning service sector. Third, large-scale capital
has increasingly participated in a gender-segmented labour market –
a consequence of high-wage competition for male labour. A manager
of a large plant in the industrial competitive zone, employing predominantly women, outlined the impact of high-wage competition on his
non-unionized work-force: 'We're right in between two major companies and ... they're probably the highest paying in Nova Scotia, so
it is very difficult for us to maintain any real super employees in our
company ... If they are good, they know they're good, and they know
they're wanted by other companies. They're going to go there because
they know they can probably make twice as much as what we can afford
to pay ... So we're in a bad location for labour force.'

MAP 8 Fish processing labour markets

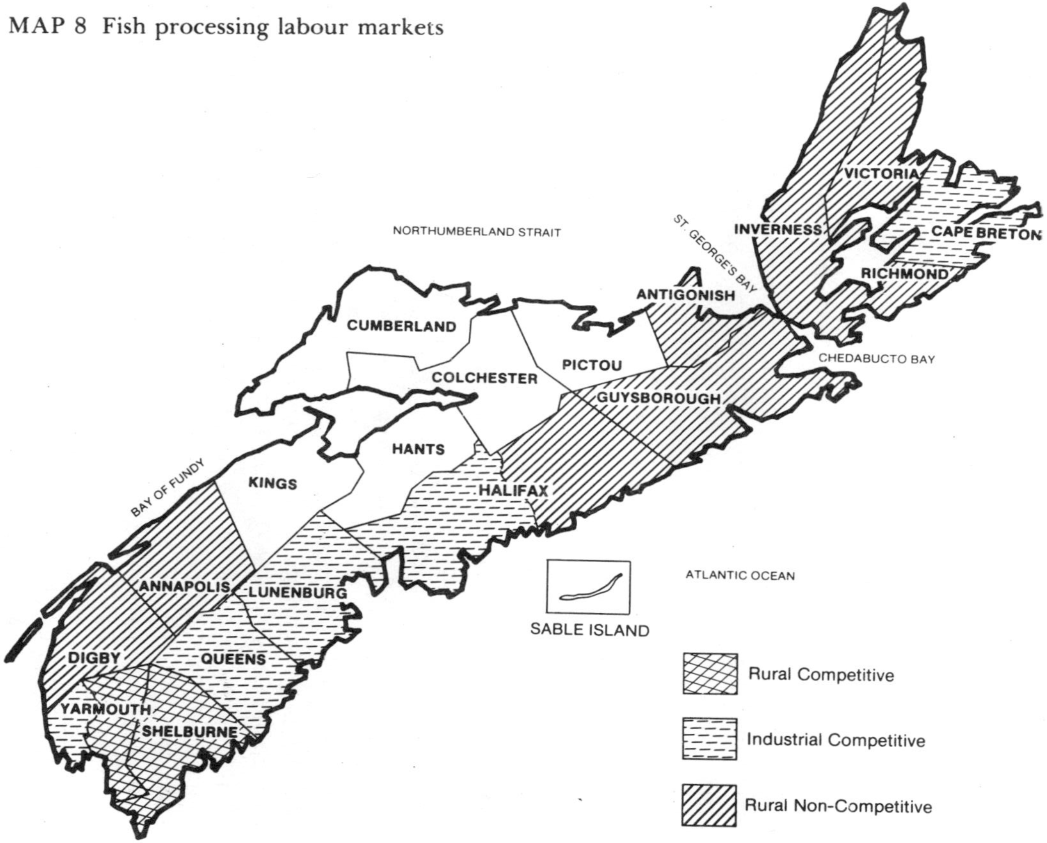

Rural Competitive

Industrial Competitive

Rural Non-Competitive

ATLANTIC OCEAN

SABLE ISLAND

CHEDABUCTO BAY

NORTHUMBERLAND STRAIT

ST GEORGE'S BAY

BAY OF FUNDY

VICTORIA

CAPE BRETON

RICHMOND

INVERNESS

ANTIGONISH

GUYSBOROUGH

PICTOU

COLCHESTER

CUMBERLAND

HANTS

HALIFAX

KINGS

LUNENBURG

ANNAPOLIS

QUEENS

DIGBY

SHELBURNE

YARMOUTH

The rural competitive zone is relatively small in area but was the centre of post-war expansion by competitive-scale fish processing and the coastal-zone fishery discussed in chapter 3. It is distinctive because of its large number of plants and the intense competition for fish, labour, and markets. For example, 78 per cent of workers indicated that they had had three or more employers in recent years. A further 74 per cent felt that they could easily get the same type of job in the area if they wanted to.

The rural non-competitive zone is not simply a residual category. It incorporates marginalized rural areas that have experienced dramatic decline in rural farms and contraction in both the fish and forest industries, which formed the basis of the traditional plural economy. Unemployment remains high, as does dependence on single industries and out-migration. This zone has three separate Acadian districts that are all heavily involved in the fishing industry.

An early assessment of secondary data seemed to indicate a simple rural–urban division in the province's labour markets. As Table 48 indicates, the industrial zone[17] had rates of male and female participation in the labour force of 78.1 per cent and 51.9 per cent, respectively. These figures ran five points above the provincial average for males and six points above for women. These figures represented real employment levels, as mean employment income was 11.1 per cent above the provincial level for males and 10.4 per cent above for females. Unemployment levels were below the provincial average, by 2.4 per cent for men and 3.0 per cent for women. Fishery-generated employment levels, as a proportion of the total work-force, were the lowest in this zone, at 7.8 per cent, or 3.4 per cent below the provincial average.

Labour-force participation rates for both rural areas were well below the industrial zone's average. Male participation in rural non-competitive areas was 10.5 per cent below the industrial area, while the female labour force was 12.4 per cent below. Trends in mean employment income for both rural zones ran in similar directions. Men earned 24.2 per cent less than their counterparts in the industrial area; females 25.1 per cent less. This discrepancy reflected high underemployment as well as the relatively high proportion of unskilled jobs. Unskilled workers comprised 30.8 per cent of the work-force in the competitive rural zone, as opposed to 19.8 per cent in the competitive industrial zone. In the rural non-competitive zone, 24 per cent of the labour force was employed in jobs dependent on the fishery, while 100 per cent of the

TABLE 48
Labour market characteristics in Nova Scotia fish processing

	Rural competitive	Industrial competitive	Rural non-competitive	Total province
A. *1981 participation, unemployment, and income levels*				
Labour force participation rate (%)				
Females	37.3	51.9	39.5	45.3
Males	70.5	78.1	67.6	73.2
Unemployment rate (%)				
Females	17.1	8.9	16.6	1.9
Males	7.9	6.3	12.2	8.7
Mean employment income ($)				
Females	5,216	8,087	6,061	7,323
Males	13,006	15,676	11,883	14,088
B. *1981 work-force composition*† (percentages of the area's work-force)				
Skilled workers	21.8	27.6	24.5	27.0
Semi-skilled workers	28.4	28.0	29.3	28.7
Unskilled workers	30.8	19.8	24.8	20.9
Captains and fishers	22.5	1.3	5.9	2.3
Plant workers	19.3	1.7	4.2	2.2
Fisheries-generated employment‡	100.0§	7.8	23.7	11.2
	(5,655)	(219,905)	(49,355)	(366,065)

* Calculations based on information from Statistics Canada (Canada, Statistics Canada, 1983)
† Figures based on a special table run by Statistics Canada for the authors from data on the experienced labour force aged 15 years and over for all census divisions and subdivisions in Nova Scotia for 1981. Definitions are based on Pineo, Porter, and McRoberts (1977). 'Skilled workers': skilled clerical, sales, service, crafts, and trades workers, as well as supervisors and foremen; captains and foremen of fishing vessels (code 7311) are included among foremen. 'Semi-skilled workers': semi-skilled clerical, sales, service, crafts, and trades workers; fishers (code 7313) are included among semi-skilled crafts and trades. 'Unskilled workers': unskilled clerical, sales, service, crafts, and trades workers; fish plant workers (codes 8217 and 8228) are included among unskilled crafts and trades.
‡ Fisheries-generated employment is calculated by multiplying captain and fisher employment by a factor of 0.7 and plant worker employment by a factor of 2.3 to account for fisheries-dependent employment (Canada, Environment Canada, 1976a).
§ The actual calculated proportion is 101.2 per cent.

employment in the rural competitive zone was ultimately calculated to be dependent on the fishers.

This description suggested two possibilities: a bifurcated wage structure in fish processing that reflected the industrial-rural split or a uniformly low wage structure resulting from unskilled work. The plant-survey data suggested neither pattern. Wages in the industrial zone were higher only than those in the rural non-competitive zone. Wage levels in the competitive rural area were as high as those in the industrial zone.

The industrial labour market had relatively many large plants in the 1984 plant survey. These in turn accounted for a very high proportion (79 per cent of 1,541 workers) of the zone's total fish-plant employment. The rural competitive market was dominated by competitive-plants, with seventeen of the ninety-three plants in this category. Competitive operations had 37 per cent of the area's fish-plant employment (1,091 workers in total). The rural non-competitive market accounted for 53.9 per cent and had an even distribution by plant type.

The 1984 survey revealed that the first two zones had comparable wage distributions, with a plurality of workers getting $6.00 to $6.99 per hour. The rural non-competitive market had decidedly lower wages: 59 per cent of 2,929 workers made less than $6.00 an hour in 1983, with wages especially low in the Acadian district of Clare in Digby County. The worker survey corroborated these findings: 71 per cent of workers in the rural non-competitive zone earned less than $7.00 per hour, while only 30.3 per cent of workers in the rural competitive and 59.2 per cent in the industrial competitive zone earned more than $7.00 per hour.

Two factors seemed to explain the high-wage zones: unionization and large operations in the industrial area and independent ownership, competitive-scale operations, and a predominantly male labour force in the rural competitive zone. Most unionized fish plants were large (seven of the nine unionized plants were large) and had a high proportion of workers in this area. The independently owned and operated plants in the rural competitive market (thirteen of the seventeen competitive plants) acted as wage leaders in that area. Regardless of surplus labour, demand for labour made wage levels for non-union men quite high. The only unionized plant had a 'negative' demonstration effect. One local manager remarked: 'We found that the union at [plant] was usually cheaper than what most plants were ... [The workers] thought the union would help them with wages but it didn't ... The other plants looked at it by saying that 'if we pay 10 cents more an hour than the union is getting, our workers will never want to see a union.' And this is what has happened.

Plant workers are largely unable to realize any sizeable benefits from high pay scales. Other factors appeared to be as important as labour costs in assessing the significance of surplus labour. Workers in the rural competitive zone (see Table 49), while earning relatively high wages, had the most seasonal unemployment and intermittent work and the least regular employment.[18] This reflects the high proportion of employment in competitive plants and those firms' dependence on the coastal-zone fishery. Further, employees were considerably more likely to work over forty hours a week, when they worked, than those elsewhere. A small, but substantial percentage were more likely to work over forty hours than were people in the industrial competitive zone, where the predominantly female work-force was more likely to be on the job less than forty hours per week.

Income from plant employment mirrored the employment distribution just described. Fifty-one per cent of workers in the rural competitive zone earned between $4,000 and $8,999 per year from plant employment. In the industrial competitive zone, 47.2 per cent of the female segment earned less than $4,000; 38.5 per cent of the male segment more than $9,000. In the rural non-competitive zone, 26.7 per cent of women earned less than $4,000, while 53.8 per cent of men earned $9,000 or more. Therefore, despite higher hourly wages, workers in the rural competitive and industrial competitive zones have incomes typical of a province-wide 'plant worker' syndrome of intermittent and seasonal work. In competitive plants, this syndrome appears to be a structural characteristic of their operations. In large plants, it seems to result from a managerial decision to reduce labour costs.

To increase flexibility during peak production periods, some competitive plants contract out portions of their salt fish production. According to several managers, this phenomenon – almost exclusive to this zone – has increased dramatically since the 1960s. Some firms had up to seven pieceworkers on regular contract. One manager explained:

In the plant you lose on percentage recovery because the people don't know what they were doing or don't give you 100 per cent [effort] when your back is turned. It is cheaper to put it out to these people because they are out of the way. You don't have to tie up a forklift or machinery to cater to the people who are cutting while you are splitting fresh fish. We have been exceptionally

TABLE 49
Seasonal employment (weeks worked) in fish plants by labour market location, 1986 worker survey (percentages, with numbers in parentheses)

Month (1986)	Rural competitive				Industrial competitive				Rural non-competitive				Total			
	0	1–3	4	(N)	0	1–3	4	(N)	0	1–3	4	(N)	0	1–3	4	(N)
January	89.5	8.8	1.7	(57)	75.9	11.1	13.0	(54)	40.0	5.9	54.1	(85)	64.3	8.2	27.5	(196)
February	80.7	8.8	10.5	(57)	74.1	11.1	14.8	(54)	35.0	8.4	56.6	(83)	59.3	9.3	31.4	(194)
March	70.2	14.2	15.8	(57)	75.4	5.7	18.9	(53)	31.3	6.2	62.5	(80)	55.3	8.4	36.3	(190)
April	60.0	16.4	23.6	(55)	64.2	11.3	24.5	(53)	21.0	16.0	63.0	(81)	44.5	14.8	40.3	(189)
May	42.3	32.7	25.0	(52)	62.2	9.5	28.3	(53)	22.8	7.6	69.6	(79)	39.7	15.2	45.1	(184)
June	40.4	28.8	30.8	(52)	57.7	9.6	32.7	(52)	20.0	17.5	62.5	(80)	36.4	18.5	45.1	(184)
July	30.8	32.7	36.5	(52)	37.7	13.2	49.1	(53)	16.0	22.7	61.3	(75)	26.7	22.7	50.6	(180)
August	7.7	36.5	55.8	(52)	9.6	36.5	53.9	(52)	12.2	24.3	63.5	(74)	10.1	31.5	58.4	(178)
September	17.3	30.8	51.9	(52)	26.9	16.9	46.2	(52)	11.7	22.1	66.2	(77)	17.7	26.0	56.3	(181)
October	35.8	34.0	30.2	(53)	48.1	21.1	30.8	(53)	20.8	13.0	66.2	(77)	33.0	21.4	45.6	(182)
November	58.9	21.4	19.7	(56)	57.4	22.2	20.4	(54)	31.3	20.5	48.2	(83)	46.6	21.3	32.1	(193)
December	73.2	14.3	12.5	(56)	64.8	14.8	20.4	(54)	34.9	14.5	50.6	(83)	54.4	14.5	31.1	(193)

busy because we have been getting a lot of fresh fish, and we really didn't have the area or the equipment to do that kind of volume. It was easier for us to put it out to these people because they bought their own skinning machines, their own electricity, and looked after picking the fish up.

Independent interviews with nine out-workers showed final annual production to range from 50,000 pounds for a part-time worker working only in summer and fall to 400,000 pounds for a family operation working from March to December, where the husband put in as many as seventy hours per week. Most out-workers were husband-and-wife operations, where the male split, boned, and skinned while the female finned the fish. In the summer of 1984, piece wages averaged forty cents per pound for the finished product. Exploiting cheap labour is clearly not a primary motivation, as all workers reported earning between $350 and $400 per week. In only two of eighteen cases were family members not paid for their labour. Annual incomes ranged from $5,000 for part-time work to $16,000 for full time. While some piece-workers said that they would telephone around every season to check on prices being paid, there was such a shortage of fish in 1984 that most took whatever work was available. Only two indicated splitting for more than one processor at a time, and most claimed a long-standing relationship with their current supplier. While the majority valued self-employment and unsupervised work, some indicated inability to get regular plant employment.

In one area in the rural non-competitive market, contracting out involves both formal and informal integration among small and large firms. Smaller operations handle the unstable components of market demand and recruit the more temporary work-force. This form is characterized by universally low wages.

The largest group of such feeder plants was made up of affiliates of a large operation in one of the province's Acadian districts. The main plant processed the high-value shellfish, as well as herring and groundfish, while the larger feeders did herring and groundfish. The relatively low value of the herring, as well as its seasonal character (particularly for herring roe), probably explains the feeders' decidedly lower wages. The owners and managers of the main plant are regarded highly as business people by plant managers outside their area. However, they may take advantage of local labour surpluses. As one manager outside the area puts it: '[A large company on the French Shore] is the best operation around. They are fair, but they control everyone. Their wages are probably not as high as they should be; they are lower than around here.' In addition, a more conservative social environment, reinforced by religious, ethnic, and linguistic ties, probably facilities labour control. One manager in this district commented: 'Of course, in [the Acadian district] it might be a little different than many places. Everybody knows everybody. Here, a person can get a reputation, but it's quite easy to get at the facts also.'

However, these bonds also limit the use of cultural similarities for personal advantage. The owners of the main plant are expected to take community and individual welfare into account in their business dealings, sometimes showing paternalistic concern for the well-being of employees. For example, one owner of the main plant stated: 'We have two guys from [town]. Every fall they seem to have financial problems of some sort. For example, their car breaks down, and they ask, 'Can I have a thousand dollars, please? Can you deduct $25 a week next year?' We've come to the conclusion that they don't need it, but they're worried about their jobs for next year. They're very good workers, they're no problem. We've had them here for a long time.'

Conclusion

This chapter has explored some unresolved issues in the sociological analysis of labour markets. Using data gathered in two surveys, it has investigated the institutional factors that condition and constrain the supply of and demand for labour in one industrial setting. Conceptually the study addressed the interaction between capital and labour in terms of industrial structure and location. The differentiated structure of the industry, and the locational configurations that correspond to it, while consequences primarily of resources and markets, have important secondary causes stemming from the conditioning effect of surplus labour. Three observations are salient in this regard. First, the industry is overwhelmingly located in rural areas with surplus labour. Second, capital has shown marked reluctance to locate in the metropolitan-industrial areas of Halifax and Lunenburg counties. Third, large-scale capital has avoided or has been singularly unsuccessful in locating in the Shelburne/Barrington/Argyle competitive labour market. Except in small-capital and subcontracting competitive plants, low wages do not much affect locational decisions. However, a pliable work-force with few alternatives is vital to the industry's existence, particularly for large plants.

Surplus labour influences capital's choice of technique. Resource and product peculiarities make the entire industry seasonal and labour-intensive. Choice of technique is a function primarily of market considerations, availability of state subsidies, and specific labour-market assumptions. Competitive operations, for example, choose labour-intensive strategies to maximize flexibility and to minimize overhead costs, based on a clear expectation that a skilled and experienced labour force can be drawn on as the market dictates. Historically, this labour force has been primarily male. Large operations choose more capital-intensive strategies, primarily to maximize productivity. The resultant job tasks are more routine, monotonous, and intermittent. In those areas with the greatest labour-market competition for males, female workers provide a conducive environment.

In a labour market where capital might be assumed to have a clear advantage, wage rates did not uniformly reflect such dominance. First, unions significantly improve wages in large plants, particularly for women. Second, high levels of competition among independently owned, competitive plants in Shelburne/Barrington/Argyle have increased wages for predominantly male workers. Third, large-scale capital tries to reduce labour costs – either by using part-time and seasonal workers or by contracting out to low-wage feeder operations – and competitive-scale capital contracts out labour-intensive aspects of salt fish production, in an effort to increase production flexibility while keeping plant overhead low.

11 Capital and work-force adaptation in Clare

Marie Giasson

Situated in Digby County, the municipality of Clare comprises about thirty villages and hamlets. The most populated ones stretch along the shore, the remainder being scattered over the area. Clare is the heart of Acadian culture in Nova Scotia, with a total population of nearly 9,675 inhabitants, close to 80 per cent of whom are francophone. Since the first settlers arrived on the shores of Baie Sainte-Marie after the deportation in 1755, to the present-day inhabitants of the Ville Française,[1] the community has remained rural and characterized by a great capacity for adaptation. For more than a century, forestry, fishing, and subsistence farming have dominated the local economy. Economic development remains fragile, depending on outside markets. Despite periods of intense activity followed by periods of recession, Clare has shown a certain prosperity in the last few decades, accompanied by slow but constant population increase[2] and relative ethnolinguistic stability.

This chapter examines the fish processing industry in the Acadian region of Clare, first sketching in the historical context and then considering in detail the structure of the industry, capital's organization and control of the work-force, and the social reproduction of workplace relations.

Historical context

Since the 1950s, fish processing has dominated Clare's economy. This resulted not only from the increase and diversification of fishing effort but also from the growth of local entrepreneurship (see chapter 3). In the last three decades, the number of fish plants has increased from seven to more than twenty, and two or three new businesses start up each year. The seasonal character of the fishery provides access to unemployment insurance and has also led to incorporation of women into the work-force. This situation modifies strategies of adaptation not only within the family unit but for the whole of the work-force.

Fishing-related activities in the regional economy of St Mary's Bay traditionally meant the inshore fishery: groundfish (cod, haddock, and hake), herring, mackerel, and lobster.[3] While the bulk of the fish was exported dried, salted, or smoked to the West Indies, fresh lobster was shipped to the Boston market. Canning factories were also started up in Cape Ste-Marie, Meteghan, Comeauville, and Le Bas de la Rivière (New Edinburgh). Introduction of the Cape Island style of boat increased fishing and the division of labour (see chapter 3). Wholesale fish merchants, generally ex-fishers, appeared and soon became entrepreneurs. They set up processing plants, specializing in herring (fillets, smoked, dried, or bottled) and salted and dried groundfish.

The economic crisis of the 1930s lowered prices and saw partial loss of the West Indies market. Not until the 1950s and deep-sea scallop fishing did Clare's fishing industry see further expansion and new companies. One venture initiated scallop harvesting and processing: 'Bay Seafoods' (a pseudonym) was set up by two local brothers and ex-lobster fishers. It equipped itself with a dozen 65-foot scallopers, as well as a plant specializing in the refrigeration and packaging of scallops, and began with smoked and salted herring. Today Bay Seafoods has several plants in Clare and elsewhere, processing scallops, herring, or groundfish, and is the largest employer between Digby and Yarmouth. Vertical integration, diversification, proximity to the American market, and the availability of manpower explain the economic success of fishing activity in the area since the Second World War.

Between 1945 and 1960, other fish plants also began, generally by specializing in fresh fillet production. Some were absorbed by Bay Seafoods, while others remain among the region's most important firms. While stimulating every aspect of the local economy – from vessel construction to related services (engineering, plumbing, electrical and mechanical work, carpentry, and the like) – the expanding fishing industry also opened a new job market and increased demand for qualified labour. According to MacInnis (1983: 15), nearly half of the active labour in Clare in the early 1980s was linked to the fishery sector.

In response to Japanese demand, the extraction of herring roe was integrated into fish plant operation about fifteen years ago and has increased enormously in the last six or seven years. All of the new fish processors in Clare have added it (see Table 50); many would have difficulty in surviving without it. Some small plants produce roe specifically, operating only for eight or nine weeks per year. This seasonal activity demands mobilization of considerable labour, and the industry must turn to other areas for additional workers.

Overall the expansion of fishing and fishing-related activities as well as the resulting increases in house construction, service firms, and wholesale and retail businesses and a proliferation of mink and fox ranching has had dramatic social repercussions. Development of fish processing has massively increased participation of women in the workforce.[4] Young people tend now to stay in the region rather than migrate to larger centres in search of work; even more surprising, more are returning, after having worked in Ontario or the west. These trends are reflected in an appreciable increase in Clare's population – 8,993 in 1976 to 9,675 in 1986 – and in vitality in housing construction.

Structure of the industry

Clare currently has more than twenty private companies, large and small, processing sea products. Production of fresh or frozen fish fillets remains the basic activity of most plants, except for a few that specialize in salt or dried fish, in smoked or pickled herring, or in shellfish. Distributed among many of the coastal villages, more than half of the plants are family businesses; others were formed by two, three, or four associates.

Bay Seafoods has the greatest production capacity and the most employees; it also offers a variety of marketed products, of which one, the scallop, has great value. Bay Seafoods influences the region's processing industry, particularly in division of labour and preferential bonds between certain companies. Thus, a few of the plants dealing in groundfish 'feed' fresh fillets to Bay Seafoods, which has intermediate and final operations, such as preparation, packaging, freezing, and marketing. Similar agreements are now taking place for herring roe production; plants with a good capacity for freezing and storing are thus becoming the intermediary between less well-equipped plants and potential buyers (see chapter 4).

Some companies can supply themselves partially or totally with catches from their own fishing fleet; a few own boats and fishing licences,[5] which assures them a minimum quantity of processable fish. Among

TABLE 50
Fish processors, products, and equipment in Clare district

Plant	Starting year	Products	Boat owners	Freezers
A (Bay Seafoods) (two plant sites)	1946	Groundfish Herring (roe and fillets) Scallops Smoked salmon Fishmeal	Twelve scallopers (offshore) Three herring seiners (one offshore) (one carrier)	Yes
B	1948	Groundfish Herring roe	–	–
C	1948	Groundfish Herring roe Fresh lobster Scallops	–	Yes
D	1981	Groundfish Herring roe Fresh lobster	–	–
E	1981	Groundfish Herring roe	–	–
F	1982	Groundfish Herring roe	Three boats (40', 60', 65')	–
G	1980	Groundfish Herring roe	One boat (55')	Yes
H	1984	Groundfish Herring roe	–	–
I	1980	Kippers Herring roe	–	–
J	1955	Groundfish	–	–
K	1912	Salt fish Herring roe	–	–

the eleven companies surveyed (see Table 50), two fished for ground-fish themselves. Plant G has a 55-foot boat, while plant F operated three boats: of 45 feet, 60 feet, and 65 feet. The latter firm, which began in 1982, is representative of a few fishers who reorganized themselves to form a fish processing enterprise. In the context of lower quotas and more plants, this proves to be a major asset. However, in most cases, independent fishers provide the basic product to the processors.[6]

The forces that structure the industry flow chiefly from idiosyncrasies in the raw material related to quantity and quality of a species. However, plants' technological capabilities, work organization, operations performed, products turned out, and market characteristics (labour as much as product) also play a part (see chapter 1). The main species fished in the region are groundfish,[7] herring, scallops, and lobster. The first two categories are the backbone of nine of the eleven plants in this study and the exclusive products of four of six new firms started since 1980. While the availability of the resource and a strong market explain initial growth, there is a growing problem of supply associated with the proliferation of plants and a decrease in quotas. Current processing capacity exceeds what total fishing effort can supply, and so processors are seeking suppliers from as far away as Cape Breton and New Brunswick – adding to transportation costs.[8]

Some fishers regroup to form a company and build their own plant. One manager whose plant has no fishing licence said: 'It's the boats that have the plant, and not the plant that owns the boats!' Some plant owners who have been in the fishing industry for a long time are unable to obtain a licence to procure and operate their own fleet. Competition is therefore high when buying fish, which helps independent fishers (see chapter 7). To guarantee supply, there are generally verbal agreements between these fishers and the plants. But processors with their own fleets have the definite advantage.

Local fish processors compete for groundfish and herring stocks. Despite substantial efforts during summer, which intensify during the spawning season for herring, actual catches (in the thousands of tons) are immediately absorbed by the local industry. A medium-sized enterprise can process a hundred tons of herring per day – if it only extracts roe. However, the industry then has to get rid of herring remains – thousands of tons of waste is liable to become a major source of pollution. Two options are available: to bury the waste at a specially designated site – only one exists in Clare – or to reduce the waste into fishmeal – only Bay Seafoods has the necessary technology. Signs of pollution have appeared, and the municipality declared that it would permit use of the burial site only for 1987. Many enterprises are therefore searching for other inexpensive options.

Difficulties in obtaining raw material – groundfish or herring – and in eliminating herring waste link plants in the region. Technological capabilities, however, differ in mechanical equipment and type of processing and in freezing capacity. A system of driving belts and conveyors carries the product to processing tables staffed by approximately twenty workers. Once the heads are removed and the fish are cut along the backbone, fillets are then directed toward the skinning machine. Fillets, depending on the plant's capabilities, will either end up on another production line, where they are trimmed, weighed, and packaged, or be put directly in bulk containers (fifty to sixty pounds) to be sent to another plant or to fresh fish markets. The whole process needs a good supply of fresh water, and some plants in Clare have problems of quantity and pressure. The quality of the water is otherwise very good.

A plant specializing in production of fresh fillets can make do with a cold storage room, while for another plant, aiming to make a frozen, packaged product, freezers are a necessity. Clare has four plants with refrigeration installations, but only two plants use them in groundfish processing. The other two use them to freeze herring roe[9] but foresee, over the short or mid-term, producing frozen fillets once they are able to acquire more fish.

Among the nine plants that process groundfish, only one regularly carries out the complete chain of production. By contrast, another plant sends off 80 per cent of its product fresh, round, or dressed; it makes fillets only if the market for round fish is weak. However, most plants producing fish filets have agreements with larger plants that are equipped to make a finished product, frozen blocks, or packaged and frozen fillets. Bay Seafoods, which buys and processes fillets for export absorbs production from four other plants,[10] three of them local. This division of work is generally satisfactory. As one plant manager said: 'They [Bay Seafoods] are there to find markets to sell to, while we're in the market to buy [from fishers].'

The same horizontal integration is found also in herring processing, with plants having refrigeration capabilities absorbing fresh herring roe produced by other plants. Of the eleven plants surveyed, ten produce herring roe. Four have freezer capacity which allows them to stock their own production and buy from others and gives them direct access to the export market – essentially Japan. One of them absorbs production from five smaller producers, having special arrangements (a type of consortium) with two of them. Bay Seafoods operates two of its own production plants for herring roe and buys from two others. One of these latter plants is already a feeder of Bay Seafoods in groundfish (fresh fillets), while the other acts as a subprocessor, also specializing in lightly salted kippers and smoked herring.[11] The other two processors with freezer capacity are Plant G, a new company selling directly on the Japanese market, and Plant C, the second biggest processor in Clare, all species considered.

In herring processing, 'mother companies' usually supply the smaller plants, fulfilling agreements with fishing boat captains for purchase and distribution. Bay Seafoods owns three herring seiners, which ensure a minimum quantity of fish to process. This supply is supplemented by purchases from independent fishers. Only Bay Seafoods can reduce herring waste into fishmeal. Other plants[12] send it their herring remains for processing.

Lobster and scallops are the other main species landed. Of the companies surveyed, two sell fresh lobster on the Canadian and US markets. Bay Seafoods holds the initiative regionally in the scallop fishery, operating a fleet of twelve scallopers and processing the scallops.[13] Scallop has a high market value, ranking first in value, and third in volume, in the firm's total production. Pacific salmon is also an important product for Bay Seafoods. Equipped with the most modern installations, the smoking plant operates year-round with about ten employees, processing salmon and herring.

Given frequency of landings, species fished (many of them seasonal), and types of processing carried out, the fish plants of Clare tend to be very active during summer and to slow down during winter. Almost all plants would prefer to operate for the entire year and are attempting to extend work periods. Ten of the eleven plants visited operate year-round as a rule, but production lines often halt for two or three weeks at a stretch during January, February, or March. Obviously the larger, more diversified plants operate the longest. But vertical and horizontal integration requires that Bay Seafoods start production as soon as fish is unloaded at any of the four plants supplying it with fresh fillets. Also, because Bay Seafoods uses herring stocks even before the females mature, it can start up as soon as the herring arrives in June – several weeks before the other plants. This situation, linked to other factors, has a positive effect on hiring and wage-setting by Bay Seafoods, especially during the herring roe season, when activity is intense in all the plants and fifty-to-seventy-hour work weeks are common. Because the company processes scallops,[14] as well as herring and groundfish, it often operates at peak capacity during summer, employing the firm's entire work-force.

The major processors (those that buy the production of smaller ones and offer a finished – packaged and frozen – product) have developed outside markets. Here again, Bay Seafoods leads by sheer quantity and diversity. However, market competition among major producers, occurs on national and international levels, not regionally (see chapters 4 and 6). With increasing demand, the problem is to gain the best possible prices while keeping production costs low. Being close to New England[15] is an advantage: the Clare region exports 75 to 80 per cent of its total groundfish production to the United States.

Certain small firms sell a bit of their production directly on the market. But fish is quickly perishable; most fresh fillet is sold to other companies in the region and reaches market packaged and frozen. For groundfish, a few companies in Clare (one specializes in salt and dried fish) deal directly with outside market agents. Bay Seafoods owns installations and a sales office on the US west coast and deals with distributors in Montreal and Boston, which handle the Canadian and the northeastern and southeastern US markets respectively. These distributors (brokers not exclusive to Bay Seafoods) see to product promotion, while Bay Seafoods makes constant efforts to maintain and improve product quality. Daily contacts with brokers and markets permit exchange of information on demand, prices, and production. Besides the North American markets for groundfish and scallop, Bay Seafoods has developed others, including Germany for herring fillets and Greece and the Caribbean for smoked herring.

All herring roe production is sold to Japan. For such enterprises, freezing capacity is essential. Four plants in Clare, including Bay Seafoods, have this capacity. Direct access to the Japanese market results in high profitability. As one plant manager emphasizes: 'Herring is the cherry on the cake – it makes a big difference if we market it [herring roe] ourselves – the profit we make from it we don't have to give to someone else for operations like freezing, packaging, and storing.' More tenuous outfits usually deal with a more powerful neighbour, rather than going directly to the market, ensuring faster sale and necessary working capital.

All these links between companies rely on oral agreements. As one manager stated: 'It's a matter of one's word. They're going to need us and vice-versa. In general, it works well.' In most cases, the same relations exist in companies' ties with independent fishers. Nevertheless, processors will advance money for purchase of new boats in return for the fisher's tacit obligation to sell all fish caught to the company. Such informal agreements rely on strong community ties and a tightly knit social fabric: 'In Clare, everyone knows each other.' Such arrangements have existed between certain companies for many years. New firms have to compete to buy the fish and tend to deal with more partners.'[16] Arrangements often extend into neighbouring counties.

Clearly larger plants with diversified production, such as Bay Seafoods, have more manoeuvring room and bargaining power as to type of agreements and price. But they generally do not threaten smaller firms, because they complement each other. According to one plant manager, 'It's not necessarily the larger one eating up the smaller one; it's often the smaller ones attacking each other.' Many plants have difficulty maintaining adequate production and adaptation to the stops and starts of an industry where supply can vary drastically from month to month. Extensions to buildings, construction of cold storage rooms, and addition of new production lines are all familiar strategies to extend processing capacity. But with declining quotas for many species, and market fluctuations, risks remain high. In the mid-1980s, however, increase in demand and fine performance of the fishery sector created confidence. The processing industry in southwest Nova Scotia in general and in Clare particularly performed well. Of the two or three new plants that intended to open in 1987, one was to be a sizeable herring processor employing more than a hundred people.[17]

Despite performance, however, anxiety was beginning to spread in 1987 – from plant owner to worker – and focused on one question: in view of quotas, would there be enough fish for all the plants to survive? Groundfish is the basis of Clare's processing industry, and most plants get supplies from inshore fishers who were demanding an increase in the total allowable catch (TAC). The key concern for plant owners, as well as employees, is to operate at full capacity and for the longest possible time during the year. As we shall see, an important factor for adaptation strategies of industry and work-force is the number of weeks of activity.

To summarize, interdependence and ties of complementarity among firms at the processing and market levels typify the fishing industry, even though Clare's processing firms are owned privately (by families or associates). Because competition is high at the buying level, companies owning their own fishing boats hold the advantage. Vertical and horizontal integration is characterized by division of labour among plants and by the dominant position held by Bay Seafoods because of its processing capacity, its diverse operations, and its share of the market.

The second part of this chapter will focus primarily on Bay Seafoods in an analysis of the organization and control of the work-force and labour market. The third part looks at social reproduction.

Organization and control of the work-force

Distributed unevenly among about twenty plants, most members of the fish processing labour force have seasonal, casual, and part-time jobs. Incomes are generally supplemented by financial assistance from the state. Plant production is at the mercy of the hazards involved in fishing and the annual exploitation cycle of different species; thus availability of workers becomes a key element in the production process (see chapter 10). Organization of work and distribution of jobs by gender imply substantial differences in each person's job time, or the number of active weeks that allow the worker to qualify for UI. Organization and control of the work-force rest as much on access to transfer payments (which in turn are tied to a certain amount of job security) as on wage or work conditions.

Labour force characteristics

Clare's fish processing industry employs more than 750 workers seasonally – and almost twice as many during the herring roe season. A little less than half of seasonal but regular workers are women (Table 51). Among casual or part-time workers, women count for close to 80 per cent. Thus women constitute the largest numbers in the work-force held in reserve (Connelly and MacDonald 1983; Ilcan 1986). During herring spawning season most plants get students to work at minimum cost. Casual workers, both men and women, are generally younger and more mobile than regular workers. Some come from as far away as Cape Breton. The majority of regular employees come from Clare or the immediate area. In the largest plants, particularly Bay Seafoods, places of residence are more diverse, and some workers travel up to 25 miles to work.

The average age of plant workers in fish processing depends on many factors: the firm's age, types of processing carried out, and the specific job. In general, new plants have relatively younger workers: the av-

TABLE 51

Composition of labour force by gender and employment category in Clare's fish processing plants in 1986*

Plants	Regular employees			Casual employees			Grand total
	Men	Women	Total	Men	Women	Total	
A†	128	160	288	20	80	100‡	388
B	35	5	40	25	60	85	125
C	30	50	80	10	30	40	120
D	34	16	50	10	55	65	115
E	25	25	50	40	110	150	200
F	32	20	52	5	25	30	82
G	18	17	35	5	30	35	70
H	25	10	35	5	30	35	70
I	5	18	23	5	20	25	48
J	11	4	15	–	–	–	15
K	26	4	30	5	70	75	105
Total	369	329	698	130	510	640	1,338

* Approximate figures, based on employee lists from each plant and data provided in interviews with plant owners or managers. They include office workers (secretaries, accountants, managers, and the like).
† Bay Seafoods (two plant sites).
‡ In the case of Bay Seafoods, casuals are mostly those hired by the plant, which operates only during the herring roe season.

erage age is twenty five to thirty years. Certain jobs (e.g. handling, working at the herring-filleting machines, working in a fish reduction plant) are physically more demanding. Younger personnel, are capable of such sustained work. Older female workers, from forty to fifty-five-years, are more likely to be packaging fillets or sorting scallop. The level of education of younger workers is higher than that of older ones, even though most of the foremen have not finished high school. A few female production-line workers have either secretarial or nursing-related training and could not find work in their fields or wished to stay in the region.

Work organization and job distribution by gender

The many tasks linked to production range from handling to salting, smoking, or freezing, through the different processing steps of gutting and splitting fish, skinning, trimming, weighing, and packaging fillets.

While certain jobs are generally carried out by men (i.e. unloading boats, loading trucks, storing in either cold storage or freezer, machinery operation, and maintenance), female workers form the majority at processing tables. Their many tasks generally enter into the category of unqualified work and can be learned in a few days.

The percentage of female personnel is a function primarily of the type of firm and its operations. The more finished the product (e.g. packaged frozen fillets or canned), the higher the number of female workers. Conversely, male workers are more important in plants where processing operations are basic or limited. The two most important local firms (Bay Seafoods and Plant C) that aim for the export market have a work-force of approximately 60 per cent women. Plants B, D, H, and J, which produce mostly fresh fillets, have a majority of male employees (see Table 51).

The proportion of women at Bay Seafoods is particularly high in scallop and groundfish processing, while in other sectors (e.g. refrigeration or fish reduction plants), only men are hired (Table 52). More and more women are hired to cut fish in new plants (E, F, and G), because of the difficulty in recruiting new cutters and because some women are learning and becoming qualified in a job traditionally occupied by men. This segmentation of the labour market according to task and gender has, as we shall later see, noticeable effects on the conditions, work time, salaries, and horizontal and vertical mobility of all workers.

The numerous production tasks carried out by women are similar from one plant to another. The work is repetitive and boring ('It's not interesting work, it's always the same thing'). Co-workers stand side by side, enduring standing, cold hands and feet, noise, absence of windows, and odours from fish and sometimes ammonia vapours.[18] These conditions demand physical stamina, particularly during peak activity: 'Even though I like working overtime, at the end of the summer I'm tired. If I can keep up the rhythm it's because it's only a few months a year.'

Some jobs are plainly more demanding than others,[19] and some people are more sensitive than others to particular conditions. Most plants allow employees to talk while working, as long as the production pace is maintained. Work teams are often formed, which seems to satisfy both supervisors and workers. However, organization depends initially on production demands, and often staff members are transferred around the plant. As a female packer from Plant C stated: 'Some

244 Emptying their nets

TABLE 52
Number of employees at Bay Seafoods according to sex* and production sector

Production sector	Men	Women	Total
Herring	34	61	95
Groundfish	19	54	73
Scallops	8	38	46
Smoking (herring and salmon)	5	4	9
Refrigeration	28	0	28
Fish reduction	11	0	11
Total	105	157	262

* Excluding supervisors, who are all male workers

days our work changes many times, and other days we do the same job from 8:00 to 5:00. It depends on what the plant has in fish.' Such movement initially involves less-qualified employees – more than 80 per cent of whom are women: trimmers, packagers, sorters, and weighers. In larger firms, male workers can rise to jobs on specialized work teams: machine operators, mechanical maintenance attendants, carpenters, and the like.

Bay Seafoods organizes its work-force somewhat differently because of its diverse operations and sheer volume. Teams of workers are assigned to various production sectors, and employees in a scallop processing plant will rarely be used in another sector, except during overtime. Thus, the firm ensures a minimum number of employees in each sector when all production lines are in operation.

'On call': job structure and overtime

Work-force organization with division of tasks according to sex, affects employees' work time. The majority of jobs in fish processing plants are seasonal or casual. Plants – especially smaller and more specialized ones – are almost deserted from December through February. This generally means five to six months of complete inactivity for the worker. The largest plants as well as the more diversified ones will attempt, however, to operate all year long. For the majority of plant workers, the number of weeks worked in a year can vary from approximately sixteen to twenty-eight. At Bay Seafoods, women employed by the groundfish plant work longer than women at the herring plant. The ten or so employees assigned to smoking salmon and herring, machinists, and refrigeration workers are active almost year-round. Availability for work and seniority also affect work time.

Crucial to being hired is availability, which, under present technological conditions, is tied to the idiosyncrasies of fishing. Plants are unable to predict accurately the rhythm and regularity of production, so managers must have a labour reserve that can be mobilized quickly by telephone calls[20] the moment fish arrives. This system demands constant adaptation by the employee and effectively controls the workforce. Although most workers accept this method, a few female informants find it difficult to fit family life to their job, especially if they have young children: 'You can't plan your day. Sometimes in the morning you don't know if you should start a wash, go to the grocery store ... You don't dare go too far from the house in case you're called.' The choice of workers to be called is based on seniority, availability, or proximity of residence to the workplace.[21]

Some companies, particularly Bay Seafoods, will help workers complete the minimum number of work weeks required for UI. Workers are called more frequently and offered various jobs. Thus, a female employee of the firm's scallop plant might work one or two weeks in groundfish processing in order to 'get her stamps.' A handler will be offered a few weeks of extra work in general maintenance, cleaning, and painting. This practice suits most workers and helps the employer in recruiting and retaining staff. Because of its diversity and its prolonged operations, Bay Seafoods can stretch work time by switching employees among production sectors or by giving them priority on the 'on-call' lists.

Nonetheless, mobilization and use of the work-force remain entirely at the employer's discretion. For example, two persons hired for the same task in the same plant worked twenty-one and twenty-eight weeks, respectively, in the same year. According to a woman employed by Bay Seafoods for the past thirty or so years, seniority and greater availability explain most differences: 'I'm like a senior and I can do all the jobs, so I have more work [weeks of work] than the others. They [the firm] know that I want to work.' It is thus harder for new recruits to 'get their stamps,' especially since UI laws require a minimum of twenty weeks of work to receive benefits the first year. The requirements are lower for employees who have been on the labour market for a few years.

Overtime is related also to availability of labour and the special character of the fisheries.[22] Since landings are irregular and fish need to be processed quickly, work schedules are often extended. Extra hours are generally in the evenings after 5 p.m. and on Saturdays, but rarely on Sundays. Work weeks of fifty to sixty hours are not uncommon during summer, especially in herring spawning season. Employees are generally expected to do extra hours in the job they do most regularly, but can refuse if the sector or job is unfamiliar. Such opportunities increase income, since some sectors need longer working days than others, particularly at the herring plant, where daily landings in season can amount to several hundred tons.

In general, male employees work more weeks during the year and put in more overtime. Owners and many workers still regard men as breadwinners. This division of labour directly affects the annual incomes of both men and women.

Wage conditions and transfer payments

The basic salary for unskilled workers is generally fixed according to regional averages. In the winter of 1987, the hourly rate was between $5.30 and $6.00. New employees of certain companies (such as Bay Seafoods) are subject to a three-month probation period at provincial minimum wages. In principle, all production-line workers in the same plant – except cutters, who are considered qualified – receive the same hourly wage, notwithstanding experience and years of service. In practice, however, wages are decided by the employer. Although seniority is not systematically recognized, especially among female workers, it often affects remuneration, priority on 'on-call' lists, and the limited forms of job security. The most available or most assiduous workers sometimes receive a bonus at the end of the year.

According to plant management at Bay Seafoods: 'All women get the same salary because otherwise it would create problems.' Supposedly seniority does not guarantee faster or better work, and competitiveness and jealousy might cause disputes. With men, however, seniority, stability, and personal situation are considered: 'If a guy has five or eight kids and you know he's a good worker, you're going to try to keep him. He's not the type who usually goes and shows his pay cheque to other people. You need him and he needs you. That's where we have the possibility of doing what we want.'

Above all, wage policy is focused on retaining competent, dependable workers. In a non-union shop, where everything between employer (manager or supervisor) and employee is negotiated individually, it is difficult for workers to establish trust and solidarity. Wage differences and sexual division of labour keep workers divided. Since annual wage increases over the last few years have only been 25 cents per hour, employees often plead their own case to their employer. Competition for workers increases as plants grow in number, and it is getting more difficult to keep workers. In early 1987, Bay Seafoods raised its wage by 65 cents per hour (from $5.35 to $6.00) for production-line workers, thereby matching some local firms and surpassing the rates in others.

Plants producing fresh fillets, which hire teams of cutters, offer an average hourly wage of $7.00.[23] Because of a shortage of qualified workers in this traditional male activity, women have become more and more numerous, especially in the newer plants. Some employers offer them a slightly inferior wage, claiming that larger fish must be cut by men. Women cutters, however, manage very well. In the opinion of many managers: 'In groundfish, women do as good a job as men, and are maybe a bit more stable.' As with contracting out, in the intermediate term, the use of women may be a way for capital to downgrade the trade of fish-cutting (see chapter 10).

At Bay Seafoods, the diversity and range of operations and the broader occupational structure – with managers (manager, accountants, supervisors), skilled workers (mechanics, electricians, machinists, carpenters), and support staff (secretaries and general maintenance attendants) – imply a wide salary range. Almost all these jobs, except secretarial, are occupied by men, and wage disparity therefore becomes linked to the gender division of labour. Workers in the fish reduction plant are among the better paid, making an average of $8.00 per hour. Tedious conditions – such as noise, strong odours, and intense heat, as well as extended and shift-work schedules (twelve hours per day, and twelve hours per night) – make recruiting personnel for that plant difficult. They also explain the higher salary for the dozen or so workers – usually men – who work there: 'They're among the best paid salaries in Clare!' declares a young man whose father makes only 25 cents per hour more than he does, even after thirty years of service in a different part of the same firm.

Evaluation of wage conditions would not be complete without discussion of overtime and transfer payments. Overtime hours are generally worth one and one-half times the basic hourly wage. Here again, slight differences exist among plants, depending on task and gender. For cutters, overtime is paid after a forty-hour week,[24] regardless of number of days. Extra hours on a Saturday afternoon or a Sunday – rare except during herring season – are paid at time and a half and double-time respectively. At Bay Seafoods, where there is a more marked gender division of labour, overtime is paid differently for men and women. Women are paid time and a half after eight hours of work per day, while men are subject to the forty-hour-per-week system. Since this practice was established a number of years ago, many of the positions occupied by men have become more regular, while women's positions are often casual.

Strong competition for workers during herring season has recently led to modifications of wage structure in many plants, with 'piecework' being introduced for herring roe extraction. Pieceworkers are paid approximately $8.00 for preparing a twenty-pound box.[25] Plants are currently outbidding one another to attract workers for the eight or nine weeks of activity. For some, like Bay Seafoods, this system creates a major wage disparity within the same work unit between female workers extracting roe and those filleting herring. However, Bay Seafoods keeps its workers, because it uses staff in diverse tasks, offering them at least the minimum number of weeks necessary to collect UI benefits. In any case, a female worker who leaves to do piecework risks losing her initial job once the herring season is over. For most of the people interviewed, length of work time is more important than wage conditions. Workers prefer more weeks of work to better wages – partly a result of the region's labour surplus and even more a strategy to gain access to transfer payments.

UI is an integral part of annual income for seasonal jobs in fish processing. Except for some regular employees (managers, supervisors, and office personnel), almost all interviewees depend on state assistance for part of the year. Informants indicated annual income (including overtime, UI, and social assistance) before deductions of $8,000 to $15,000 for a single person and $18,00 to $25,000 for families where both husband and wife work in a plant. Although many agree that they are poorly paid ('They [Bay Seafoods] could pay us more'), most female workers are relatively satisfied, particularly since the last increase, and recognize that they would not earn more elsewhere in the Clare region. The increase in the number of plants has certainly affected remuneration, but 'getting one's stamps' is a constant preoccupation and helps determine labour availability. Those who do not complete the required number of insurable weeks in fish processing – mostly younger workers and those new to the job – will sometimes work on mink and fox farms during the slaughtering season.

In most cases, UI represents more than one-third of annual income. Benefits vary according to number of insurable weeks[26] and average number of hours worked per week. The method of accounting for these hours is crucial. During a low-activity period, some employers group into one week work hours that were staggered over a two-week period. This tactic favours individuals who have already qualified for UI, by increasing their average income per week, and can penalize others.

Recruitment and labour mobilization

Given Clare's labour surplus, it is easy for the employer to obtain a work-force. Hiring is generally done informally and often favours people who have a relative already working for the firm. At Bay Seafoods, one out of three employees has a close relative working there (often in the same plant). A member of the community, with a good reputation socially and as a worker, will be favoured over an outsider. These practices also apply to students – sons and daughters of employees – hired for the summer.[27] Also, most workers are francophone: 'We're among our own. We all speak French.' However, such affirmation does not presume conflicts with the few anglophone workers coming from neighbouring areas. Rather, it helps keep the work-force in place.

Though the work-time relation appears the means by which some larger plants have maintained their work-force, there is some horizontal mobility. Younger and qualified workers, particularly men, tend to shift plants according to wage conditions or work offered, or for other, more personal reasons. However, female production-line workers who have quit their first job must sometimes wait several months before being rehired elsewhere.

Generally, women workers tend to be more stable than male employees. Some people have worked at Bay Seafoods for fifteen to twenty years and even for thirty to thirty-five years. As elsewhere, the average length of service is generally higher for women. Women and men, respectively, average seven and six years for each main activity sector (groundfish, herring, and scallop plants). This work-force stability, though appreciated, is not formally recognized by payment or job security. The company ladder has no more than two or three rungs (called assistant foreman, foreman, and section chief), with little opportunity for upward mobility. There is 'nothing to look forward to, no stimulation ... you feel that you're not getting anywhere.' The young are conscious of limitations of the plant work: 'It's a hard job, boring and repetitive; you have no chance of advancement; I do it because there's nothing else, not many work possibilities in the region.'

Bay Seafoods usually goes outside the Clare region to hire supervisors and fishing captains. It prefers to hire people unknown to the plant workers. Some informants deplore this practice: 'They [Bay Seafoods] should give more opportunities to the young. It would be more interesting to work for them, because we'd know we could advance. Since competition for manpower is starting to be felt it could be a way for them to get young people they could count on.'

Outside fish processing, job opportunities within the region are limited and future job prospects are not particularly promising.[28] As for employment in commerce and services, especially in jobs traditionally held by women (as clerks, secretaries, or nurses), one would require either training or 'production' – a parent or close friend who could 'open the door for you.'

Work relations and class consciousness

Employers and employees agreed that work in fish processing is not easy, even if conditions have improved. Workers, especially women on the production line, have almost no formal control over their work process. Decisions concerning distribution of tasks, operations carried out, and pace of production all rest entirely with management. A few direct and the rest take orders, particularly in large enterprises employing mass workers. In some plants, however, the supervisor, manager, or even owner actively participates in production.

The separation at Bay Seafoods is emphasized by the size of the plant and multiplicity of its operations. Indeed, top management has new offices completely independent from the plants, mirroring the ever-growing distance in social and work relations: 'The bosses are on one side. We don't see them too often. Sometimes they make the rounds and if you're not working they'll tell you off! It's changed [relations with the bosses]. You'd say it was closer before. When we had parties, people were together more.'

Though the majority of workers agree that relations between staff and management are relatively good, 'it's bosses on one side, employees on the other. I've been working there for fifteen years. It's always been like that.' Some attribute the changes to bureaucratization: each sector now has a manager, a general foreman, foreman, and assistant foreman. 'Before, we had more freedom. It's a lot stricter since we've had all these foremen.' Some women even say that the supervisors are more demanding and harder on female than on male workers: 'They're tougher on us women. The men have it easy ... They don't get bawled out like we do.'

Men have a lot more freedom to move around within the plant and are not constantly watched by supervisors, because of the types of tasks they are given. Women are subjected to tighter control, as much from co-workers as from supervisors. As fatigue and stress develop, the climate becomes tense and verbal clashes occur with supervisors and sometimes even between workers: 'We're like a big family at the beginning of the week, but on Friday we take care not to step on anyone's toes: everybody's tired.'

Nonetheless, most employees in the same plant share a feeling of belonging to the immediate group: 'It's changed. Before we were all together. Now there's a little group here, another one there. It's no longer like a family. Three-quarters of the people, we don't even know them.' At Bay Seafoods, the familial atmosphere exists mainly among workers in the same production sector. Shifting part of the work-force from one plant to another sometimes creates friction: 'All the women in herring get along. It's not like in the fish [plant]. I don't like the women in fish. You'd think it bothers them when we go to help them. They look at us as though we've come to take over their job.' 'In scallops, I got the impression that a lot of the women acted as if they owned the plant. When we go over to help, some of them think we're taking away their work from them.'

Because the feeling of belonging is so narrowly focused, individual manifestations of annoyance or discontent rarely end up in collective action. Problems relating to organization of work are generally settled on the spot by the supervisor. However, working conditions have given rise to collective protests at one plant in Bay Seafoods, which went as far as sporadic work stoppages. Since these demands were concerned mostly with workplace conditions (e.g. lack of heating, strong odour of ammonia), they were quickly and at least partly resolved.

Arguments about wages gave rise many years ago to organized meetings at Bay Seafoods: 'The girl who spoke got fired.' Threats of layoffs, and even of closing down the plant, impeded group solidarity. One female worker recalled: 'Once when we went to see the boss about raising our wages, he told us: "I'm only doing this to give you work. I could leave this tomorrow." And we know that it's true, so ... what can we do? Once we went on strike to get more heat. The boss arrived and told us: "I'm going to fire all of you and hire other people." And all the employees went back to work. So you don't have a chance – there are always those who back down.'

There is no local tradition of trade unionism, and paternalism is more or less accepted as the traditional 'way.' Most people find it normal that the bosses should decide arbitrarily who will work for them and for how long; tasks, wages, and other working conditions; and terms of relations between employers and employees. Workers scarcely see how they can influence these matters, except by 'going to see the boss' on an individual basis. Even though modern management techniques are being used in some fish plants (Bay Seafoods, for example),[29] paternalistic rapport permits better control of the work-force and restrains the development of class consciousness and formulation of group demands. Therefore, there are no formal negotiations based on a well-defined balance of power: 'They [management] tell us: "If you're not satisfied, quit!" But I don't really think they mean it, I think it's just talk. For sure, there are others to take the place but if there's no one left? They couldn't replace everybody!'

Awareness of potential strength does not, for several reasons, create a move toward unionization. Many employees have several years' seniority. Some of them, among the oldest employees – men and women alike – have known the company since its beginnings and respect and admire the initiators, two local fishers who 'worked hard to get what they have, and they're not going to give it away.' Moreover, for a majority of women aged thirty-five and up, fish processing constitutes their first and only paid work experience. They have no basis for comparison with other production sectors and no training for skilled, better-paid jobs.[30] Further, arbitrary hiring practices and a closeknit social fabric obscure the real balance of power. Workers' demands thus do not lead to a more structured employee organization.

Many interviewees do not see how unionization could improve working conditions: 'I think it's OK as it is; I don't think the union would change things.' Others clearly fear closure: 'We don't want a union. They [the firm] would close their doors if the union got in. That's a lot of jobs! But they [the owners] are the ones with the money. They could live without it. They wouldn't be worse off, it would be the workers.' In the event of a major work conflict – a strike, for example – some believe that 'the plant could still ship out the fish without processing it.'[31] According to a female worker, this threat has already been expressed by Bay Seafoods management. Therefore, fear of lay-offs or of job cutbacks affects work relations. The industry's recent expansion, however, encourages workers to develop a collective or class consciousness.

In summary, the gender division of labour, the work-time structure, and access to transfer payments allow the industry to organize and control its work-force, develop hiring strategies, and determine working conditions. Seasonal production and labour surplus keep the balance of power uneven. Capital is sustained by good economic circumstances and favourable marketing policies,[32] and labour is often divided over short-term objectives.

Social reproduction

The work structure of Clare's fish processing plants moulds the social structure of the community. Large-scale participation by women modifies the socioeconomic situations of family units and influences their life-styles and futures. The assignment of gender roles and the feeling of belonging are starting to change. The development of the industry, while allowing a marked cultural identity to be preserved, leads to greater economic disparity. Social and spatial polarization is heightening socio-political awareness.

Working women and family strategies

Jobs in fish plants are the principal source of employment for women in the region. Their arrival on the job market is thus tied closely to the development of the processing industry and to its seasonal nature. The working woman contributes greatly to family well-being. An additional salary and the accompanying transfer payments constitute an extra income and frequently cover all or part of a family's essential needs. Such is the case of single-parent families, where the mother almost always has custody and care of her children, and for other informants whose husbands are ill and incapable of working regularly. Even when a husband has a regular job, many women interviewed claimed that the family situation would be very difficult if they had to live on one salary only. Children's education, mortgage payments, the rise in the general cost of living, and the increase in needs to be met push women onto the job market. For younger women, either single or living with a partner, wage work in fish processing often represents the only means to obtain financial autonomy and individual and social independence. For a young couple, buying a house is a priority, and it often becomes a necessity for the woman to have a job.

Even for women aged forty and over, who have raised their children and for whom plant work represents mainly a chance to get out of the house, wage work brings financial and personal independence. Though their roles as wives and mothers remain a priority, they also become 'productive.' Women employed by a plant become part of the 'working world.'[33] Women's paid jobs thus affect individual and familial strategies.[34]

Despite some task sharing – 'the kids help around the house, and my husband starts dinner when he arrives before me' – meal preparation almost always falls to women – 'at night, I prepare the meals for the next day.' Wives are also chiefly responsible for their children's education. Many women who work in plants admit not having any – or enough – time to plant a vegetable garden, especially since working weeks in summer are often fifty to sixty hours long. Wood heating, while often combined with oil or electric heating, is a habit that has persisted, especially since energy prices continue rising. Off-months allow many workers to exploit nearby woodlots cheaply, and firewood becomes an important economic contribution in some cases.

Socioeconomic reproduction of the family unit now depends almost exclusively on wage work and transfer payments. These, in turn, have changed life-style and adaptation strategies. Women who work in fish processing are usually not fishers' wives or daughters (see chapters 10 and 12).[35] In many cases, both husband and wife work in a plant, which implies special organization of the household. Fish plant jobs mould the household's rhythm, with summer's intense and demanding activity being succeeded by a winter break, with house construction or interior repairs and more intense social and family life, including visits to relatives and friends, outings to the local Legion Hall to play bingo and to the Social Club, and greater participation in organizations, associations, clubs, or committees.

Women working in fish plants turn often to three main social and leisure interests. Most are involved in handicrafts (quilting, knitting, cloth-doll-making) at home, and some get together once a week at each other's homes to share these activities and exchange ideas and patterns.[36] Further, 'Frenchy's' (or 'Smithy's') used clothing stores are also a favourite meeting place: many women go there not necessarily to buy, but rather to 'have a quick look in' and a chat. These businesses partially replace the general store, with some retired men frequenting them for the same reasons. Finally, bingo attracts mostly older people; the younger ones prefer bars or restaurants.

Adaptation strategies are first of all economic: ensuring sufficient good work weeks to collect UI benefits. Further, a good season in the plant affects workers' level of consumption, and therefore local and regional commercial activities. The pulse of the community beats to the pace of production in the fishing industry and, in particular, the processing plants. Though this economic concentration produces more and more pronounced dependence, it is toward this sector that hopes of expanding the region's job market are directed.

Many deplore this situation, primarily because of the lack of work opportunity for younger, educated people. Most plant workers encourage their children to continue their studies: 'I encourage my kids to study. I wouldn't want them to work in fish – it's slave labour. I do it because there's nothing else – but it's not only a question of salary. It's boring, always the same work and there's no advancement.' With no certificate or diploma themselves, these workers often focus aspirations for their children on the few technical courses offered at high school or college: nursing, secretarial skills, or hairdressing for the girls, mechanics or electronics for the boys. University studies are still the privilege of a very small number; starting up your own company remains, in popular ideology, the most obvious and accessible route to material well-being and social prestige (see chapter 14).

The recent development of entrepreneurship – in fish processing and elsewhere – has accentuated social cleavages. Several families stand out by virtue of their homes and their buying power. As one informant emphasized, this differentiation is more and more noticeable in residential zones: 'I live in the centre of the village where the plant workers live. On that side of the village, older people live there – they are richer, inherited from their parents, etc. ... and further on, it's the rich people. We live in the least rich part of the village.'

Even though most interviewees consider themselves middle class, their incomes place them in the lower middle class, and some families (e.g. with a single parent or health problems) find themselves rather underprivileged, drawing welfare cheques along with UI and wages earned from the plant. Private property is more accessible than in a city, and apartments are quite difficult to find, so that most family units (even young couples) either own a house or are on the way to doing so. Mutual help among families and aid within the community still play a big role. The building of a home mobilizes family and friends: 'Around here people help each other out. If someone has a need, we're there. To build a house, for example. Around here, we're together. People help each other out in sickness, all kinds of things. They stick to each other.' This mutual aid is linked to a strong community identity which persists in spite of recent socioeconomic changes.

Social fabric and community identity

'We feel good here. We've never thought of living elsewhere.' Most interviewees confess that they are quite satisfied to live in Clare and would not want to live elsewhere. Almost all originate from the immediate region, and most have parents living within a few miles, if not next door or across the way. Everyone knows his or her neighbours, and use of community services, like the Town Hall or health facilities, increases chances to meet. Even the youth tend to stay, although it means getting a poorer-paying job – in this case, in the fish processing plant. This stability, coupled with the length of Acadian settlement in the region and, more recently, the concentration of jobs in fish processing, contributes to a tightly woven social fabric.

Family bonds remain strong. Visits among relations are fewer since the advent of television and women working ('You have less time to visit if you work'), but Sundays continue to be devoted to family gatherings. Even now, young couples tend, when possible, to build a home close to the parental home.[37] People help a child, sibling, or cousin build his or her home; they ask a mother, mother-in-law, or sister to mind the children; an uncle or aunt or cousin may help them get a job at the fish plant or elsewhere. At Bay Seafoods, in spite of a high number of employees, often five or six members of a family work practically side by side.

Working in the same economic sector, often in the same plant, creates common interests and a certain class consciousness. Although this consciousness is still faint in work relations, it manifests itself mostly through a feeling of equality and belonging to a 'big family' or 'the same gang.' As one informant put it so well: 'We're all dressed the same: boots on our feet and hairnets on our heads. We're all equal. Nobody's trying for a beauty contest here.' The people referred to here are on the production lines, not office workers. Feeling on equal footing weaves workers together and helps integrate newcomers (men and women), even those coming from a neighbouring community or a village outside the region. Most already know a few people in the fish plant before starting to work there, and relatives and friends sometimes find themselves on the same team.[38]

As for people, mostly anglophones, originating outside the Clare area, their language and culture do not seem to present many problems – at least in the workplace. One female employee at Bay Seafoods said: 'Everybody there works in English. They're French but they talk English, no problem.' Although most of Clare's Acadians express themselves just as well, if not better, in English, using an Acadian dialect and belonging to the 'francophone' community still shape community identity. Linguistic conscience is not absent in the workplace, at least for some: 'I'm happy to work in French. I don't like speaking English at all.' One worker said: 'In the fish, it's getting worse. The foreman is an English person among French. I don't know if it's because he's English but it seems like he knows everything better than others!' Others are proud to work for employers from the region: 'For the workers, it's not really important that they be from the area. But for the owners, yes, we're proud that they're from here.'

Despite greater socioeconomic differentiation, Acadian origin, language, and religion,[39] as well as place of residence, remain the roots of Clare's identity and feeling of belonging and underlie social and political cohesion and people's attachment to region and village. The municipality of Clare is a political entity that is almost totally francophone. There are representatives of Acadian stock at all levels of government. And while most female informants are not politically active, many personally know their representatives, having first met them on school-boards or through other local and regional activities: 'They're people from here.' Community-based cultural, athletic, and religious clubs, associations, groups, and events abound. People enjoy dances organized by the Royal Canadian Legion or the Fire Department, parish bingo, cultural or local talent shows, and the Festival Acadien. Social involvement appears to be no more and no less evident among plant workers than among others.

The processing industry participates little in local social life, except for sponsoring some sports teams. 'Apart from giving jobs, [company owncrs] don't do much in the community!' Many firms contribute to develop community services[40] or participate in public collections.

However, increasing work opportunities in fish processing allow Clare's inhabitants to stay in the area, which is what most of them want: 'I am proud to have found myself a job here.' Many – mainly married women who are not the major family support – would rather withdraw from the job market than leave the area to work.[41] Those who would accept 'exile' would do so only out of necessity, not by choice. Family ties, the community's rural character, and the living space and feeling of freedom it creates keep people there. But as for working in a fish plant: 'It's only a job!'

Adaptation strategies: labour market and cultural values

People in Clare are conscious of the exploitative system to which they are subjected. Class structure and class relations are part of everyday life in the fish plant, but workers' consciousness has not yet developed an explicit political and social dimension. Faced constantly with the problems of surplus labour, seasonal jobs, and access to transfer payments, workers feel, and are, very dependent on the industry and try to make the best of their situation (see chapter 14).

Paternalism still prevails in most plants. It can be seen as a means by which capital increases control over the work-force, but it is also used by the individual to secure a job and qualify for UI in a highly competitive labour market. As Barrett and Apostle (1987: 203) put it: 'The main plant owners are expected to take community and individual welfare into account in their business dealings.' The majority of plant owners in Clare district are – as is the case for Bay Seafoods – from the region, and people on the French Shore are proud of their local entrepreneurs and glad to be among their own. In this perspective, the presence of a union may appear to many as a threat to community unity and traditional values.

The 'one big family' pattern that still exists in the fish plants is also applicable to community life. Ethnic, linguistic, religious, and family ties constitute the backbone of Acadian culture and identity. The sense of belonging is also linked to territory. In order to remain in a community they feel part of, and to preserve an identity they value, fish plant workers, especially women, accept difficult working conditions. Ethnic identity and intense community life make up for dissatisfaction in the workplace. The observations presented here suggest that retention of traditional values gives the people of Clare security in a changing and more complex economic environment.

Though remaining a rural area, Clare district, through development of the fish processing industry, is becoming more and more integrated in a broader capitalist production system. Increasing numbers of jobs, participation of women, changing relations between owners and workers (through modern management techniques), and dependence on transfer payments all indicate a more polarized society, where production is the meeting-point between capital and labour.

Conclusion

Clare's fish processing industry is a young economic sector, in full expansion, and represents the most dynamic element of regional development. Efforts to diversify production and create fish plants specializing in specific products further augment the increasing number of companies. Although activities are seasonal, one person out of four gains the major part of his/her income from work in a fish plant. Use of a female labour force and the gender division of labour facilitate low production costs. Because of the organization of the work-force and the structure of work time, government must assume, during much of the year, the costs of social and economic reproduction. Transfer payments keep the work-force in place. Without this financial support, workers would have to seek jobs elsewhere, probably in another sector. This situation is not unique to the Clare region and is apparent throughout the Atlantic provinces. It exists as much because of the nature of the fisheries industry as because of the labour surplus. Further, workers have little bargaining power because paternalistic rapport still prevails in most of the firms studied, often obscuring the real balance of power. A tight-knit social fabric and pronounced community identity make owners, directors, or plant managers influential people whose material success creates respect and admiration, if not envy.

A certain class consciousness is developing, however, as social cleavages become more pronounced. Desire to remain in the area and attachment to family and culture keep most people accepting of low salaries and difficult working conditions. Lack of education and relatively high regional and provincial unemployment rates greatly limit social and spatial mobility. Paid work done by women, however, allows for socioeconomic reproduction of family units within the narrow framework of the village, where mutual aid and community organization compensate partly for the job market's structural weaknesses and limitations. In short, plant-type work for the people of Clare is a means to make a living while preserving one's culture, values, and identity. It thus become a life-style, insofar as social reproduction of the community become more and more conditioned by development of the fish processing industry.

12 Communities and their social economy

Richard Apostle
Gene Barrett

Industrial and locational differentiation in the fish processing industry cannot be understood without reference to community structure. As is observed in chapter 11, the industry not only reinforces decentralized patterns of rural settlement but thrives on the social conditions that these community structures regenerate. This reciprocal tie represents the crux of the social economy of fish processing in Nova Scotia. This chapter explores these connections in more detail by analysing community attachment from three different viewpoints: those of plant managers, fishing boat captains, and fish plant workers.

We found heavy reliance by capital on communities as well as strong personal satisfaction among fishing boat captains which derives from living and working in an environment in which work and community are closely linked. Plant workers were less satisfied and confirmed a typical dilemma of rural community life. Apart from disenchantment with the intrinsic nature of plant labour, workers' dissatisfaction is related also to increasing stratification in fishing communities between families of fishers and those of plant workers. More important, it is linked to the patriarchal structure of the household. The problem is most acute for plant workers, most of whom are women, since women's work forms the unacknowledged basis of community life.

Plant managers[1]

A major dimension of the relationship between fishers and fish buyers concerns the link between fish plants and the communities within which they operate. A large percentage of plant operators (owners and/or managers) have lived in their community since birth. Moreover, a majority of the firms surveyed – particularly competitive plants – obtain their supply of fish within 15 miles of the plant (see Tables 53 and 54). As might be expected, larger fish plants obtain significant portions from distant locations, but even the largest obtain over half of their raw material from points within 15 miles of the plant gate (Table 53). The connectedness of fish plants to communities is emphasized in the geographical location of firms. As the data in Table 54 show, 85.8 per cent of Yarmouth County's plants, 75.9 per cent of those in Shelburne County, 55.5 per cent of those on the South Shore, and 64.3 per cent in Cape Breton and eastern and northern Nova Scotia obtain all of their supply from with 15 miles of the plant. Only firms in Digby/Annapolis do not conform to this pattern. There, special conditions – such as a decline in local groundfish and scallop landings and intensive herring processing – have compelled processors to obtain supply from more distant locations (see chapter 8).

As in Gangen Harbour (chapter 9), such intense connectedness between plants and supply, coupled with processors' personal links with their communities, translates into great social familiarity between fishers and buyers. They often grow up in the same communities, attend the same schools, and participate in the same organizations and activities.

As we have seen, wage rates in the fishing industry are generally below those of the food and beverage and the manufacturing sectors in Nova Scotia. Fish plant workers often constitute a 'captive labour force.' Regardless of plant size or geographical location, operators reported that 80 per cent or more of their employees live within 10 miles of the plant gate. Most of these plants are rooted in communities, drawing on local labour. They represent, especially for women, the primary, if not the only, source of local employment.

Being part of a captive labour force has advantages as well as disadvantages. When asked about hiring decisions, 52 per cent of managers surveyed indicated that prospective employees should come from the surrounding community; many felt an obligation to offer available work first to local residents. Fully 83 per cent noted that the reputation of the applicant was significant. Workers who have a history of transgressing conditions of employment (rules, job specifications, and so on) are quickly known and subject to informal 'black-listing.' Given employment dependence on the plant, this condition most likely compels conformity.

Another indicator of labour dependence is stability of employment.

TABLE 53
Distance from plant within which fish is purchased, by plant type (percentages)

Plant type	Less than 1 mile	1–15 miles	Greater than 15 miles	(N)
Small	39.3	25.0	35.7	(28)
Competitive	51.4	20.0	28.6	(35)
Large	33.3	22.2	44.4	(27)
Total	42.2	22.2	35.6	(90)

TABLE 54
Distance from plant within which fish is purchased, by geographical location of fish plants (percentages)

Location	Less than 1 mile	1–15 miles	Greater than 15 miles	(N)
Yarmouth	42.9	42.9	14.2	(14)
Digby/Annapolis	33.3	8.3	58.3	(24)
Shelburne	55.2	20.7	24.1	(29)
South Shore	33.3	22.2	44.5	(9)
Cape Breton, eastern and northern Nova Scotia	33.7	28.6	35.7	(14)
Total	42.2	22.2	35.6	(90)

When asked what proportion of their workers had been employed for five or more years, 26 per cent of plant operators with businesses over five years old replied over 76 per cent, 44 per cent responded over 51 per cent, and 64 per cent stated 26 per cent or more. A substantial core of the labour force has remained continuously employed in those plants. This pattern was particularly strong in the centre of competitive capital in the industry. Fifty per cent of operators in Shelburne County reported that 76 per cent or more of their staff had been employed for five or more years, while 65 per cent claimed that over 50 per cent of workers had remained for at least five years. No other area of the province reported figures approaching these. As noted earlier, Shelburne County operators also pay among the best hourly rates.

When asked about difficulties in hiring labour, 81 per cent of managers replied in the negative. Eighty-nine per cent of the operators in the management survey indicated no unionization. Many argue that it is their prerogative to establish wage rates, working conditions, and the like, since they have invested capital and take the attendant risks. They suggest that, while their business must realize positive results in order to remain operating, they are not insensitive to employees' needs. One owner noted: 'I live here too, you know. I was raised here. I've known these people all my life. We do our best for them. Got to. Otherwise how could you walk down the street and look anyone in the eye.' Unions would only disrupt what they saw as reasonable relations with their employees.

Fishing boat captains

The social life of Nova Scotia fish captains is characterized by strong attachment to community, substantial intergenerational occupational succession, and comparatively high satisfaction with work. These interrelated factors form the social core of post-war rural reconstruction. The captains in this study exhibited very high work satisfaction; 85 per cent said that they would go into fishing if they had their lives to live over – a proportion found normally in professional groups and indicative of real satisfaction with work conditions.

These captains were also considerably more likely (Table 55) than other private-sector work groups to label their current job the best they had ever had (55 per cent v. 25 per cent, respectively). A factor analysis (not shown here) of a twenty-six-item list of questions about work satisfaction demonstrates that captains value their work in terms of preference for inshore work, sense of independence, earnings, and time for family activities.[2] Answers to specific questions, which are reasonably consistent with patterns identified elsewhere (Apostle, Kasdan, and Hanson 1985), indicate the significance that fishing captains[3] in Nova Scotia attach to the survival of a viable inshore fishery which will permit them to retain such freedom.[4]

Sixty-five per cent of the captains still live in their communities of birth. Over 72 per cent have fathers who are or were fishers, and many currently fish with relatives. (Where the median crew size is 2.10, the median number of relatives included in these crews is 1.20. Captains exhibit high integration within their communities: over 85 per cent of them are married, and over 89 per cent own their homes.[5]

Homeownership is one of the main assets and bonds in Maritime culture.[6] Land for building is more readily available, and one can obtain materials and other building assistance through kin and friends.

TABLE 55
Job satisfaction levels among different Maritime samples* (percentages)

Possible responses†	(1) Fish plant worker's survey (1987)	(2) Captain's survey (1986)	(3) Southwest Nova Scotia fishers (1982)	(4) General Segmentation Sample (1979)	(5) Marginal Work World Sample (1977–8)
The best job you've ever had	15.5	55.5	41.4	25.1	13.1
One of the best jobs you've ever had	28.9	22.4	18.7	25.2	21.1
Above average job	45.7	7.2	9.8	15.0	17.3
Average job	9.9	14.1	26.8	30.8	38.9
Below average job	0.0	0.4	1.7	3.0	6.1
One of the worst jobs you've ever had	0.0	0.0	1.0	0.5	2.1
The worst job you've ever had	0.0	0.4	0.7	0.4	1.4
Number in sample	(232)	(403)	(418)	(930)	(427)

* Columns 1 and 2 present figures from this research project's two surveys. The survey in column 3 covered 597 fishers (both captains and crew) in the five western counties; the sample covers all fishing communities from Northwest Cove in Lunenburg County to Westport on the southern tip of Digby Neck (Apostle, Kasdan, and Hanson 1985). The surveys in column 5 covered 587 private-sector low-wage workers in these provinces (Apostle, Clairmont, and MacDonald 1978). Column 4 presents figures from a 1979 survey of 1,513 workers employed throughout the private sector in the three Maritime provinces (Apostle, Clairmont, and Osberg 1980).
† For columns 1, 4, and 5, the question reads: 'Overall, how satisfied are you working there?' For columns 2 and 3, the question reads: 'Overall, how satisfied are you with fishing?' The responses for columns 2, 4, and 5 include only those who have held more than one job, and the answers in column 4 include only those who have had more than one employer.

Over 59 per cent of the captains report that they built their current homes. Typical for small communities, 51.9 per cent of the captains know all their neighbours well, 32.0 per cent most of them well, 10.2 per cent some of them well, and only 5.9 per cent a few or none well. Correspondingly, 70.0 per cent think that they 'really belong' in their community, and the other 30.0 per cent think they 'belong,' while none reports feeling belonging not 'very much' or 'not at all.'[7]

Plant workers

In contrast to fishing boat captains, workers in fish plants have less work satisfaction and community attachment. Both captains and fishers generally manifest high overall job satisfaction. Over 55 per cent of the captains in one study, and over 41 per cent of a more general sample of Nova Scotia fishers in comparison to 25.1 per cent of private-sector workers in the Maritimes, thought their work the best they had ever had. Plant workers in the study closely parallel other marginal employees, as only 15.5 per cent of the former and 13.1 per cent of the latter made such a claim.

In general, plant workers have fewer integrative ties with their communities than fishers and demonstrate less attachment. Although the differences are attributable partly to age, plant workers have low marriage and homeownership rates, very similar to those of other low-income workers in the Maritimes. Over 85 and 89 per cent, respectively, of captains are married or own homes, compared with less than 60 per cent of plant workers. Proportionately fewer (about 10 per cent fewer) plant workers know their neighbours well or feel that they 'really belong' in their communities.[9]

The post-war era has clearly witnessed increasing stratification in fishing communities between 'fishing' families and 'plant worker' families. Connelly and MacDonald (1983) point out, for example, that, with professionalization of the inshore fishery through limited-entry regulations, fisher's incomes improved and their wives left fish plant work to pursue household work full-time.[10] Fishers tend to follow the occupation pursued by their fathers and to marry into fishing families (Table 56). Over 73 per cent of captains have fathers who were fishers and over 42 per cent of their fathers-in-law were fishers. Further, about 70 per cent of brothers and 64 per cent of sons were fishers. Approximately one-third of fish plant workers came from fishing families, or married into them, while only 17 per cent have spouses who fish. Twenty-seven per cent had spouses who were also fish plant workers. Over 24 per cent of all general household members, save for sons, are themselves plant workers.[11]

Household work

A limited amount of research has been done on women in Canada's Atlantic fishery, and Marxist-feminist theories have been most evident.

TABLE 56
Kinship occupational affiliations (percentages, with numbers in parentheses)

Affiliation	Fishing captains		Plant workers	
Fishers				
Spouse	6.5	(170)*	17.4	(172)
Father	73.3	(396)	34.9	(284)
Father-in-law	42.1	(314)	38.6	(167)
Son	64.0	(75)	13.9	(36)
Brother	69.6	(23)	37.5	(24)
Most common household members				
Father	40.0	(5)	24.1	(58)
Daughter	5.0	(40)	0.0	(23)
Sister	0.0	(0)	0.0	(10)
Mother	0.0	(1)	0.0	(54)
Plant workers				
Spouse	16.5	(170)	27.4	(172)
Father	3.0	(396)	10.9	(284)
Father-in-law	4.8	(314)	7.6	(145)
Son	5.3	(75)	13.9	(36)
Brother	0.0	(0)	25.0	(24)
Most common household members				
Father	0.0	(5)	24.1	(58)
Daughter	15.0	(40)	43.5	(23)
Sister	0.0	(0)	60.0	(10)
Mother	0.0	(1)	31.5	(54)

* All eleven women who are categorized as 'fishers' by the exclusively male captains in the survey act as crew members on boats owned by their husbands. As Allison (1988: 255) points out, this pattern means that while 'women are working in an occupation considered non-traditional for women,' they are on board in customary male-female 'relationships.' Most of the women match their husbands' full-time efforts in fishing ('Wife fishes alongside her husband in the boat,' 'Fish together'), but some do have part-time involvement ('Fishes in June'). Nine of the eleven work on smaller direct-producer operations, and nine are located in the more economically marginal Cape Breton and eastern Nova Scotia area. In a majority of cases, these women make an important contribution to household income, at least partially through their UI earnings.

As Porter (1987) points out, both liberal and Marxist social scientists doing fisheries research in the Atlantic region have neglected women's issues. Government publications, like the Kirby Report (Canada, Task Force on the Atlantic Fisheries, 1982), and the substantial body of academic literature on fish harvesting have little, or no, mention of women's concerns. The recent literature on women in the Atlantic region (particularly Nova Scotia) has looked at their role as a labour reserve. Women accept more seasonal, part-time jobs with lower wage levels and thereby subsidize the fish plants and their communities (E. Antler 1977; Connelly and MacDonald 1983; Ilcan 1986).

Porter (1987: 44, 45, 52) suggests three areas that require more investigation. First, marginality can be used to analyse the overlapping experiences of women and their communities. Second, women's activities inside and outside the household can reveal the amount of subordination that is based on gender relations. Third, women are actively involved in family and community: social reproduction does not occur automatically or passively; women are key actors in the cultural life of their communities.

In a wide variety of social contexts,[12] women do a large proportion (perhaps 80 per cent) of household work. Further, there is strong gender-related division in tasks: women concentrate on the time-consuming work of cleaning, meal preparation, and child care and men on repairs, garbage removal, and tasks outside the house (Miller and Garrison 1982). Men are beginning to perform some traditional women's tasks when women take up employment outside the home – usually they assist women in tasks for which they still have primary responsibility. Seccombe (1987) refers to this as 'helping her out.'[13] Households tend to revert to a more traditional division of labour when a woman leaves outside employment (Thiessen 1987).

Analysis of the data on division of household labour among plant workers confirms many of the patterns found in other, similar settings.[14] Women assume major responsibility for the more onerous tasks, like laundry, house cleaning, dish washing, and child care, and usually take care of the medical needs of children, shop, attend parent-teacher meetings, and run family finances. Men are more inclined to do traditional men's tasks of auto, appliance, and home repair, as well as wood-cutting. They are also slightly more likely to do the gardening. Among younger workers (whose parents still do most of the work in the home) female respondents are consistently more likely than men to say that they do some of the tasks traditionally defined as women's work (77 per cent v. 47 per cent, respectively, sometimes wash dishes). Similarly, younger males say that they do repairs, wood-cutting, and gardening more frequently.

Women with paid work get little relief (except in the minds of their spouses)[15] from more demanding tasks (Table 57). First, they do ap-

TABLE 57
Self-reports on the division of household tasks (percentages, with numbers in parentheses)

Task*	Husband – with wife doing paid work (44)			Husband – with wife not doing paid work (29)			Wife – with husband doing paid work (71)			Wife – with husband not doing paid work (21)		
	Never	Some-times	Always (N)	Never	Some-times	Always (N)	Never	Some-times	Always (N)	Never	Some-times	Always (N)
House cleaning												
Self	39	61	0	65	28	7	1	51	48	4	58	38
Spouse	0	75	25	7	45	48	59	39	2	48	42	0
Food preparation												
Self	27	71	2	52	45	3	1	65	34	5	67	29
Spouse	2	84	14	3	55	42	37	62	1	43	57	0
Child care												
Self	4	92	4 (24)	22	78	0 (18)	0	85	15 (39)	10	70	20 (10)
Spouse	4	92	4 (24)	6	78	17 (18)	16	84	0 (39)	20	60	10 (10)
Home repairs												
Self	5	61	34	11	43	46	44	48	8	63	37	0
Spouse	50	50	0	57	39	4	11	56	33	11	47	42

* The question was: 'There are many household tasks which people do on a regular basis. For each of the tasks mentioned below, please tell me who did it last year, and how often. Never/Some of the Time/Always/Not Applicable.'

proximately the same house cleaning (48 per cent always) as wives of male plant workers who are not doing paid work (38 per cent always). Second, they report little change in workload when their husbands are unemployed. In these two groups, there is little variation in 'women's tasks' and only a modest inclination for men to do more of the 'men's tasks' when they were unemployed.

As Porter suggests, women plant workers are indeed active in community affairs. Women are slightly more likely than men (43.9 per cent v. 93.7 per cent) to belong to voluntary organizations, which J.M. Acheson (1988: 26–9) identifies as one set of key institutions in fisheries communities in central Maine. Women are significantly more likely to be involved in educational organizations and Christmas clubs; men, in fraternal organizations and volunteer fire departments.[16] There is virtually no formal political activity, with only two plant workers (one woman, one man) indicating that they belong to a political association. Women are significantly more likely (24.7 per cent v. 10.2 per cent) to report that they attend church at least once a week. Community involvement is shaped by gender, with (fisher)men being more politically active (see chapter 14 below). Women are even more active, chiefly in educational and religious activities, and help maintain a number of key social institutions.

Conclusion

This chapter suggests the dense social and economic connections among plants, fishers, and plant workers. Fishers are tied in a variety of ways to the processors to which they sell fish; plant workers are frequently dependent on plants for any employment, and the rural communities in which all these people work have their futures closely linked to the future of the fishing industry. The social bonds and subjective commitments typically cited as positive elements in smaller, more rural settings are not uniformly distributed. Whereas coastal-zone fish harvesting is associated with some of the more rewarding social and economic aspects of life in maritime communities, fish processing resembles marginal employment in other industrial sectors and more urban settings in producing fewer economic or cultural rewards. These various patterns within rural communities clearly affect differently the viability of these communities and their life-styles.

Plant workers, particularly women, do much of the work necessary to ensure the survival of these communities and receive little recognition. Their contributions take three forms: paid work in fish plants, unpaid work in the household, particularly in raising children, and voluntary community work. Without this work, the fishing industry and rural social structure would collapse.

13 Household and workplace strategies in 'Northfield'

Pauline Barber

This chapter's discussion of structural change and sectoral linkages in the regional fishery focuses on industrial northeast Cape Breton, a unique labour market placement for the Nova Scotia fishery. It examines historical and cultural conditions affecting interrelationships among households, labour markets, and processing plants. The main purpose is to demonstrate the vitality of cultural commitments associated with social relations in various plants.

Although this processing sector is located in an industrialized labour market, the area has long had a labour surplus. At the turn of the century urban settlement expanded to serve the prosperous coal and steel industries of Cape Breton. Coastal communities dependent on these industries have faced chronic economic uncertainty. Class conflicts, including periods of labour militancy unprecedented in Canada, underscore the region's industrial history (McKay 1983). Decades of state intervention have failed to reverse the deindustrialization that has made the region peripheral.

Under these unstable conditions, Cape Breton has displayed tenacious commitment to locality and way of life. Routines of everyday life and strategies for contending with hardship reveal contradictions generated by the configurations of capital, community, and culture.

Economic restructuring has caused ripples of concern and sometimes provoked conflict throughout the fish processing industry. However, unlike fish plant workers elsewhere in Nova Scotia, those in Cape Breton live in communities with a history of proletarian wage labour.[1] Cape Breton thus provides a singular illustration of the regional fishery situated in an industrialized class culture.

Historical and cultural processes are represented differentially in the routine of everyday life, at home and in the workplace. This discussion is concerned primarily with a processing-plant workplace employing more women than men. Historically, male-dominated mining and steel industries have experienced a strong union movement and workplace traditions.[2] However, class culture cannot be read off solely from the predominantly male work worlds and labour history of mining and steel – nor indeed from fish processing.

As recent feminist theories have maintained, social relations of the workplace and of the family household are brought together or conflated within the household. The inherent dualist and productionist bias in Marxist theories of labour neglected domestic activities and ideological structures which (as feminist critiques have noted) are associated with processes of social reproduction.[3] Ideologies about work, about class and gender, and about family and community persist over time. The complex attitudes and types of behaviour that give form to ideology and culture are nurtured in the family household and in the workplace. But it is the household that displays the fullest extension and the intersection of these ways of making sense of one's life and circumstances.

More concretely, issues of history, culture, and community surface at the household level when workplace experiences are described, interpreted, and planned for by family members. As well, household-derived labour converts wages into necessary goods and services. As well, in industrial Cape Breton, self-provisioning and maintenance activities result in a highly developed informal economic sector. And in the household we can examine the consequences of processing's wage and cyclical work structure in concert with household ties to other industries and inputs from the state. As with labour markets, however, ideologies of gender constrain household divisions of labour and may render allocation of resources and income less than consensual, occasioning political negotiations according to gender and sometimes age.

Economic strategies for 'making ends meet' are constructed within households according to local conventions, using available resources of labour. These strategies ensure reproduction of households to the locally established, class-related standards. Householders participate in labour markets that are in turn structured by regional, national, and international processes of political economy.[4]

Evidence of this historically constituted class context is abundant in plant workplace traditions, experience, and the basic shaping of production politics, especially escalation or containment of conflicts be-

TABLE 58
Fishing fleet in industrial Cape Breton–Northfield area

	No. of vessels	% of Nova Scotia total
Offshore (over 65')	31	17
Nearshore (45'–65')	13	6
Inshore (less than 45')	641	11

Source: Canada, Employment and Immigration Canada (1987)

TABLE 59
Processing plants in industrial Cape Breton–Northfield area

	No. of employees
Ocean Products (processing fresh and frozen)	180–200
O'Brien's (processing fresh and frozen)	50–60
Harbour Lobster (retail/wholesale, some processing fresh)	9–15
Feldberg's (retail/wholesale, some processing fresh)	6–12
International Seafoods (buying station only)	6–12

Source: Field notes

tween management and workers. Local variations in the organization of the labour process reveal how different management strategies manipulate social forms within communities, such as kinship networks and familial ideologies, created in part by the domination of mining capital earlier in community history.

What will be argued here is that within industrial Cape Breton culturally mediated relations of kinship and community provide both labour and management with workplace strategies facilitating their divergent interests. This occurs under 'benevolent' or 'authoritarian' styles of paternalistic management and under the now familiar corporate ideologies.[5] And it is precisely in analysing production politics and the ongoing articulation of workers' resistance that historical and cultural perspectives become most compelling.

In 1987, Cape Breton supported a viable inshore-midshore fishery (see Table 58), despite widespread tendencies toward industrialization, capital concentration, and vertical integration (Barrett 1984).[6] Federal initiatives for restructuring the processing sector had accorded clear provincial advantage to National Sea Products (NSP) by the late 1980s. But the processing sector continued to support small and medium-sized enterprises as well (Table 59). Clearly, this diversity further confirms the need for closer scrutiny (Apostle and Barrett 1984; Barrett and Apostle 1985).

TABLE 60
Major employers in Northfield area

Employer	No. of employees
Northfield Community Hospital	257
Cape Breton Development Corporation (DEVCO)	240
Northfield General Hospital	238
Ocean Products Fish Processing Plant	180
Sea Haven Nursing Home	75
O'Brien's Processing Plant	50
Blacks Hardware and Auto Repair	40
Sobeys Supermarket	35
Metropolitan General Store	32
White's Corner Construction	30
IGA Supermarket	30
Bank of Nova Scotia	29

Source: Canada, Employment and Immigration Canada (1987)

The industrial community studied, herein referred to as Northfield (a pseudonym), is sizeable by the standards of many Nova Scotia fishing towns.[7] The current population is around 20,000. Coal and steel have been the principal industries; the fishery and to a lesser extent forestry have played secondary roles. In the 1980s, service occupations and mining absorbed the biggest portion of the labour force, followed by construction and fish processing (Table 60).

This study is based on participant-observation research incorporating approximately forty-five semi-structured open-ended interviews with workers in fish processing plants, family members, and plant managers. Informants were selected on the basis of gender, age, family status, job classification and history, and household involvements with wage work. Informants were chosen also to cover the five processing and/or marketing enterprises in the area. As well, a wide range of key informants were contacted regarding community characteristics and issues. Two-thirds of interviewees were women, in keeping with the sex ratio in the largest plants.

The household economy

Household income

The primary sample in this study covered a wide variety of household forms in terms of both composition and economic practices. Sources of wage income for households represented a range of sectors. Female plant workers in this core sample occupied two-income households with miners, engineers, general labourers, retail workers, and so on. Miners and engineers earn relatively good wages by Northfield standards, and approximately one-third of the households studied received these larger incomes in addition to plant wages. Annual wage incomes for households started in the $5,000–$7,000 range for those few households depending solely on a processing-plant wage from trimming and packing; these are typically women's wages. Dual-income households receive $15,000–$30,000 annually in wages. The upper levels here reflect the earnings of miners and engineers. None of the households in the sample was solely dependent on the fishery, although such households do exist. In cases of reliance on a single plant wage, there were also other contributions such as mining pensions. In summary, the fishery does not display single-sector dependence, and households are not readily defined by other socioeconomic characteristics outside of plant employment.

Plant managers confirmed households' attachments to a range of economic sectors. The overall household income profile suggests inter- and intra-class affiliations, especially when gender is considered. However, while management at the largest plant still suggested that plant work was employment of 'last resort,' it appears that such is not the case. Economic decline, resulting in persistent un/underemployment, combined with community and cultural practices, distorts typical patterns of labour market segmentation. Here, except for a gender structure that follows the typical pattern of male advantage, marginality becomes difficult to identify, even in relative terms. Further, all plant workers in the sample described themselves and their households as reliant on the plant wage, regardless of other sources of income. Processing plants pay more than retail or unskilled hospital jobs, the other likely sources of women's employment.

Household economics and spending patterns

In households where total income exceeded $25,000 (one-third of the primary sample), the processing-plant wage was put toward mortgage payments, home renovations, car payments, university education, and supplementing cash resources of adult children. These expenditures were made possible by the plant wage, modest as it might seem by the standards of non-seasonal, unskilled labour markets in many other urban centres.

As expected, households with incomes under $25,000 were less likely to have debts and more likely to apply the income toward basic necessities of food, clothing, and shelter. Debt was to be avoided if at all possible, and savings were purposeful. For example, instead of using credit to purchase a freezer or a VCR, plant workers might set aside whatever could be 'squeezed out' of the weekly household budget. Both VCRs and freezers were seen as reducing expenditures on leisure and groceries, respectively.[8] Irregular cash additions, such as the child tax credit, supplemented savings from wages for these larger purchases, or for things that children needed or wanted – clothing, bicycles, ghetto blasters, jewellery, and so on.

In several households, money was being set aside in March for substantial 'grading' presents (for completing a grade), such as bikes, stylish brand-name clothing, and walkman tape recorders, which would be given at the end of the school year in June, apparently regardless of academic performance. For children, 'grading,' birthdays, and Christmas brought costly unbudgeted gifts. Parents, especially mothers who organized these purchases, could momentarily stop 'stretching the dollar to make ends meet.' It was a point of pride to be able to satisfy a child's needs/wishes on ritualized (and therefore clearly circumscribed) occasions. On a daily basis, however, even modest requests for small treats costing less than a dollar were often denied. And many household items, including children's clothes, entered the household through exchanges with kin.

While presents and appliances might suggest that the plant wage is expendable, there was no evidence of the male-breadwinner ideology that holds that women work only for 'pin money,' or to buy extras (although men did feel that they should support their families). In context, it is remarkable that the family budget could be stretched to afford some discretionary commitment to consumer goods when balanced by non-cash exchanges in other areas of the household economy.

A typical household

John and Mary Brody (pseudonyms), both in their mid-thirties, are recently married and live with Mary's three daughters (aged eight, ten, and thirteen) from a previous marriage. Mary's sister, Charlene, who is completing high school, lives with them some of the time in an effort to reduce confrontations with her parents, who object to her current boyfriend, Ned. Charlene and Ned eat weekend meals with the Brodys, and Charlene often eats lunch with Mary's daughters on school days. John complains that Ned drinks more than his share of beer without reciprocating, but Ned does provide male companionship in an otherwise female household. John is one of ten siblings, eight of whom live in Northfield. Mary has a second sister and a brother, both married and living in Northfield. Mary works in the largest fish plant, Ocean Products, and has worked in fish processing since she was seventeen. John is a supervisor in a small landscape and construction company. His crew is laid off each year when the weather becomes prohibitive or when business is slow. When weather and contracts allow, he can work ten-to-twelve-hour days, but he earns no more than his wife – $7.20 per hour. Wage income to the Brodys is unpredictable, but in 1986 they made approximately $17,000.

Child care is often a major expense for households with two parents working. Mary spent the five years prior to her present marriage as a single parent and therefore bore the brunt of child care arrangements. Like most plant workers, she put together a patchwork of babysitters, calling on female relatives for assistance. In return she would help them out – for example by donating prepared food for a family anniversary celebration hosted by her aunt. Sometimes cash was exchanged for child care. Mary's youngest child had a subsidized placement in a local day care centre, but the hours did not coincide with extended or weekend shift work. If Mary had paid full-time child care expenses, little of her wages would have been left for the family budget. But now her eldest child was old enough to supervise the younger girls when Mary was called in to work, although Mary worried about leaving them unsupervised during her long hours of summer shift work.

The Brody family was living in a two-bedroom apartment which was small for their needs but conveniently located and affordable. They paid rent of $375 per month, a figure Mary described as on the low side because both she and John helped maintain the apartment and yard. Occupying the top floor of a house owned by a miner, the apartment was heated by a coal furnace and the relatively low heating cost was incorporated into the rent. The modest household furnishings included basic bedroom, lounge, and kitchen furniture. The fridge and stove were provided by the landlord, but the Brodys owned an older-model freezer stocked with fish bought at discount prices from

Mary's plant, a washing machine (but no dryer), a colour television set, a stereo, and a VCR, all purchased by Mary.

Neither John nor Mary owned a car, but John was negotiating with his mother-in-law's brother to buy an older-model station-wagon for several hundred dollars. After the purchase price was set, John would save the amount over several pay periods. He would then save to cover insurance and other expenses associated with putting the vehicle on the road. John planned to work on the body and engine with his brother and nephew. Because this was a contract between kin, extremely bad feelings would result if the car were turned over to someone else. However, John recognized that even the idea of extra cash might make Mary's uncle eager to complete the deal. Besides, having made the commitment, the family was keen to use the car over the summer. John felt pressed and nervous about whether he could pull it off. The car was a constant topic of conversation in the household.

Mary was also saving for grading presents and for a bicycle. She was thinking about trading her old bike and a ghetto blaster with a friend in return for a better-quality bike. The bike would enable her to exercise, travel to the beach, and visit friends and relatives during summer, without the 'hassles' required to deal with expensive and infrequent public transportation. To replace her ghetto blaster she was going to have her radio repaired. One of her male workmates needed some new curtains stitched, and in return he would fix her radio. She could use her sister's sewing-machine to fix the curtains. Mary was also able to make a little extra cash by perming hair for five dollars plus the cost of the perm solutions. This money enabled her and her daughters to buy what they required for their hairdressing needs, since hair care was a matter of some pride with them and many other Northfield women.

The youngest daughter was physically frail and required frequent costly medication. This was supposed to be paid for by weekly child support payments for the girls from their father, directed to Mary through a social service agency. Often the payments were not made on time, and Mary was sometimes worn out by having to pursue the cheques. Nonetheless, she always managed to budget the necessary money (equivalent to the family allowance payment paid by Health and Welfare Canada, around eight dollars per child per week).

Around June, Mary and the girls would begin shopping for school clothes in local discount clothing stores. The girls would each arrange to buy several pairs of pants or jeans, blouses, and sweatshirt tops, to be paid for on a lay-away plan: each week Mary would pay a small amount of the cost owing, using the family allowance payments. Lay-away, common in Northfield stores, served interests of both companies and consumers. If payments could not be met, the deposit could be applied to some other, less expensive purchase.

Mary bought groceries at a discount store, according to the week's cash flow. When cash income was down, Mary would try and stretch food purchased the previous week to last over two, sometimes three weeks. A reserve of non-perishable food built up during better weeks could be spread over 'leaner times,' although Mary did not like to eat out of cans or go without fresh fruit and vegetables for too long. Between $75 and $100 was being spent on food and groceries per week, although expenditures were seldom as regulated as this. Like other plant workers, Mary proudly considered her contributions to the domestic budget as twofold: the wages were essential to the household economy, and less obvious, but equally significant, Mary had an 'eye for a bargain' and was skilled at 'stretching the dollar to make ends meet.'

Entertainment expenses were modest. Several videos were rented most weekends, and John would usually buy a case of beer to share with friends who might drop by. Mary might drink some beer, but considerably less than John and most of their guests. John considered his beer purchase as his right – something to which he was entitled at the week's end. Mary too saw the weekly purchase of beer as an item of male privilege not to be challenged and made it a top priority in weekly budgeting. Her own personal expenses, such as cosmetic to counteract the harshness of the plant, came below food in the weekly budget. Outings to restaurants and/or movies were exceedingly rare, perhaps once a year. The Brodys seldom travelled to Sydney, the large metropolitan centre a short bus or car ride away. More common entertainments were local dances, where one could spend as little or as much as one chose on beer. Neither John nor Mary smoked cigarettes, although most other informants did.

Vacations were scarcely even considered. But like many Northfielders, the Brodys did get into 'the country' for several weekends a year – on a group expedition to one of several picturesque sites not far away. Relatives and friends set up a camp where they pool resources and collectively administer eating, drinking, and child care. Often these excursions are family reunions, for which relatives who had migrated from Northfield might return. Indeed, a family reunion drew the Brodys to the country for their only weekend of vacation in the summer of 1987.

And ironically, in those few summer weeks, when both John and Mary were working extended hours in six-day-week shifts, it became apparent to all the Brodys that such a hectic pace was tolerable only because it was seasonal. Aside from causing physical strain, expanded work routines jeopardized social contacts and 'investments' in the acquisition of knowledge necessary to make the most of informal exchange possibilities. Many informants felt the same way. For example, Mary's friend Paulette said: 'I need the money but I can't keep this up for long. The house is a mess, there's nothing in the cupboards and Ma's expecting a visit. It's alright for him [her husband]. When he gets in after work he can take it easy. I've got the supper, the laundry and the kids to take care of.'

To summarize, a blending of cash income, domestic labour, self-provisioning, and non-cash exchanges constitutes the Brody household's economic strategies. The Brodys demonstrate how plant wages provide crucial cash resources to a household – supplemented by other inputs requiring considerable effort and social knowledge to organize. These inputs are overwhelmingly provided through women's labour. In general, plant wages, along with state disbursements, do enable modest manipulation of cash resources to facilitate some planning for short term in a generally unstable and unpredictable political economy.

Coping with uncertainty

Families know that their financial inputs depend on unreliable sources – namely, fluctuating work and wages and a politicized agenda for state payments. Northfield plant workers are sufficiently cynical about their political economy and experienced enough with marginal work and wage patterns to disdain the increased consumer spending and debt typical of less peripheralized urban class cultures (Mingione 1983). Instead, they view consumer products and the 'needs versus wants' dialectic through a culture where 'getting by' calls on social currency to supplement cash incomes. And, most important, as Mary Brody's example indicates, women's labour and skills are applied to both earning wages and 'stretching the dollar.'

Furthermore, like Mary, other interviewees had considerable experience in plant work. Conversations with management confirmed stable labour forces, despite inherent insecurities. If plant work was once the last place to seek employment, or a place for only the unskilled to secure a job, often temporary, this seemed not to be true in the 1980s. In summary, most explanations given for entering the plant labour force understand the current importance of the job. Most workers expressed commitment and emphasized the necessity of the wage. Tensions over new management policies for production procedures at the largest plant sorely tested job commitment, but most informants felt economically compelled to remain and 'tough it out.'

Living in the community over a nine-month period allowed the author to participate in social exchanges with various households and to observe negotiations between kin. Sustained contact also revealed the cycles of stress in workers' lives, compounded by their ties to the fishery. For all who rely on the fishery, uncertainties arise from tensions between quotas (set each season by the federal Department of Fisheries according to its demographic projections for various species) and markets for fish. Even for smaller plants in Cape Breton, markets are global in nature. Small plants are defined as those relying on family labour, supplemented in peak periods by some casual workers – for example, high school or university students. Lobster might be shipped to Halifax, Toronto, New York or even England, salt fish to Portugal, and some species even to Australia. Boston, however, is a major hub of market activity.

Plant workers are far removed from the locus of market-oriented decisions and fisheries policy. They must attempt to make sense of rumours and of explanations from managers about matters of vital concern – how much work to expect, when the season will peak, and why the season looks the way it does. Even the quality of the catch being processed is compared with information about the season. How much income will the plant worker net in the season? How will the season's work pattern affect household activities and family members' daily routines?

These concerns are shared among processing plant workers throughout the region, although a household's dependence on the fishery may affect access to knowledge and perceptions of efficacy. Additional and uniquely derived strains, however, emanate from workplace social relationships. It is the argument here that the very nature of the politics of production is shaped by the historical and cultural context, examined next, within which women and men work and seek to make sense of their lives.

History and culture

The coalfields

Like many coal districts, Northfield has stark visual qualities. Contiguous clusters of wooden houses hug a coastline interspersed with exposed marshy fields long since denuded of the trees found there by the first settlers some 300 years ago. Housing clusters are oriented to each other, bespeaking previous ties to the sea-facing pitheads.

Successive waves of immigrants arrived in response to developments in coal production from the 1890s to the 1920s. In the 1890s, a period of sporadic petty production was superseded by capital concentration in one major firm, the Dominion Coal Co. Despite structural weaknesses in corporate organization, Dominion Coal orchestrated rapid expansion. New mines were brought into operation, along with three-shift year-round production.

The swelling population of miners strained facilities. Temporary, military-style accommodation was provided as the firm geared up to build the inadequate row and semi-detached company housing that dominates some neighbourhoods even today. It was a buyers' market for labour and a taxing existence for the new arrivals. Epidemics of smallpox and typhoid ravaged overcrowded, unsanitary, poorly serviced mining communities. Infant mortality was substantially higher than elsewhere in Canada (Macgillivray 1973). One older resident of Northfield, Mabel MacDonald, who taught miners' offspring during the 1920s, spoke of hungry, poorly clothed children too miserable to concentrate on lessons.

By 1910, 52.5 per cent of the population of the major mining town had arrived in the previous ten years, most of them Canadian- or New-foundland-born (80 per cent) – from local subsistence farming and fishing communities and other parts of the Maritime provinces and Newfoundland. The remaining one-fifth came from the British Isles and, to a lesser extent, from continental Europe. However, 44 per cent of Northfielders were of Scottish descent: the earlier group came from the Highland Clearances, while more recent arrivals were chiefly experienced industrial wage labourers. D. Frank (1974, 1979) argues that the latter group brought with them lessons of 'militant working class consciousness' from nineteenth-century Scottish coalfields. The Cape Breton coalfields also nurtured some militant trade union leaders (MacEwan 1976; McKay 1983; Mellor 1983; Harrop 1987; Mac-

Eachern 1987; D. Frank 1988). Shared traditions, combined with the harsh way of life, contributed to a sense of community demonstrated during several periods of bitter conflict (Muise 1980). Striking miners reacted against the company's attempts to control their union while profits were being increased (or at the very least maintained) through cutbacks in miners' hours and wage rates.

Collaboration between provincial and dominion governments revealed the state's commitment to the goals of capital. The military was called out several times between 1909 and 1922 to subdue, control, threaten, and provoke striking miners (Macgillivray 1980; McKay 1983). The company used every means at its disposal to achieve its interests, including cuts in food supplies from its stores and evictions from its houses. By the 1920s, a secure, albeit brutal, industry based on substantial undersea coal reserves was slumping. Short-lived intensive exploitation of the resource was followed by stagnation and decline, brought on by overreliance on unstable markets and failure to diversify (D. Frank 1979).

The fishery

Despite lack of documentation, it is clear that the fishery had been a steady secondary presence. Early accounts report two- and three-masted fishing vessels using the harbour even as the area was developing its cod-based economy. Between 1925 and the early 1950s, a significant swordfishery operated from Northfield. One plant manager recalled the harbour bustling with several hundred boats every mid-August, the peak of the season. In 1949, the majority of vessels came from other ports, but twenty-five to thirty local boats were participating the swordfishery, catching between 1,000 and 4,000 each year. Other species caught included cod, herring, mackerel, and haddock. Lobster was also plentiful. Two of the area's three companies dealing in fish also processed the stock in waterfront plants. Up to 500 people worked seasonally in fish handling and packing, and a new breakwater and wharf were constructed between 1949 and 1951 (Pearson 1966).

The decline of the swordfishery coincided with new market and policy conditions imposed after Newfoundland joined Confederation in 1949. Smaller craft were undermined by policies favouring larger draggers, and the present range of small to medium-sized firms dealing in groundfish and lobster was established. Some large draggers were built locally, but there were misfortunes. With mining's use of the harbour for dumping, waste build-up changed harbour levels and flow, but in the 1960s, with mining in decline, an urban-renewal planning study identified solutions (Pearson 1966). Fish processing has always been dominated by family-based capital, in contrast to the corporatized, unionized mining sector.

Northfield has limped along since the 1920s, tied to unpredictable market conditions in mining and the fishery. Many offspring from families who arrived during the initial boom remain attached to the area. Extraordinarily cheap housing (first rented, then bought at nominal prices) facilitated this commitment. As late as 1966 these dwellings still represented over half of Northfield's housing stock (Pearson 1966).

Today, for most people in Northfield, the pithead-focused neighbourhoods signify identity and personal support in hard times. For example, Doris, a single-parent plant worker, was worried about medical treatments her son required in Halifax. Co-workers suggested a benefit dance in the community hall of the area where she had been raised, where people knew her and would help. The cost of the benefit she held included a fee for hiring a hall and a disc jockey's services. A small cover charge plus money from the sale of drinks covered these costs and allowed for modest fund-raising. Locality or community base continues to have efficacy both socially and economically.

Indeed, the commitment people described was one of community, kin, and way of life. Most informants could recall stories told by older relatives about enduring and 'getting by.' Many themselves talk about hard times – the daily struggles, not just the major battles. History touches the present through conflicts in the relations of production and in areas of social reproduction that have been manipulated by capital. 'Getting by no matter what' – a pragmatic approach – is a cultural expression shaped by the history of 'relations of appropriation.'[9]

Kinship and households

Not surprising, active horizontal and vertical kinship ties are a striking feature of community life. Very few of the 100 or so persons contacted had lived elsewhere for any length of time. People who had left and remained away, especially younger ones, tended to have more education or an exceptionally good job. Marriage to a partner from outside could result in a seemingly more permanent move, although people did return to live after extended periods elsewhere. The long tradition of Cape Breton out-migration – 'going down the road' – occurs now, as always, in a circular process. Nonetheless, the Northfield area has experienced a population loss of around 5,000 over four decades.

Northfielders, and other Cape Bretoners, leave home seeking work elsewhere within a political and economic context mediated by cultural conventions and expectations. In the nineteenth century, rural Cape Bretoners went to Boston (deRoche 1985). Northfield itself represented an employment centre at the turn of the century. When mining and steel failed, people left for central Ontario or later tried the western 'oil patch' during its brief boom of the 1970s. Once again, migrants relied on knowledge and networks from kin and community to arrange employment, accommodation, and orientation. Young people still 'try their luck' in big cities elsewhere but need to get a job with wages great enough to cover the high cost of living there. Success now depends more on education and skills; good connections and support networks are no longer sufficient.

Three young men aged between eighteen and twenty-six attached to households in this sample were struggling for 'a break' in unskilled labour markets in Halifax. Family members predicted their imminent return and extended economic and emotional support as required. For example, Bonnie spoke with her son in Halifax by telephone several times a week. He reported only sporadic employment as a day-labourer in a construction crew.

Shelley went farther afield. As a high-school graduate she had strong ambitions but probably unrealistic expectations. She had just arrived in Toronto, where she was seeking work and perhaps training related to computers. Until she could become self-supporting, she was staying with her mother's sister in the suburbs. Shelley's calls to her mother in Northfield communicated increasing discouragement and frustration, as she began to appreciate the limits of a high-school graduation certificate when measured against the cost of living and competition for such basic computer jobs as word processing.

Unsuccessful excursions away further reinforce attachments to the local community as a safe, comfortable, familiar, and manageable haven. Hard times can be faced more easily at home when people in other places are cold, uncaring, and inhospitable. By contrast, community and kinship links can be mobilized for a wide variety of economic and social supports and exchanges.

Household composition and informal exchanges

Housing is a key resource in Northfield's economic strategies. People move between households giving and receiving aid in response to changes in their own or others' circumstances. Only a small percentage of households surveyed remained stable through the nine-month research period. Residence patterns indicated the protracted dependence of children on their parents for the provision of shelter. Requests to return to a parental home arose from adult children in mid-life. Other kin also participated in shared residence arrangements. The following examples of informal household exchange typify the Northfield experience.

The Bradleys: Roger, a handicapped boarder, moves in with the family; the second eldest son leaves, seeking work in Halifax; the eldest son returns from Halifax after several months of struggling to live on a minimum wage.

Betty White: As a female single parent she moves out of her parents' home into her own newly purchased house, her first independent residence. Earlier, her teenaged marriage to a miner had been cut short by his death in a mining accident, and she and her infant son had continued living in her parents' house. Betty's twelve-year-old son prefers his grandparents' neighbourhood; his wishes are accommodated without much conflict, and for the immediate present he remains in their care.

The Leightons: They are a married couple with three children who move in with the wife's parents while searching for a house to purchase.

Sheila Jones: She is a single parent with a toddler, who moves out of her parents' house to live with her childless aunt and uncle. Sheila had been experiencing difficulties in her natal household, especially as the cycle of shift work intensified through the summer. She felt a special affinity with her aunt, who also worked in a fish processing plant. Both were proud of their skill and confident that the two of them (working different shifts), her uncle and her mother, could manage child care.

The Franks: They are a young married couple who move into the newly renovated basement of his parents' house to live rent free. Two married brothers and their wives and children live in adjacent housing. Each couple had taken its turn living with the parents.

Agnes Murray: She is a seventy-four-year-old miner's widow. She lives with her forty-year-old daughter, Valerie, one of eight children (she also raised two children of relatives). Valerie, a single parent, had recently broken off with her male partner and moved in with her mother, bringing two of her three children, a daughter aged twelve and a seven-month-old son. Agnes provides care for the two children and a baby girl born to her daughter's own daughter, while the mother works. Agnes has shared the home with other adult children and their spouses, a situation that has produced mutual advantages for all, as well as family frictions.

These examples, which typify the extremely complex patterns of household occupancy in Northfield, represent dramatic forms of kinship mobilization, as does the substantial informal sector (see the Brody household above). Patterns of household occupancy and informal deals between kin and neighbours reveal culturally based structures overlooked by official statistics. Men and women use skills, learned because of necessity, for their families and in labour markets, where flexibility and good networks secured the best jobs. Labour and/or below-market-value cash exchanges replace otherwise major expenditures in construction trades and auto and home maintenance. Women's skills are applied to home decorating, catering, child care, hairdressing, and the customary major annual house cleaning. This informal sector is so vital that it undermines the potential for a wide range of entrepreneurial activities found in less economically stressed urban industrial centres.[10] This finding runs contrary to a view of the informal sector as a training ground for formal-sector activities (Pahl 1985).[11] As suggested earlier, such transfers in commitment probably occurred prior to this current phase of economic restructuring.[12]

Patriarchy and social reproduction

Domestic labour helps to clarify the context of plant politics. Women perform a great deal of domestic labour. High standards of household cleanliness were maintained at all times. Routine and extraordinary annual cleaning and interior painting are seen as exclusively women's work. Cooking and child care were domestic tasks generally shared with a male partner if he arrived home earlier. Even when performed by men, meal service tended to involve preparation by women. These chores were still perceived as women's work, regardless of who performed them – a discrepancy between ideology and practice that went unnoticed by informants. This pattern is consistent with descriptions of the domestic division of labour in comparable settings (Morris 1985). What is remarkable is the tenacity of the ideology in this context, where persistent un-/underemployment and multiple-income strategies severely test notions of the male breadwinner.

Comparison with practices within miners' households provides evidence that historical pressures still hold sway. Divisions of labour in mining households were more pronounced, showing even less discrepancy between gender ideology and practice. Miners' wives were expected to provide hot meals when the men came off shift and to prepare 'lunches' for the next shift, regardless of their own domestic and wage work routines. It was said many times that miners would not be 'caught dead in the public eye' doing women's work. Only extreme circumstances might provoke deviation from this pattern.

While mining is hard and dangerous work, fish processing also confers its forms of hardship and risk. For example, cutters stand on concrete floors for hours on end in a cold plant. They must handle chilled fish with a sharp knife, performing split-second precision-cutting techniques under constant pressure of daily quotas and 'keeping their numbers up.' Aside from the more predictable health conditions exacerbated by the damp cold environment (such as backache, arthritis, and chest complaints), plant workers mentioned their ever-present fear of skin cuts. Even minor lesions are liable to serious bacterial infection despite application of industrial health and safety standards intended to reduce fish-plant bacterial counts. Risks aside, physical fatigue was the most often emphasized problem. And yet, for the women, unlike most of the men, leaving the plant and arriving home signalled the resumption of a further set of tasks.

Aside from long-standing socialization to gendered work assignment, the organization of mining production helps maintain an occupationally based and rigid division of domestic labour. As Cockburn (1983, 1986) has demonstrated, segregated male workplaces tend to promote a contrastive ideology of gender, stressing sex differences and objectifying women (though not 'my wife and my sisters') as a group. Such expressions of contradictory attitudes to women support a self-serving ideology outside the workplace. Social reproductive practices and processes that are patriarchal (seeking to control women's lives and labour) are clearly connected to relations in production.[13] Indeed, as we shall see, there are also patriarchal processes in the organization of production and in the imposition of new technologies in Northfield's fish plants.

Management and workers

Three styles of management were in effect at different times in North-field's two main processing plants, both of which were family businesses, capitalized and managed by family members over several genera-tions.[14] 'Benevolent' paternalism prevailed at O'Brien's, the smaller plant, where up to fifty employees worked when the plant operated at full capacity. According to informants who had worked in both plants, the management régime at Ocean Products prior to modernization could be called 'authoritarian' paternalism. 'Impersonal corporatism' was ushered in with modernization of Ocean Products, which installed 'state of the art' production equipment, including computerized scales. The labour force there numbered approximately 180, working two shifts.

Benevolent paternalism

Benevolent paternalism is illustrated in the organization of the smaller plant, O'Brien's. Approximately forty inshore boats fished under long-standing arrangements with the company. January through April, when weather and quota restrictions limit fishing, the plant may be supplied by one small local dragger. Four of seven siblings are involved in the company started by their father in a derelict plant in Northfield. This family group owns and operates three plants in the Cape Breton fishery, giving the Northfield plant some flexibility in access to fish and infor-mation. This plant operates on the basis of market patterns and can process variously fresh and frozen fish and, especially during winter, salt cod.

The total labour force involves twenty women and thirty men. A circular belt system is used for processing fresh and frozen species: trimmers, cutters, and packers work face to face, with ample oppor-tunity for verbal and visual contact. Salt fish processing occurs primarily in another area of the plant. Trimming and packing (performed by women, and men where necessary), and general labour (men only), are paid approximately $7.00 per hour. Cutters, both male and female, earn $7.25 for hours spent cutting, which is a work task, not a job designation. Therefore, a day's wage often involves two rates of pay. Trimmers and packers have the least flexibility in job routines, followed by the few female cutters, who can earn slightly more but must maintain speeds comparable to those of the male cutters. For a female cutter to be shifted from cutting to trimming and/or packing while cutting work remains is a disappointment. Management perceived the male cutters as more skilful, despite contrary evaluations from female cutters. Skill is, of course, a social construct in this case, readily associated with (and of benefit to) male cutters, and it may involve self-fulfilling prophecy for female cutters: more stress, less skill; less skill, less opportunity to work in a skilful manner.

Management also determined men to be physically more capable of general labour jobs, which affects gendered wage levels. With only two wage rates in existence, a form of equal pay is in practice. However, female workers take home less annual pay (between $5,000 and $7,000) than men, who can earn $20,000. Probably the greater application of male labour results in longer hours, and male workers seem to be called back before females after layoffs. Some of the twenty women are employed seasonally, out of choice or because there is no room for them in the small work-force employed on salt cod and in the erratic production of the slower months. Workers who choose not to join the crew processing salt cod may do so without penalty, but a woman who declines this work may implicitly signal lack of need for the wage. Management was very well informed about households' commitment to the plant wage. Probably a notion of family wage applies to allocation of male workers and, to some extent, to female labour as well. Women seen as main contributors to household incomes were the most likely to be offered work outside the full production schedule. But, given the greater differentiation of men's jobs, even these women would receive less work than men.

Recruitment is arranged through negotiations with plant workers. If there is a space at the plant someone will come forward with the name of a relative or neighbour who needs work. At least eleven of the twenty women at the plant worked alongside siblings. Once a decision has been made to hire someone, it is taken for granted that the choice is a good one. Management believes that all workers are trainable, and there is no trial period attached to employment. The manager and supervisor actually work on the production line, maintaining close contact with employees' production performances. There is no other form of supervision or quality control, and there is no individualized monitoring of levels of production. Management wants 'dependability' in new workers – reliable attendance and hard work. Management felt that all employees did work hard; moreover, it was a point of pride that there had been no labour conflict at the plant. It saw its hiring strategies as more suited to the company's need than the alternative of seeking employees through the Canada Employment and Immigration offices.

The apparent success of this benevolent paternalism hinges on the company's use of kinship as a strategy for recruitment. Workers are grateful to the manager for giving their relative a job. New employees want to represent their kin well. The hiring of kin helps to ensure the dependability of workers and their relatives. New workers are more likely to be women than men because 'women's work' has higher turnover. Further, having sets of relatives in the plant makes for congeniality and familiarity, which management encourages. Loyalty to manager and company inspires better work. As in many processing plants, management may also provide economic assistance to help employees through difficult times. It is a flexible system. All workers spoke highly of the fairness and personal disposition of the manager.

Some potential tensions were revealed during the wind-up to full production in spring. Call-back for regular female employees was later than anticipated, and rumours, jealousies, and general stress surfaced over what the season would be like. While this competition might spur on those called back to work hard, it could also produce some form of organized protest. For the time being a union was considered unnecessary – a violation of paternal trust that would bring more trouble than any perceived advantages. Many employees had worked earlier at Ocean Products or knew about organizational changes, new management practices, and labour conflicts in the larger plant.

Authoritarian paternalism

The larger plant, Ocean Products, had been laid out and run by the father of the present manager according to a style of management practice defined here as authoritarian paternalism. Under this regime, organized but contained protests occurred quite regularly. Recruitment was also facilitated through kinship. Production took place on a belt system, with little initial formal training. Supervisors were described as being more agreeable than those in the modernized plant. The 'boss' owner/manager, however, used extremely harsh and often sexist language with female employees. Indeed, stories of his vulgarity toward female staff were numerous and consistent. So too were accounts of his loathing for trade unions, a subject which, it was said, could provoke physical demonstrations of anger.

Most employees with the seniority to be called back to the modern plant were said to prefer this old way of doing things: 'You knew where you stood and what the rules were.' The rules were fewer than in the modern plant and were, at least according to present management, not well enforced; production was 'poorly managed and ad hoc' compared with the new corporate standards. Though 'supervisors kept an eye on things,' there was no way of assessing and controlling individual output.

In the event of disputes (usually over wages), there would be a walk-out followed by a meeting of employees. A delegation would then be dispatched to inform management of the demands. Tempers would flare and cool down. Explanation of the company's poor financial circumstances, necessitating a delay in the wage increase, would be offered. Promises would be made, and work would resume.

Workers in Northfield understand hard times and have some sympathy with such explanations. The sequence of events and their outcome would be predictable to all involved. The work was hard, the pay was poor and seasonal, the hours were sometimes long – but at work, women trimmers and packers were with friends. 'Supervisors were not breathing down your neck all day.' Further, the style and pacing of work were left up to the individual, under certain general conditions. If one occasionally escaped to the bathroom for a smoke, for example, the risk was humiliation from a paternalistic verbal scolding – often derisive of women in general and the employee in particular. One would smart and grumble, but ultimately recover.

Impersonal corporatism

The present management took over the old plant of Ocean Products, operating it for several months before embarking on rebuilding and modernization. A two-year hiatus in makeshift facilities further strained an already difficult situation. Like his father before him, the new manager had a university education, but the son had strong corporate aspirations to expand production in new directions. On taking over the company, he initiated a company concept and corporate identity. The management level was expanded, with new appointments to production and personnel. While several members of the family continued in administration, new appointments were not necessarily based on experience in the fishery but rather followed the corporate logic of preferring management experience. This was a source of some consternation to employees. A frequent complaint about a supervisor might be that 'he doesn't know one end of a fish from another.' Along with receiving uniforms in the new corporate colours, employees were issued with policy books outlining the company's concept and rules.

'Participatory management' was also part of the new style, and management initiated an employees' committee to facilitate 'democratic' communications. In retrospect, most informants saw the committee as less than democratic, the equivalent of a management tool. Employees felt that management exerted control over the agenda through its presence and because the process was at its initiative. Management also found the committee frustrating: many employees appeared reluctant to participate and did not understand the new approach. Very discrepant definitions of the plant as a workplace were surfacing.

When the new plant went into operation, management, disappointed by the responses of plant workers to new concepts and practices, set about implementing a program of 'human resource management.' Beyond obtaining technical efficiency (including use of a computer for measuring production output), the trimming labour process, predictably, became the prime focus of attention. Technology is supposed to streamline production and increase efficiency. Trimming, in contrast, is like a 'human bottleneck in production,' where managers seek to make up for the capital costs of modernization (MacDonald and Connelly 1986). Workers were oriented to the new physical plant, where individual trimmers' work stations stood in tidy lines to reduce social interaction. Retraining programs impressed on trimmers the need for efficiency, economy of effort, and conformity with company policy. Computerized scales weighed individually tagged pans of processed fish and waste, enabling measurement of both quantity and quality of output. These changes placed substantial pressure on all, but especially trimmers. Management was concerned also about a series of delays in achieving a smoothly running production unit.

Gender and technology

The fact that nearly all trimmers are women reveals a further complexity in the articulation of structures of gender in the plant. First, predictably, job assignments and wage rates at Ocean Products are more spread and gendered than those at O'Brien's. Trimmers' and packers' wages are lowest of all the rates.[15] Thus women are confined to the lowest-paying jobs. Second, as is typical of many East Coast plants, the supervising staff is male, although some training and quality control are performed by several women. However, for most of Ocean Products' history, men have supervised women's work. This division of labour, however disguised by paternalism, entails a power relation that becomes more obviously gendered to supervisors and workers, both female and male, as supervision becomes more aggressively oriented to increasing individual production. Women workers expressed humiliation at being harassed by male supervisors about their inability to adopt new techniques. Many trimmers indicated that their workplace experiences reminded them of their school days – of personal humiliation through subordination. For example, Linda recalled in great detail her anger and distress at being ordered to stand and trim at a table in a section where new employees were trained, for committing the double transgressions of chewing gum and 'answering back' when reprimanded. She was placed at the end of the section by herself, out of the vision and hearing of her usual companions. Not only was Linda being socially isolated and humiliated, but being placed next to trainees implied that her work was being compared to theirs. Linda recognized that she was being publicly humiliated for a trivial rule violation; more important, her sense of competence suffered in the aftermath.

Thus in fish plants, as in other industries, women operate under new technological regimes maintained and managed by men. The expression of power differential accords with gender patterns found in other sets of social relations. Issues of technological competence and incompetence are correlated with notions about masculinity and feminity, respectively (Cockburn 1986). There is an ideology of gender attached to a gendered technological division of labour. Female workers are, therefore, doubly disadvantaged by technological interventions in their relations with capital. At Ocean Products, trimmers lose autonomy in their labour process, compounded by a gender ideology which holds that women are incapable of understanding either the workings of the new technologies or the logic behind them.

Male workers expressed regret about the plight of female trimmers, recognizing that their own jobs as general labourers, mechanics, and engineers ensured them a degree of precious mobility and work autonomy. In contrast to trimmers' concerns about losing their jobs because of inability to maintain new production levels despite attempting to 'do my best,' men saw themselves as much less affected by the changes. In fact, if management's accounts of its difficulties in implementing new technologies and work regimes are correct, it would appear that male workers had and took more opportunities than trimmers to resist the imposition of change. Management attributed equipment break-downs and damage prevalent throughout the modernization phase – a time of nascent conflict – to the lack of ability of the male labour force. An equally plausible explanation would be to view these actions as workers' resistance.

Resistance and retaliation

Two firings pushed the plant's labour force to take action beyond the ritualized walk-out. Two supervisors were, in the manager's language, 'terminated.' One was several years from retirement and had been with the company for most of his working life. He had hired and trained most of the plant's employees and appeared to be widely respected. The other was younger, a father of several small children. Rumours circulated that these firings, generally perceived as unjust, represented a hidden agenda to 'get rid of all the old workers' (i.e. those on staff prior to modernization). Concern over the injustice was fuelled by ties of kinship extending throughout the plant's labour force. The family status of these men was also an issue – a man close to retirement, 'where would he find work now? – and the other man with young children to support.'

Management was devising a system for allocating desired jobs at the scales. It was also studying capabilities within the female labour force with a view to constructing a new category of skill. 'Working the scales' was not a better-paid job, but it could be distributed as a reward to favoured employees. Workers holding less than grade 10 education – most of the labour force – were required to write a grade 10 math test, but some refused. Grade 10 education was also made a require-ment for new employees, further reinforcing fears for job security.

The logic of having an educational standard imposed on production work was not clear to employees. For example, Dolly, an older and more successful trimmer who was able to satisfy, and often exceed, new production demands, pointed out that while she had only a grade 4 education, she had grown up 'around fish.' She had learned from her father how to process fish when he was marketing fish in a rural community. Dolly was quite scornful of management's association of skill in fish trimming with years of formal education. (Management practices in fact associate education levels with the worker's receptivity to corporate ideology and discipline.) After months of attempting to justify the new procedures, management was unable to convince workers. Its negative view of the capacities of its labour force was further confirmed by the resistance to and fear of hidden motives behind the testing. A self-fulfilling prophecy ensued: management found justification for its policies in the reactions of individual employees.

The firing of the second supervisor provoked a walk-out. In the course of events outside the plant, a delegation of employees spoke with an officer from a local union, and procedures were immediately implemented toward a certification vote. Management's impersonal and technical corporate strategies had violated previously acceptable paternalistic codes of control deeply rooted in this close-knit community. Plant employees needed no lengthy organizational process to reach the consensus necessary to form a union. When management's practices were collectively recognized as unjust and threatening, knowledge of unions as protection precipitated certification. The mining sector had, after all, been unionized early in its development.

This is a story that is unfinished. The collective agreement was signed after a period of extremely stressful production politics for all concerned. Management reported declines in plant production despite aggressive supervision. Production losses are always a risk in technological upgrading (Thompson 1983), and in this case managers and workers alike reported them to be severe. Some employees resigned as a direct result of increased stress. New patterns of recruitment and training were implemented. There was talk of employing men at the trimming tables – breadwinners who really needed jobs – and of recruiting new trainees from different localities to break down the internal patterns of kin and neighbourhood alliances. Other rumours reported an inability to recruit and hold trainees because of management's unreasonable expectations and for reasons of community loyalty.

Kinship relations, and empathy with their culturally understood meaning, appear to hold greater force than relations based on a potentially patriarchal sex/gender structure under the extreme pressures applied to labour during modernization and restructuring. As one respondent said, 'You take care of your own and your buddies.' In Northfield, the concept of a 'buddy' derives from the partner system – the cornerstone of coal production – which made two miners mutually dependent for personal safety and the productivity necessary to achieve meagre wage levels. The bond between partners and buddies was tied to life itself. Family, workmates, buddies, neighbours, and union all resonate with the idea of standing together. Male plant workers took risks and collaborated with female co-workers at a time when, as a gender-based group, they might have sought to strengthen their privileged position.

Also, the role of the state in this context is contradictory and not simply in the interests of capital. State theorists have long argued that the interests of capital are served through state programs that subsidize the costs of labour. A lower wage rate can be paid, but the labour force is still able to maintain or reproduce itself over time. However, the state may not control the terms under which such subsidies are awarded. In the case of Ocean Products, management contested the awarding of UI stamps to union officials receiving disciplinary layoffs ('time on the grass') for rule violations supposedly unrelated to union activities. Such intervention in UI administration was unprecedented, since company managers appreciate the state's willingness to contribute to employees' incomes. Further, state officials, as members of neighbour and kin networks, may bring those loyalties to an interpretation of policy guidelines and disputes, resulting in a second order of contradictions surrounding the relationship between capital and the state.

Conclusion

The culture of Northfield has developed out of the collective experience of surviving adversity. Historically, people were forced to rely on kin and neighbours for basic survival under brutal conditions imposed by the domination of mining capital and later entrenched in a peripheralized political economy. They are precedents for collective action, given sufficient provocation, but, paradoxically, support networks and 'making do' can diminish the impact of stressful changes and compensate for the breaking apart of kin and community relations by capital. The culture of 'making do' represents resignation, a form of adaptation to structures of domination and the vagaries of political economy.

Alternatively, history and recent events suggest that more militant responses remain a possibility. The case of fish processing has been particularly relevant in this context. Resistance became reconstructed as protest, protest was sustained into union organization, and union organization triggered management strategies reinforcing the concept of a union; yet none of these events would necessarily follow in another setting that presented a different configuration of capital, culture, and community.

The use of kinship and neighbourhood networks for job recruitment is revealing. When used as a management strategy, control over the labour process may be enhanced through the appearance of providing favours to employees, by maintaining a familial tone in the workplace, and through recourse to the social paternalism, benevolent or authoritarian, that has characterized management in fish processing plants. Alternatively, these social networks may be mobilized for workers' resistance under conditions – such as impersonal corporatism – collectively recognized as threatening accepted codes of control which ideologically reproduce familial idioms such as loyalty and commitment. The unionization of Ocean Products in 1987 occurred in response to such violations, despite the fact that capital should have the upper hand in such depressed labour markets. Forming a union in such conditions constitutes extraordinary risk-taking, explainable in part by the historical structuring of relations of 'kin and community.'

14 Populism and alienation

Richard Apostle
Gene Barrett

As discussed at length in chapter 7, recent studies of Nova Scotia's fishing industry by political economists have focused on the emergence of monopoly capital and the proletarianization of fishers. In what amounts to a fairly reductionistic argument, these increasingly asymmetric class relations are thought to be following predictable, organized, class-based forms. In the context of a highly state-regulated industry and regional underdevelopment, the political economy of the fishing industry has assumed central importance in what is seen as an emergent radicalism.

The analysis here suggests a more cautious approach. This chapter examines a number of unanswered questions concerning the political economy of the fishing industry that, for many reasons, have been considered unimportant. It focuses on class differentiation and political response among capital and among direct producers (fishers) and on the impact of marginality on fish plant workers. Of particular concern is the relationship between class and right-wing populism.

Capital: differentiation and right-wing populism

As we have seen in chapters 3 and 4, corporate capital in the fishing industry has been thought to be monopolistic and of direct mercantile and industrial lineage. If the strength of large-scale capital has not been overestimated, then its significance certainly has. Clement (1986: 53), for example, wrote: 'Viewed from the perspective of the workers, most large processing plants are industrial factories. There are still many small, seasonal plants (especially in Atlantic Canada) employing only a few workers who are often unorganized. In general, however, the industry has been subject to centralization, which has meant the closing of smaller plants and shifting of processing to larger, more populous locales.' This perspective has led to disregard of the small- and competitive-scale sectors and the unwarranted assumption that monopoly capitalism has brought producers in the industry to the brink of proletarian radicalism.

Williams (1987: 80) has recently analysed restructuring in the industry more subtly: 'The so-called "independents", the small- to medium-sized fish plants, didn't know what they wanted. Some called for letting Nickerson/National go bankrupt. Others insisted on a return to private enterprise, even though competition with the huge company had put many of them out of business. The content mattered little – the main point was to enthusiastically endorse private enterprise and condemn government control.'

While Williams belittles these private-enterprise sentiments, they may represent the roots of a right-wing populism that is the dominant ideology of much of rural Nova Scotia. The differentiated structure of the fish processing sector provides the key to explaining the prevalence of this viewpoint. Despite extremely high political alienation, the industry and much of rural Nova Scotia value individualism, believe unshakeably in free enterprise, and intensely dislike big government and big companies. This view has close affinity with analyses of ideology and the small-capital sector put forward by recent studies in the differentiation perspective (Curran and Burrows 1986).

This project's 1984 survey of plant managers uncovered general apprehension about the potentially adverse effects of continued government intervention on behalf of large-scale processors in the region. Aside from plants that were part of National Sea Products (NSP) operations, processors widely regarded government intervention as a 'bailout' for organizations that were being given unfair competitive advantages.

Small- and intermediate-scale processors thought that the large companies would eventually be forced, because of their inherent inefficiencies, to return for more government support. They repeatedly argued that the largest companies (NSP being the obvious case) were unable, because of their size and bureaucratic structure, to respond effectively to substantial variations in supply and demand.[1] The ideal production units were small or intermediate-size, family-owned and -managed establishments; such operations could react more quickly and provide the direct supervision necessary to run a profitable company in a labour-intensive industry.[2] One manager of an intermediate-size plant observed: 'The fish business is not a nine-to-five operation. There are times when you have to work ... I have here Thursday night until 2:30 at night, I was here Friday night until 12:00 at night, I was up at 6:30, drove to Barrington to get some salt, and stayed here until 8:00 Saturday night. [Executives of large fish companies], you see, can they do that? I've got my rubber boots on, that might tell you a little bit.'

Hostility to 'big' government tended to be both personalized and universal. Dissatisfaction focused on arbitrary regulatory activity and on the anticipated negative effects of refinancing the large processors. Fisheries officers were frequently described as being ignorant about the technical issues at stake. The ever-present realities of patronage politics contributed to cynicism about 'big' government. The awarding of some government salt fish contracts is clearly facilitated by political connections, and one intermediate-sized processor who is widely regarded by his competitors as inept survives through government support. However, being affiliated with a party out of power can have negative consequences. One large processor with Conservative inclinations needed repairs to the local wharf during the 1970s and early 1980s, and other processors attributed the delay of such repairs to his political affiliations. One small processor detected a certain irony: the area (and many of the large processor's employees) traditionally supported the Liberals.

Despite high levels of disaffection, small and competitive-scale capital has little class-based industrial organization. Less than half (45 per cent) of the interviewees belong to any 'business-related organization' or 'industry associations,' and those that do have little faith in their efficacy. Of those that belong, only 36 per cent believe that the organizations effectively represent members, and a further 51 per cent say that they 'hardly at all' represent the interests of the plant whose manager was interviewed.[3] Under 'normal' economic and political conditions, capital usually remains highly individualistic in its responses to problems; only a common threat pulls everyone together.

The structural pressures that generate political fragmentation are threefold. First, diverse product lines and market conditions, in conjunction with differences in scale and geographic location, inhibit common definitions of problems. For example, differences between small and intermediate-sized processors in the Southwest Nova Scotia Salt Fish Packers Association are based on dragger ownership and the use of dragger-caught fish. Dragger ownership implies a scale of operation generally associated with frozen fish production. Small processors accordingly suspect that intermediate-sized ones, which have been active in the leadership of this organization, will not fully represent their needs. The association was nearly destroyed by infighting over imposition of import duties on Canadian salt fish entering the US market (see chapter 6).

Second, smaller plants fear negative reactions by the large companies or by government should they make vigorous efforts to organize. Virtually all managers (96 per cent) acknowledge that 'some fish companies have more influence with government than others,' and large companies can block their supplies of fish (particularly during winter). They also fear retaliation by government officials if they try to resist orders to alter their plant's operating procedures. One manager, who has been in conflict with the Department of Fisheries and Oceans (DFO) concerning application of regulations, quotes an officer with whom he had a disagreement as saying: ' "There are a lot of things we can be tough on" ... and I have lived to see that he was right, because I have suffered the consequences of retaliation. They have been hard on this plant ... Because of my big mouth, I got myself in a mess, and now I am being squeezed harder than the other guys.'[4]

The Seafood Producers Association of Nova Scotia (SPANS) is open to all producers, but there is a perception that the largest processors both control it and use it to co-opt and deflect criticism of them. For example, one small processor, who believes that NSP should have been forced into bankruptcy and who has to compete with an NSP operation in his locale, says that he considered withdrawing from SPANS because it is dominated by the three largest companies in the province.[5]

Third, heavy competition among many smaller operations for the loyalty of their captains and for access to the marketplace prevents full-fledged support to common organizations. Low-level antagonism can flare into the open when a plant draws captains away from other plants or attempts to break into new markets. One processor, for example, admitted that he had been exporting his product to the United States without the appropriate licence. Nearby buyers notified DFO authorities in New Brunswick; an eventual seizure of his truck looked like a random spot check.

Fishers: alienation, fragmentation, and protest

Williams (1987b: 81) has indirectly identified the ideological contradictions within Nova Scotia's fishing industry. He notes the differences
between the 'narrow self-interest, parroted, unquestioned assumptions
about "free enterprise" versus government' that many 'wealthy' fishers
believe and the less conservative views of 'an inshore guy in Pictou
County who has to sell to National Sea Products.' As we saw in chapter
7, political economists like Williams and Clement have explained such
differences through a model of proletarianization: 'independent' commodity producers have lost most of their autonomy, and their organizational and political activities reflect their loss of economic independence and of autonomy as a social class (Sinclair 1984: 34–5).

In delineating organizational choices in the East and West Coast
fisheries, Clement (1984) suggests a useful distinction among small-,
intermediate-, and large-scale producers.[6] He indicates (1984: 19) that
unionization has been strongest among trawlers and small-scale fishers
who deal with industrial processors. Crew members on medium-scale
units are better off and identify more closely with their skippers and
so are less likely to unionize.

Clement (1986: 61–8, 190–6) posits organizational preferences that
vary according to differences in the scale of productive units and position in the division of labour. The captain of a large-scale vessel –
'equivalent to a manager of an onshore operation' (78) – would lean
toward an association; crew members – proletarians – would favour a
union. On intermediate-scale vessels, skippers, who range from being
supervisors to small capitalists, may prefer associations or co-operatives,
depending on the extent to which they are 'dominated by large capital'
or can 'engage in entrepreneurial activity' (78, 191). Crew members,
whose only revenue comes from the crew share, are 'clearly proletarian'
(77) and should favour unionization. However, close personal ties may
incline them in their captain's direction, particularly for East Coast
herring seiners (82). Depending on market location, individuals working on small vessels are dependent or independent commodity producers (see chapter 7). The majority are dependent, by virtue of their
weak market positions, and have 'a much greater prospect' of allying
with the working class than with independent commodity producers
(68). Unionization is most likely among fishers who sell to industrial
processors (82).

Clement's characterization of primary production raises a number of questions and issues.[7] Loss of economic power, when it occurs, may vary considerably by area and sector. Such variations presumably will have major consequences for organizational behaviour. Many scholars presume that groups with a certain type of structure will hold the 'appropriate' ideology. Some groups, particularly dependent commodity producers, may still find traditional small-business ideologies appealing. Minor threats to economic independence may stimulate defensive behaviour designed to protect or restore independence and intensify populist sentiments.

Individualism seems to be a fundamental cornerstone of this ideology, and most fishers would normally reject 'organizational' responses altogether – not through false consciousness, but rather as a key ideological response. What are the unusual circumstances that lead to 'organizational' action? Another question is virtually ignored by political economists: what factors lead to short-term, issue-based alliances among direct producers, and what factors produce more permanent organizational solutions? In particular, what factors determine whether fishers with the same social location will join a union or an association? Why do some 'dependent commodity producers' choose an association (the Eastern Fishermen's Federation), and others a union (the Maritime Fishermen's Union)?

Nova Scotia fishers have manifested particularly high levels of political alienation and spontaneous political activity[8] (Apostle, Kasdan, and Hanson 1984). Fishing captains are more likely than other occupational groups to become active in issue-based, non-electoral politics, especially in reaction to government fisheries policies and regulations. This section explores some of the factors that affect their political activity, ideology, and organizational affiliations, based on the 1986 survey of 451 captains. It investigates the effects of social class, location, income, and religious affiliation.

Among twenty-six questions about work satisfaction, only a question about the performance of federal and provincial officials elicited a the score in the dissatisfied range of responses. A majority (52.3 per cent) expressed negative opinions about the manner in which DFO enforces regulations and policies, and only 18.4 per cent thought that the federal government was doing a good job of regulating access to the fisheries.[9] Several themes emerge in the captains' comments. They were concerned about government's intrusive involvement in the fishery, lack of responsiveness to their needs, and uneven or arbitrary enforcement of regulations. 'It is a mess; they usually don't do what they say they are going to do. They leave too many loopholes in their regulations. The important policies are not enforced as much as unimportant policies.' The captains also emphasized their dislike of the dragger fleet and of government's role in facilitating its growth. There was an impressively high level of agreement that draggers destroy gear, deplete stocks, and produce low-quality fish. One lobster fisher from Digby Neck said that government 'has created a monster and there's nothing they can do about it. The draggers have ruined everything' (see chapter 8). Finally, some captains are unhappy that they will not be able to ensure a future in the fisheries for their sons. One captain commented, 'Any son of a fisherman who wants to continue in this field should automatically be able to get a fishing licence (lobster or otherwise) without having to buy it from someone else.'

Much has been made about a number of radical protests by fishers against fishery officers and DFO regulations generally (A. Davis and Kasdan 1984: Sacouman and Grady 1984). Despite an obvious class dimension, these episodes seem more akin to primitive rebellion than class conflict. Aggravation and hostility lead to an unplanned explosion which, depending on the circumstances, may draw other members of the community into action.

Although the captains may threaten to block harbours or roads in order to influence government, most draw the line at physical violence or destruction of property. In 1983, some lobster fishers burned and sank two government surveillance vessels in East Pubnico as protest against enforcement practices.[10] Of the captains in the survey 65.6 per cent strongly disapproved of this action and only 14.3 per cent approved. Lobster captains were most likely to express strong disapproval. Kearney (1986) has suggested that the majority of lobster fishers, who operate boats under 41 feet in length and generally abide by trap limits, are not sympathetic to the growth of the 'midshore' fleet of 41-to-45-foot lobster boats which break trap limits and are prepared to confront government in the attempt to expand fishing activity. One small lobster fisher on Cape Sable Island said: 'It was a crime. It costs all fishermen money. There was no need for it. Some fishermen buy big rigs and want to pay for them only through lobstering. They will ruin lobstering because of it.'[11]

Overall, the only group unwilling to condemn these boat burnings was inshore dragger captains. Twenty-eight per cent said that they approved or strongly approved. High debt loads and perpetual disputes with DFO over quota allocations incline them to be more equivocal in their answers. However, their support is hardly class-based solidarity but rather an expression of relative deprivation.

Captains in the survey differ in the types of organizations to which they belong and in the types that they think will best represent their interests. Only 24.8 per cent belong to any fisheries-related organizations. Of these 100 captains, forty-two belong to associations, thirty-eight to co-operatives, and twenty to unions. Further, 41.3 per cent of all captains rejected any organized solutions, saying that fishers could best represent themselves individually; 28.0 per cent preferred co-operatives; 19.8 per cent chose associations; and only 10.9 per cent selected unions.

The boat typology in chapter 7 distinguishes among captains in terms of types of organizations. Only 24.8 per cent of the captains in this survey belong to any fisheries-related organization (Table 61). Relatively few captains of processor-owned vessels belong to any organizations, and the few who do almost uniformly prefer associations.[12] Captains in the petty-capital category, particularly inshore dragger captains, are more likely to be involved with organizations. Inshore dragger captains are more likely than any others to belong to unions, and inshore scallop captains, to associations. Support for co-operatives is concentrated among direct producers and inshore draggers. The only significant union membership among direct producers comes from lobster captains and captains of boats using all three main types of inshore gear.[13]

Captains indicating extensive service or input ties to buyers (see chapter 7) were no more likely to join unions or associations than those without such connections. However, they were more likely to belong to co-operatives. Since producer co-operatives provide fishers with services and inputs, this relationship is true by definition. The level of competition in the port market had no appreciable effect on the probability that captains would join an organization. Captains selling in monopsonistic markets were no more likely to join unions, for example, than captains selling in competitive markets. In fact, in every area of the province, captains selling to single buyers were more likely to favour individualistic responses than were captains selling to multiple buyers.

There were some clear connections between income levels and organizational membership. Most captains with the lowest incomes (1985 fisheries income under $18,000) had no organizational affiliations, while increasingly higher income levels tended to mean a shift from unions

TABLE 61
Organizational membership by boat type (percentages, with numbers in parentheses)

Boat type	None	Association	Co-operative	Union	(N)
Director producers					
Lobster	73.2	8.3	11.3	7.2	(97)
Longline/handline	96.3	3.7	0.0	0.0	(27)
Lobster/longline-hand-line	75.9	10.1	12.7	1.3	(79)
Lobster/longline-net	75.2	10.8	9.9	4.1	(121)
Total	76.5	9.3	10.2	4.0	(324)
Petty capital					
Inshore draggers	55.6	7.4	14.8	22.2	(27)
Inshore scallopers	75.0	25.0	0.0	0.0	(12)
Longliners/seiners	75.0	16.7	8.3	0.0	(12)
Total	64.7	13.7	9.8	11.8	(51)
Processor-owned vessels					
Intermediate vessels	87.5	0.0	0.0	12.5	(8)
Offshore vessels	76.2	23.8	0.0	0.0	(21)
Total	79.3	17.2	0.0	3.5	(29)
Total, all categories	75.2	10.4	9.4	5.0	(404)

to co-operatives to associations. A plurality of captains in the three higher-income groups ($18,000 to $29,500, $30,000 to $48,000, and $49,000 to $200,000) belong successively to unions, co-operatives, and associations, suggesting a relative-deprivation explanation of organization affiliation.

Geographical specifications affected organizational affiliations and preferences. Captains in Halifax-Guysborough, Shelburne, and particularly Queens/Lunenburg were most likely to lack organizational affiliation. Those in Shelburne were especially inclined to individualistic responses, perhaps because of the inner-worldly asceticism propagated by a number of Baptist groups concentrated there. This asceticism encourages the expression of individualism and conservative forms of populist politics. Captains in Shelburne were particularly resistant to organized representation.[14]

Membership in, and preferences for, co-operatives and unions were concentrated in the west (Annapolis/Digby/Yarmouth) and particularly in Cape Breton. Again, a control for religious affiliation explained the proportions. Catholics, concentrated in western Nova Scotia and Cape Breton, were more likely than conservative or liberal Protestants to belong to any organization.[15] For historical reasons, Catholics in both areas tended to belong to unions and especially to co-operatives. Religious affiliation also helped to specify the types of organizations that captains would like to have. Both conservative and liberal Protestants expressed greatest preference for associations, second preference for co-operatives, and least for unions. Catholics, by contrast, were most likely to prefer co-operatives.

In summary, Nova Scotia fishing captains manifest great alienation from government, both generally and with respect to DFO. They are also likely to engage in issue-based, non-electoral political activity to advance their position on an individualistic basis. Like other populists, they are also ideologically committed to social and economic equality. They do not view themselves as part of a stratified class structure and clearly wish to preserve an economic environment in which each will be able to make an independent living.[16]

Social and economic forces operate simultaneously to create varying levels of activity and to generate left- and right-leaning organizational solutions to the problems that fishers face. Petty capitalists are most likely to be involved with organizations; paradoxically, they also embody the most polarized affiliations and preferences, with the inshore draggers most strongly inclined to unions and scallopers to associations. From a conventional political economy perspective, the low levels of organizational activity, particularly in unions among direct producers, and the lack of interest in such activity, are a theoretical challenge and a practical disappointment. Direct producers are the least likely to unionize, irrespective of dependence on port markets. Their connections to species-product segments, competitive-scale processing capital, and the lucrative lobster fishery underlie this tendency. Since the significant biological and economic basis for this segment of the Nova Scotia fishery is likely to persist, it is not easy to dismiss the situation as temporary or anomalous.[17]

Plant workers: marginality and rejection of mainstream political life

The Marxist and segmentation theorists considered in chapter 1 arrive, from different logics, at the same general expectations about worker

politics in the marginal economic operations so common in fish processing.[18] Marginal enterprises, by contrast to centrally situated organizations, will generate more political alienation and apathy, as well as
lower political participation, because they have more arbitrary authority systems, less job stability, and lower wages. Although there is
little research in this area, available evidence shows that marginal workers are less interested and involved in politics and more cynical about
their political representatives. They are also more likely, other considerations held constant, to favour more economic equality and oppose
favouritism toward the 'big interests' (Apostle, Clairmont, and Osberg
1986). Recent feminist literature on women in fish plants also suggests
that women may manifest even more negative political attitudes and
less political involvement (E. Antler and Faris 1979).

As shown in Table 62, plant workers interviewed generally manifest
higher alienation, or lower efficacy, scores on four standard questions[19]
than either a 1977–8 sample of low-wage private-sector workers in the
Maritimes (Marginal Work World Sample) or a 1979 cross-section of
private-sector workers in the Maritimes (General Segmentation Sample). These differences range from a low of −1.4 per cent to a high
of 24.5 per cent. However, the plant workers' scores closely parallel
those in the survey of captains and a 1982 survey of fishers in southwest
Nova Scotia.[20] But unlike captains and fishers, fish plant workers in
this study have levels of political involvement that parallel those of
other marginal workers in the Maritimes. These rates are considerably
lower (Table 63) than those for either captains or fishers in general
or for a representative 1977 cross-section of Ontario. There are no
significant gender differences, but the absence of any clear contrast
may merely reflect high overall alienation and low activity.

Plant workers' political alienation and lack of involvement are reflected in comments about community and political affairs. While a
slight majority can identify at least one person to whom they would
turn for help with community or government problems, the more positive responses typically indicate that the individual contacted will 'try'
to assist them. There are a number of reports of negative outcomes,
with some considerable scepticism about the effectiveness of even well-
intentioned leaders. As one plant worker commented, 'probably nothing' would be done by the mayor or MLA 'unless it is a serious problem
and a lot of people are behind it.'

The labour-force strategies typically employed in fish processing have
led not only to political withdrawal by workers but also to a patronage

TABLE 62
Comparison of captains with Maritime samples on political efficacy (% who agree or strongly agree; numbers in parentheses)*

Political efficacy questions†	(1) Fish plant worker survey (1987)	(2) Captain's survey (1986)	(3) Southwest Nova Scotia fishers (1982)	(4) Marginal Work World Sample (1977–8)	(5) General Segmentation Sample (1979)
1. 'Generally, those elected to Parliament soon lose touch with the people.'	85.8 (274)	88.0 (400)	89.2 (575)	80.1 (572)	77.6 (1437)
2. 'I don't think that the government cares much what people (like me) think.'	78.2 (275)	79.8 (396)	81.6 (580)	69.3 (576)	56.6 (1448)
3. 'Sometimes, politics and government seem so complicated that a person (like me) can't really understand what's going on.'	91.6 (274)	92.1 (403)	93.4 (589)	93.0 (583)	85.2 (1474)
4. 'People (like me) don't have any say about what the government does.'	82.6 (275)	78.6 (397)	83.5 (581)	67.5 (579)	58.1 (1466)

* A series of chi-square tests was run on the sample differences in item scores, and all samples by item scores are significant at the 0.001 level.
† For explanations of surveys, see note * to Table 55, above. The wording in brackets was included in the 1984 CNES, but not in the Maritime surveys. Exclusion of 'like me' may have diminished the differences between the fishers and the Marginal Work World samples, on the one hand, and the 1984 CNES, on the other: some of the former respondents have probably used a broader reference group than people like themselves.

TABLE 63

Comparison of plant workers with other samples on political behaviour (% who say yes; numbers in parentheses)

Questions*	(1) Fish plant worker survey (1987)	(2) Captain's survey (1986)	(3) Southwest Nova Scotia fishers (1982)	(4) Marginal Work World Sample (1977–8)	(5) Ontario Provincial Election Sample (1977)
'1. Signed a petition directed to some government agency?'	25.9 (282)	41.2 (403)	38.9 (592)	22.2 (562)	34.8 (1,139)
'2. Helped draft or circulate such a petition?'	5.0 (283)	20.3 (403)	10.0 (592)	9.4 (563)	8.2 (1,189)
'3. Attended a city council school board, or similar meeting to support or oppose some policy?'	14.1 (284)	35.1 (404)	24.0 (592)	18.9 (562)	22.6 (1,193)
'4. Worked with an organization of neighbours or other members of your community to persuade government (such as Department of Highways, the School Board) to do something you feel ought to be done?'	20.4 (284)	48.5 (404)	29.1 (591)	19.4 (563)	–

* Surveys 1–4 as in Table 55 (1–4). The wording begins, 'I'm going to read you a list of things people sometimes do *between* elections to keep in touch with government. Can you tell me, for each of these things whether you have ever done such a thing when there was not an election campaign going on.' These questions were initially developed by Ontario political scientists Robert Drummond and Fred Fletcher (column 5).

orientation toward government activity. Marginal operations that employ significant proportions of women at low wages are particularly prone to dependence on government support (e.g. UI and welfare) and are sensitive to government regulations on minimum wages, overtime, and holidays (Apostle, Clairmont, and Osberg 1986: 920). These strategies also affect workers, who tend to support parties in power or parties that espouse patronage politics (Apostle, Clairmont, and Osberg 1986: 923). Thus direct financial involvement by the state, as opposed to regulatory policies, has helped shape the political organization and activities of the industry.

Conclusion

The general arguments advanced in this chapter emphasize the ways in which internally generated economic processes divide particular groups and in turn, fragment their political efforts. Processors are divided by scale and production lines, and fishers by scale, type of gear, location, and cultural differences. Differentiation among capital and producers, however, provides a structural basis for right-wing populism in a buoyant economic climate – 'free enterprise' – or as an alternative to closely aligned interests between government, corporations, or unions. Individualistic political responses to problems typified both capital and producers in the mid-1980s. This pattern, particularly in southwest Nova Scotia, found strong cultural support in fundamentalist religion. While plant workers are not internally differentiated in their orientations to conventional political activity, comparative analysis suggests that work conditions in economic enterprises like fish plants encourage marginality and political passivity.[21] The prevailing conservatism of rural political life, especially in southwest Nova Scotia, probably reinforces the marginality of fish plant workers; for female workers, a patriarchal structure in household and family makes class-based protest and solidarity difficult and exacerbates this marginality (see chapter 13).

Conclusion

The case of the fishing industry has offered a vivid example of industrial differentiation and the resilience of small capital. This volume began with a discussion of three major perspectives that guide much of the research now being done in this area and concludes with some comments abut possible contributions that the current work makes to the ongoing debates. These investigations also have added to the growing body of fisheries social science, and this chapter suggests some ways in which the theoretical concerns expressed here raise questions not only about industrial sociology but also about potential lines of inquiry for future fisheries research.

A body of theory emerged in the late 1970s arguing that the capital accumulation process is indeterminate and that a number of centrifugal forces reconstitute small capital on a continuous, even expanding basis. Differentiation theory points out that in certain industrial sectors, structural or conjunctural patterns – typified by competitive product markets, low entry costs, or production cycles resistant to mass production – generate a small-capital structure with decentralized organization and independent ownership and control. This sector is distinguished by the artisanate origins of owner-operators, family-centred organization, flexible management based on personal control, paternal (if not fraternal) employer-employee relations, low-cost but highly adaptive technology, a versatile labour force, and intimate community ties.

Playing down the prospects for such autonomous development, the dependency perspective stresses an accelerating tendency for large-scale capital to control labour costs and increase efficiency by decentralizing production, through subcontracting strategies. As our research shows, such developments are a separate tendency in the modern capitalist economy. Significant differences separate the 'dependants' from the 'independents,' such as lack of managerial autonomy and control, singular product markets, financial vulnerability, and, most important, the centrality of labour costs.

This study illustrates the relevance of both dependency and differentiation theories, confirming many observations and adding a number of insights. Resource-related idiosyncrasies in the fishing industry extend the limits of independent commodity production, while inhibiting the 'normal' concentration and centralization of capital. We saw, for example, the difficulty of privatizing sea tenure, the production risks associated with an uncultivated resource, and the penchant industrial harvesting techniques have for overfishing. Successful adaptations of capital are seen to be based on decentralized production, while the attempts of large-scale capital to pursue centralized industrial models in land-based processing facilities and integrated harvesting facilities are more problematic.

The resource basis of the industry also represents a source of locational differentiation. While technological change in harvesting is leading to increasing county-level convergence in Nova Scotia, the southwest's continuing comparative geographical advantage is magnified by proximity to the New England market. Locational differentiation is seen to affect entrepreneurship and growth patterns of small capital, port market relations, the class character of the harvesting sector, and the structure of local labour markets.

In historical terms, this study identifies two relatively successful forms of small capital through an analysis of long-term trends in the accumulation process. An incipient, highly decentralized, but *archaic* merchant's capital successfully adapted itself to the fishery between the family harvesting and processing unit and a fragmented wholesale marketing structure. A monocultural salt fish market was characterized by instability and low prices, social costs that the family fishery was made to bear through debt dependence. Post-1945 modernization broke a vicious circle of poverty and debt, increasing competition, incomes, and prices. An artisanate flourished and generated a *modern* small-capital sector which, while sharing similar resource-related adaptations with merchant's capital, was in no sense a simple vestige of the earlier era.

Attempts by large-scale capital to 'mature' were truncated in their earliest phase and thwarted in their later phase. While showing a marked tendency toward corporate concentration, the centralization of processing and harvesting capacities has unleased a technological imperative, making large-scale capital especially vulnerable to crisis and stagnation. State intervention has repeatedly insulated capital from collapse. In the past two decades, corporate capital has used scientific management techniques to trim costs and make production more efficient through improved resource targeting, intra-firm supply allocation, worker production quotas, and subcontracting. A particular stratum in the small capital sector is now a dependency of corporate capital, characterized by the most acutely seasonal production, the lowest wages, but the highest relative profits. Although productive decentralization has not yet become generalized in the industry,[1] this case seems a classic example of a basic tendency postulated by dependency theory.

Our study has shed most light on the internal dynamics of an independent, competitive stratum of small capital and its external ties to product markets, port markets, labour markets, and community. Like patterns found in the knitwear industry in Italy, strong artisanate ties – skills and experience, personal connections, family and community ties, and the like – typify this group. This background is a vital aspect of the flexibility needed to adjust to the vagaries of the fishery and a highly competitive market. While product innovation has been a general hallmark of this process since the war, managerial flexibility and resourcefulness are more significant short-term criteria of success. Some of the out-workers interviewed were 'worker-entrepreneurs.' Unlike large capital's concern to lower wage costs by subcontracting, competitive capital seeks to control overhead costs while maximizing production flexibility. Indeed, out-workers' income appears to be comparable to that of plant labour. Operating efficiency rather than profitability characterizes the competitive sector, and the 'livelihood principle' seems an important qualifier. Plant managers are concerned above all with plant viability, decent standards of living, and community survival – at the expense of labour productivity and low profit margins.[2] Such considerations may be considered extravagant in the dependent world of small subcontractors.

Overall, this analysis reveals the complex levels of integration in the competitive segment of the industry. Managers are tied to their communities, fishers to processors, processors to each other, and fish companies to brokers. Communication and exchange among these actors generate unparalleled interdependence and awareness. This is the substance of the linkage notion behind 'flexible specialization.' Such connections also support Weiss's notion of community collectivism, but they would suggest two qualifications. First, plant workers, especially women, are increasingly marginalized from the political process of rural development while remaining stalwarts in family, community, and economic life. Second, such community collectivism is most clearly evident during expansionary phases and, as is argued below, breaks down quickly during recessions. The fragile competitive-co-operative dynamic of the first stage gives way to atomization and factionalism in the second, and if collectivism re-emerges it is on a completely different basis.

The study of fish brokers reveals that independent brokers are tied to discrete product market segments in the US northeast. These brokers and traders provide competitive and large independent processors in southwest Nova Scotia with a multiplicity of outlets and minimize vulnerability in individual markets. In turn, managers have instituted flexible production schedules that respond to seasonal and market changes. The discussion of port markets in Nova Scotia isolates key variations in the formal and informal ties between fishers and processors and in the socio-political context. Small-scale fishers and petty capital in the harvesting sector occupy a series of product-specific niches which animate differentiation at all levels of the industry. The data presented in this study demonstrate the central role of the lobster fishery in sustaining a coastal-zone fishery. These fisher-processor relationships are especially important for competitive port markets in the southwest. The myriad informal ties between processors and fishers no longer reflect the subsumption of fishers by capital but rather measure the extent to which the power relation has shifted in favour of fishers, who are able to compel processors to share in the risks involved in harvesting. This is a very impressive feature of competitive capital – direct producer relations.

The study of the social and cultural dimensions of port market relations reveals a series of non-market factors that explain the community organization of competitive port markets. The family provides a hidden network for communicating and acting on information about catch and markets. The study also illustrates the comparative significance of the 'Emilian' model for southwest Nova Scotia. The relative proximity of the New England market has led to an explosion in artisanate entrepreneurship. However, as Storey (1982: 78–81) points out, educational background and the 'livelihood principle' make these ventures more circumspect and less likely to generate growth in other sectors.

In contrast to this decentralized, diffuse, and competitive structure, there is also a trend toward accelerated centralization and backward integration by a fraction of small capital. The study of Digby Neck and the Islands illuminates the origins of this tendency and shows how state policies of modernization facilitated disintegration of the diversified coastal-zone fishery, unwittingly precipitating a crisis in fish stocks and the demise of adjacent rural communities. The introduction of groundfish dragger technologies unleashed a technological imperative that led to overfishing while making fishers more dependent on processors. Processors used state financing to expand their productive capacity to maximize complementarities with the new harvesting technologies. The expansionary dynamic accentuated a tendency to local crises which, in the period 1987–9, was repeated in other areas of the province (*Halifax Mail-Star* 12 May 1989). This trend illustrates again the point taken by many differentiation theorists – that a technological imperative has to be kept in close check if small capital is to be considered qualitatively different from its large-scale cousin.

In terms of labour, this study looks at the industry from both a macroscopic, industrial-location perspective and a microscopic, labour-process perspective. The examination of labour markets assesses complex interactions based on industrial structure, surplus labour, and location. It characterizes the province as having rural competitive, industrial competitive, and rural non-competitive labour market zones. This typology provides a more detailed examination of the surplus labour concept in dependency theory. It shows that the two competitive zones (rural and industrial) have higher wage levels than the rural non-competitive zone – the industrial because of the unionization of large plants, and the rural because of high demand for both labour and fish. Flexibility of labour, far more than its cost, explains the impact of surplus labour on competitive capital. By contrast, wage levels are of paramount concern to the costly, and increasingly cost-conscious, large operations. The rural non-competitive labour market, with its monopsonistic port markets and large plants, illustrates the strengths of the dependency perspective: large capital is preoccupied with cost-control in its relations with fishers and workers.

In the case of Clare district, the segmentation perspective on captive ethnic labour markets and the concept of clientism are especially relevant in explaining industrial structure and surplus labour. The study helps to specify community-based labour market dynamics and shows in detail how one large company has constructed an inter-plant division of labour based on subcontracting to small capital. It reveals also how plants in the region use ethnic and kinship ties to recruit relatively stable work-forces in an area with few other viable sources of employment and no trade union traditions. Women are important as cheap labour for both large plants and their small subcontractors.

While employer-employee relations are generally fraternal-paternal, the fraternal connection is restricted to men and more characteristic of manager-fisher relations than of manager-worker relations. This reflects growing stratification in rural communities between fishing families and plant workers' families. Paternalism is the dominant relation in non-union plants, particularly in competitive and small ones. It is also the primary male-female dynamic and is transposed from roots in the patriarchal structure of the rural household. This situation illustrates one central weakness in both the dependency and differentiation perspectives on small capital: gender blindness. Patriarchal relations not only pervade the household but are integral organizational elements in the plant division of labour. While paternalism is part of a more congenial 'family' atmosphere in competitive plants and contributes to high worker satisfaction and low turnover, in large plants it is another aspect of labour control in management's fairly single-minded concern with increasing labour productivity.

The dilemma of small-firm workers appears to be not simply between working for low wages and being unemployed but between accepting irregular work – and low incomes – and leaving their communities. One lingering contradiction threatening the stability of the competitive fishing industry occurs in the labour market: between uneven resource supply and the social need for regular work and a decent standard of living. On the one side, competitive plants survive on fish produced by the coastal-zone fishery, which uses selective technology and is most compatible with maximizing long-term resource yield. However, the coastal-zone fishery is also most prone to fluctuations in catch levels. Plants have to orient their production time to these variations; hence work is inherently irregular. The social cost of protecting the resource and stabilizing capital accumulation has therefore been borne by plant workers and their families. On the other side, workers, fishers, and plant managers attach great importance to community survival and the ability to live and work in their communities. The very strength of these rural communities has helped sustain fishing activities and has helped residents cope with hardships inflicted on them by the environment, external society, and their own leaders.[8]

By 1989 the fishing industry was once again in crisis. In late May 1989, National Sea Products (NSP) announced permanent closure of its Lockeport operation and temporary shutdowns for four other plants in Nova Scotia. In December, it announced imminent closure of its second largest plant in Canso and of its plant in St John's and reduced operation in North Sydney. NSP lost close to $6 million in 1988 and perhaps a further $1 million in the first quarter of 1989 (*Halifax Mail-Star* 12 May 1989). Clearwater Fine Foods, as the other major corporate participant in the Nova Scotia industry, has been performing better financially, but its rapid acquisition of groundfish operations has given it a debt of about $400 million – and substantial interest payments (*Daily News* 16 March 1989). In March 1989, when Clearwater closed a newly acquired plant in Port Mouton, the president, John Risley, stated: 'There is no question there is going to be industry rationalization. Port Mouton won't be the only plant closed this summer' (*Daily News* 16 March 1989; also Canada, Fisheries and Oceans, 1989).

The strength of the competitive sector is most evident to owners, fishers, workers, families, and communities during periods of economic boom. The market – be it product, port, or labour – operates as a more or less efficient mechanism for allocating reward for work (see Canada, Fisheries and Oceans, 1989: 17–18). The ideology of individualism and of laissez-faire free enterprise, so characteristic of small- and competitive-scale managers and a large segment of the coastal-zone fishery, reflects this situation. During periods of economic bust, the structure is at its weakest, however. The market makes those who can least afford it pay for problems not of their own making and leaves people unaccustomed to collective action, isolated, and unable to plan for reconstruction. Internally generated economic processes create significant divisions within the harvesting sector which, in turn, fragment fishers' political efforts and their communities. Processors are divided by scale and production lines, and fishers by scale, type of gear, location, and cultural differences. Plant workers are extremely marginalized and represent a new rural underclass. While the unions that represent plant workers and offshore trawlers organize and protest plant closures and reduced fish quotas,[4] inshore fishers and owners of competitive plants cannot or will not support efforts to maintain large plants and their trawler fleets. Ironically, the unions are opposed by free enterprise for being too conservative.[5] Even in extreme economic crisis, collective action remains fragmented and subject to spasmodic, issue-centred orientations.

We do not offer a blueprint for resolving the confusion and conflicts that characterize the industry. This study of the industrial structure of the fishery offers vital clues for understanding the structural basis of its problems and possible starting-points and ways of proceeding in its reconstruction.[6] Three dimensions of small capitalist enterprises are crucial for a new appreciation of the industry. First, many small businesses are family undertakings, and connections between kin structures and economic activities should be a central concern for future research. Family involvement gives small capital a flexibility that is not sufficiently understood and alters basic assumptions about rationality in economic behaviour. Second, small capital also thrives within a complex and contradictory web of paternalism that facilitates both better labour relations and gender-based conflict. Third, the ties between small capital and the communities in which it is situated frequently provide social and economic advantages that are overlooked in any narrow assessment of the economic viability of small enterprises. Kinship ties, paternalism, and community networks are not easily or neatly incorporated into standard research designs, but studies that ignore these concerns will result in poor social science and even worse social policy.

Sensible long-term restructuring of the fishing industry has to be both self-sustaining and decentralized. There are three basic prerequisites: elements of competitive-scale processing, selective-gear fish-harvesting technology, and community-based organization. The first prerequisite is vital to decentralized rural development, quality-based production, and worker satisfaction. The second is crucial if the overfishing inherent in industrial harvesting is going to end. The problem of catch variability that intermediate technology faces could be addressed possibly through development of offshore longliners,[7] which would help ameliorate the worst aspects of seasonal plant employment. The third prerequisite is the most difficult to obtain, and yet the most important if the key contradiction facing community life is going to be resolved. The existing social organization of competitive-scale capital, both private and co-operative, and of household structures perpetuates sexual inequality in rural communities. Attitudes and decision-making structures have to change if the industry is going to benefit all people equally. One obvious model would be a community-based co-operative network of plants and vessels.

APPENDIX

Method for surveys

The initial outline for this project proposed three interrelated surveys in Nova Scotia, of fish plant managers, of the captains who supply these plants, and of the workers employed by these plants (Apostle et al. 1984). A 99-case survey of fish plant managers conducted in 1984 provided a representative cross-section of the approximately 225 plants operating in the province. A 451-case survey in the spring and summer of 1986 covered captains who supply 50 of the plants in the initial survey, and in the spring and summer of 1987, 292 fish plant workers employed by the same 50 plants were interviewed. This appendix explains the sampling and fieldwork procedures for the management survey and the surveys of captains and plant workers.

Management survey

Using establishments as the unit of analysis, project members interviewed 99 fish plant managers in 1984. The initial stratified random sample consisted of 113 plants chosen from a list of 225 plants operating in Nova Scotia that year.[1] The population of 225 was divided by region of the province and employment size, and a 50 per cent random sample was taken from each region (Table A.1).

The fieldwork was conducted between 7 May and 21 September 1984; most interviews were completed during June and July.[2] The bulk of the interviews in Lunenburg, Queens, and Shelburne counties were done in June; those in Digby and Annapolis counties, as well as the eastern and northern areas of the province and Cape Breton, were completed in July. The interviews in Yarmouth County were evenly divided between the two months. The mean length of the interviews

TABLE A.1
Completed interviews with fish plant managers by plant location and size (completed interviews/potential interviews)

	Plant size (no. of employees)				
	1–20	21–50	51–100	101 or more	Total
Southwest Nova Scotia					
Shelburne/Lunenburg/ Queens	26/66	6/16	3/5	4/10	49/97
Yarmouth	3/6	6/12	1/7	4/8	14/33
Digby/Annapolis	8/16	9/19	3/6	5/10	25/51
Cape Breton, eastern and northern Nova Scotia	10/22	3/11	2/3	3/8	18/44
Total	50/110	24/58	9/21	16/38	99/225

was 82 minutes. The overall completion rate was 87.6 per cent, and interviews were quite evenly distributed by size and location. Given current survey standards, these patterns indicate a high level of success in our data collection.

Surveys of captains and plant workers

Sampling

On the basis of the 1984 management interviews, it appears (Tables A.2A and A.3A) that there were 3,786 captains selling to, and 12,804 workers employed at, 225 plants in 1983. The captain estimate is in keeping with two external calculations made by local officials of the Department of Fisheries and Oceans (DFO). They estimated in 1982 that there were 5,197 fishing-licence holders in the five western counties of Nova Scotia (Apostle, Kasdan, and Hanson 1985: 257) and, in 1984, that there were 5,204 vessels in the Nova Scotia portion of the Scotia-Fundy region (personal communication). One would expect these numbers to be somewhat higher, as some licence holders or vessels sell directly to retail outlets and truckers, and some are inactive. The estimate of plant workers corresponds with a calculation made by the local DFO office. In 1983, it estimated that there were 14,273 plant employees.

For creating sampling frames, lists were obtained of all captains who

TABLE A.2
Plants, captains, and interviews with captains by plant size

	Plant size (no. of employees)				
	1–20	21–50	51–100	101 or more	Total
A. *Captains and plants*					
Captains/plants (percentage of total captains)	1,427/110 (37.7)	1,093/58 (28.9)	441/21 (11.4)	825/38 (21.8)	3,786/225 (100.0)
B. *Potential captain interviews*					
Potential captain interviews/selected plants (percentage of potential interviews)	236/27 (38.8)	189/12 (31.7)	70/4 (11.7)	105/7 (17.8)	600/50 (100.0)
C. *Completed captain interviews*					
Completed captain interviews (percentage of completed interviews)	180 (39.9)	142 (31.5)	58 (12.9)	71 (15.7)	451 (100.0)

sold fish during 1984 to, and all plant workers employed during 1986 in, 50 of the 99 plants in the management survey.[3] The lists yielded names of 899 captains and 3,147 workers, and 600 and 500 of them, respectively, were randomly selected to constitute the sample. The proportions selected closely mirror overall distribution for the province (Tables A.2B and A.3B). In addition, the number of names provided generate estimated provincial totals of 4,119 captains and 14,402 workers – reasonably close to the management survey results of 3,786 and 12,804. If one conservatively assumes a 35 per cent diminution in efficiency because of clustering, the initial samples of 600 and 500 are the equivalent of simple random samples of 516 and 427, respectively.

Fieldwork

The project attempted to conduct personal interviews with the sample of fishing captains between 23 February and 7 July 1986. Interviews with plant workers were done between 4 April and 8 June 1987.[4] Experienced interviewers were hired and individually trained in the

TABLE A.3
Plants, plant workers, and interviews with plant workers by plant size

	Plant size (no. of employees)				
	1–20	21–50	51–100	101 or more	Total
A. *Plant workers and plants*					
Plant workers/plants (percentage of total captains)	1,168/110 (9.7)	1,952/58 (15.2)	1,802/21 (14.1)	7,882/38 (61.6)	12,804/225 (100.0)
B. *Potential plant worker interviews*					
Potential plant worker interviews/selected plants (percentage of potential interviews)	49/25 (9.8)	100/13 (20.0)	64/4 (12.8)	287/8 (57.4)	500/50 (100.0)
C. *Completed plant worker interviews*					
Completed plant worker interviews (percentage of completed interviews)	29 (9.9)	22 (7.5)	36 (12.3)	205 (70.3)	292 (100.0)

use of the data collection instruments. Every effort was made to hire interviewers from the geographic regions to be surveyed. This approach reduced travel expenses and ensured that interviewers were familiar enough with their area to locate individuals for whom addresses were incomplete or telephone numbers unavailable. Interviewers were contacted once a week so that progress could be monitored and any problems resolved. As indicated in Tables A.2c and A.3c, interviews were completed with 451 captains and 292 workers. These figures represent completion rates of 75.2 and 58.4 per cent, respectively. Given the lengthy and complex face-to-face interviews carried out,[5] completion rate for captains is very good, and that for workers is adequate. These interviews are distributed reasonably by plant size, except for proportionate under-representation of workers in plants with between twenty-one and fifty employees.

Notes

Chapter 1

1 The small-business sector in the Third World is generally referred to as the informal sector. Similar patterns have been identified in that context (see Schmitz 1982; 1989; Storper 1990).

2 The development literature was dominated by the dualistic perspective in its formative period (Redfield 1947; Higgins 1956; W.A. Lewis 1963; P. Kirby 1969).

3 Corporate organization represented a dramatic departure from the orthodoxy of perfect competition that dominated the microeconomic theory of the firm. As Dewey (9174) and Adams and Brock (1986) have pointed out, the arguments that economies of scale necessitated multi-plant firms (concentration), or that monopoly profits were a precondition for technical progress and innovation (Alchian and Demsetz 1972), were largely articles of faith. For most, microeconomics consequently tended to be either irrelevant or unjustifiably conservative vis-à-vis the growing importance of corporate concentration (Dewey 1974: 4–5; Adams and Brock 1986: 15; Blair 1972: vi). This criticism represents one point of departure for small-firm economics in the post-Keynesian critique of industrial concentration.

4 Braverman's (1974: 64) characterization of labour processes in small-scale plants as essentially pre-industrial, archaic, and transitional is a further example of the linear model of industrial development and an essentially dualistic perspective that underlies classical Marxism. The post-Fordist perspective that emerged in the 1980s is a critique of this model. (See Piore and Sabel 1983; 1984; Rainnie 1984; Sabel and Zeitlin 1985; Elam 1990; Lovering 1990.)

5 Curran and Burrows (1986: 267–8) take issue with the way Marxist
theory copes with the petite bourgeoisie generally, while Bechhofer
and Elliott (1985) point to the dualistic bias in Marxist class and industrial analysis. They argue that in each case 'Marxists argue that while
the petite bourgeoisie may persist within a capitalist system it is essen
tially to be seen as an anachronism, a vestige of the simple commodity
form of production which attended the transition from feudalism to
capitalism' (185). As Rainnie (1985a: 153, 158) points out, a growing
body of evidence challenges the theory of the centralization and concentration of capital. Study of the capitalist economy cannot be based
on a priori assumptions and 'theoretical premises' about historical
epochs. The drift into theoretical abstraction (Semmler 1982; Wheelock 1983) is for some a safe haven. The Marxist–populist polemic distorts and deflects serious scholarly inquiry (Bromley 1985a: 321–2).
6 Sociologists had been drawn into development research much earlier
to account for the failure of industrial transformation in the Third
World. Various aspects of culture and personality were identified as
obstacles to modernization: attitudes to wage work, industry, and
achievement were considered relatively poor; entrepreneurship and innovation were hindered by traditionalism; and efficiency and rational
organization were constrained by patronage and particularism (Long
1977: chapter 2).
7 The simple two-case model has been made more complex (Hodson
1978, 1984), but the concern is still with gross statistical indices and
discrete levels of analysis and measurement.
8 Blair (1972: 93–4) draws attention to the 'barriers to exit' represented
by large-scale fixed capital investment. This notion is developed below
as a technological imperative in large-scale fish processing.
9 The marginality debate has paralleled the segmentation perspective in
the development literature. At issue is the relative permanence of the
traditional or competitive sector – and, in class terms, the marginal
mass – under conditions of economic disarticulation (Cardoso 1971;
Quijano 1974; Amin 1976). The concept of internal colony, developed
by Casanova (1965) in his study of native communities in Mexico, has
also been applied to blacks in South Africa (Wolpe 1975; see also
Wilkie 1977).
10 The dependency perspective developed in the Third World as critique
of both neoclassical (modernization) theory and orthodox Marxism; see
A.G. Frank (1969), LaClau (1971), Booth (1975), O'Brien (1975), and
Foster-Carter (1978). The post-Marxian debate represents a wide-rang-

ing challenge to the dependency approach (see Booth 1985; Mouzelis 1988; Corbridge 1990).

11 Blair (1972) anticipated this perspective in his discussion of corporate 'arteriosclerosis' and the centrifugal tendencies associated with new decentralizing technologies in the post-war period.

12 Murray (1983) points out that decentralization is strongly affected by product type and the degree of divisibility in the production cycle; decentralization is facilitated by modularization in industries of high divisibility, such as aeronautics, machinery, electronics, and clothing (77–8).

13 In a study of the Italian knitwear industry, Solinas (1982: 337) found that women represented the highest proportion of production-line workers in small plants and homework (also Rainnie 1985a).

14 In Marxist analysis of Third World underdevelopment, metropolitan capital subordinates pre-capitalist relations of production, small independent capitals and their workers, petty producers, domestic units, and women in a 'chain of exploitation.' Central capital 'super'exploits these groups through an indirect and unequal exchange, usually through acquisition of commodities or services in exchange for low wages. Wages are lower than their real value, or what capital would pay at the centre, because of the deliberate maintenance of pre-capitalist relations of production, such as the fallback systems provided by extended kin units, women's work in the domestic unit, and so on (LaClau 1971; Meillassoux 1972; Safa 1986).

15 Since the early 1970s, development economics working within a welfare tradition has felt that the informal sector has a large capacity to absorb labour and also provides cheap commodities and services to an impoverished population (Singer and Jolly 1973; Anderson 1982). The dependency critique emphasizes how the informal sector reproduces conditions of labour surplus in underdeveloped countries (Moser 1978; Bromley and Gerry 1979), while a third perspective focuses on household survival strategies among the poor of the Third World (Uzzell 1980; Lozano 1983). A good summary of perspectives is provided in Bienefeld and Godfrey (1978) and Bromley (1985b).

16 In Third World cities, casual labour, hawkers, petty entrepreneurs and middlemen, and small artisans are seen as helping to maintain dependence and capitalist subordination, undermining collective efforts to eliminate the structural sources of exploitation. Paradoxically, individual initiative enmeshes producers in a web of dependence while providing a seemingly viable way out for all. See Gerry (1979).

17 Oakey (1985: 136) argues that small-capital, high-technology firms assume a minimalist financial strategy vis-à-vis research and development and investment capital.

18 Despite Beverton (1953), economists and biologists have generally ignored technology in favour of numbers of fishers and their effects on fish stocks, probably because of a strong commitment to modernization: 'Progress will come from large-scale industrialization and all other techniques are backward.'

19 The 'by-catch' – unwanted species – is wasteful and disrupts the ecological balances that underlie the food chain.

20 Contrary to what most economists believe, a great deal of fishing activity is relatively sedentary. In 'bedding' species such as scallops, this is most obvious, but fishers also have preferred locations that they 'tend' on a regular basis: grounds less likely to be fished by other gear (see chapter 9, below), areas that the nearby community considers a 'fishing common' (A. Davis 1984), or inherited berths (Sinclair 1985b). See also McCay and Acheson (1987) and Berkes (1989).

21 Sinclair (1985b: 14–20, 141–8), for example, comprehensively reviews recent anthropological perspectives on the subsumption of petty producers in agriculture and explores the relevance of this work for the fishery, particularly in Newfoundland.

22 The historical relationship between technological advancement and risk is not as simple as it seems. As the 'catchability' of fishing vessels increased through improved boat design, navigational aids, communication, and fish-finding technology, overfishing increased. Fish became scarcer, and producers now carried a greater debt load and were driven to take more risks in distances, weather, and amount carried (chapters 8 and 9, below). State regulation appears to keep these risks within 'manageable' proportions.

23 While scholars (Lucas 1971; Himelfarb 1976; Bowles 1982) usually focus on single-enterprise towns, the fishing industry, with an open-access resource and low entry costs, more typically has multi-enterprise, single-industry towns – and interesting class dynamics in both the labour market and the port market under competitive conditions (see chapters 7 and 10, below).

Chapter 2

1 The secondary data used in chapters 2 and 3 to draw these conclusions did not lend themselves to clear examination of the 'contracting-out'

hypothesis put forward by dependency theory to account for capitalist decentralization. Consequently, primary data based on a survey of fish plants are presented in chapter 4 to explore this and other conceptual issues in greater depth.

2 The fishery of Atlantic Canada was central to aboriginal foraging, in which seasonal salmon, herring, and mackerel runs and the seal fishery figured centrally (McGee 1978; Upton 1979). Initial European contact upset this balance (see Martin 1974; Upton 1979).

3 Ryan distinguishes among the ship, the bank, and the bye-boat fisheries, all branches of the British migratory fishery. The latter two reached their peak in a transitional phase between the migratory and the sedentary Newfoundland family fishery (K. Matthews 1968; Ryan 1983: 35–6).

4 In the wage structure that evolved by the eighteenth century, fishers signed on for a fixed wage with planters or bye-boatkeepers resident in Newfoundland (E. Antler 1977a: 14; Sider 1976). This was a transitory phase in the emergence of the resident family fishery. Fishers were in an advantageous position vis-à-vis planter-merchants because of a shortage of labour, as contracts ensured fishers a fixed wage and shifted risk onto the merchant.

5 In a study of English emigration to Newfoundland in the 1780s and the early nineteenth century, Handcock (1977: 15–48) finds that a very high proportion of settlers came from Devon, Dorset, Hampshire, Somerset, and southeastern Ireland, in addition to a significant number from Bristol, Liverpool, London, and the Scottish Lowlands; surprisingly few came from Cornwall and the Channel Islands. Such patterns reflected old lines of communication and trade in the ship fishery and conditions of emigration-push. However, some significant 'pull' factors may have been at work also. Crew members on the fishing ships were renowned for receiving relatively high incomes.

6 This paralleled closely the two patterns that emerged in the mercantile structure of the timber trade in nineteenth-century New Brunswick (Wynn 1981).

7 Marx (1967a: 768) felt that labour control was the essence of the primitive accumulation problem, particularly in the frontier colonies with abundant public land. Marx, however, underestimated the heterogeneity of capital in this context. Kay (1975) has pointed out that the imperial system extended the life of merchants' capital in the colonies. Social and economic conditions on the frontier provided an independent basis for prolonged reproduction of merchants' capital. Wynn

(1981), for example, illustrates the role of credit in overcoming prolonged difficulties in communication and transportation.

8 Ommer (1979), in the most significant contribution since Innis's *The Cod Fisheries*, terms this expansion the 'second commercial "empire" of Jersey' (1979: 76) (the first was the ship fishery in Newfoundland). Hughes (1981) analyses the impact of these merchants on Miscou and Lameque, especially familial and corporate ties among these firms (1981: 22–3). Samson (1984) incisively studies one merchant's 'empire' in Gaspé.

9 The 'habitant-pecheur' fishery that flourished around Louisbourg between 1700 and 1758 on Isle Royale was a variant on this mercantile form. Under the Treaty of Utrecht, Isle Royale became France's only possession in the area after 1713 (Innis 1954: 138). 'Fishing stations' emerged, based on a resident 'habitant-pecheur' fishery (Balcom 1984: 14–16), and required a minimum of two shallops, which had a boat crew of three for each vessel and a shore crew, to be shared among the boats, of at least four (Balcom 1984: 59–69). Clark (1968: 313) observed a subregional pattern dependent on household and seasonal labour in Cape Breton (Balcom 1984: 15, 52–3), where labour shortages gave rise to indentured labour from France.

10 Chang (1975: 42) provides a list of forty-four Newfoundland merchants with ties to the West Country of England.

11 Remiggi (1979) describes cleric–merchant conflict in Gaspé: the impoverishment of fishers – and reductions in tithes – forced the diocese of Quebec City to encourage parish priests to side with fishers against merchant interests.

12 Populist movements of fishers through the Fishermen's Protection Union in Newfoundland and the Antigonish Movement in Nova Scotia seem the most significant pre-1945 attempts at industrial restructuring and modernization (MacInnes 1978; I. MacDonald 1980).

13 E. Antler (1981) examines the role of the colonial state in discouraging indigenous agricultural development in Newfoundland to protect import traders and the salt fishery.

14 In Newfoundland, the Harvey Commission (1894) called for separation of harvesting from production in the family fishery (E. Antler 1981: 147–9).

15 In northern Nova Scotia, Burnham and Morell owned the canneries and invested heavily in small inshore boats, gear, and traps, even importing fishers to crew their boats. Such fishers received a captive price – in one instance, 51 per cent less than the price paid to one in-

dependent fisher. In other areas, independents were used, but company collusion kept prices low, with prices often announced after fishers had 'geared up' (Canada, Parliament, 1910c: 495–686). The truck system may have been changing here because of a shortage in supply. Given increased port market competition, capital used producers' operating debt to gain access to fish rather than to enforce low prices (see chapter 7, below).

16 This did not happen in Newfoundland, and the economic crisis of the 1880s ruined the industry (Neis 1981: 132). Alexander (1980: 27) points out that between 1884 and 1911 'almost 30 percent of the labour force ... was shifted out of fishing into other occupations.'

17 One fisheries officer was killed in 1926 for attempting to enforce regulations (Canada 1927a: 12). Almost forty years earlier a fisheries official had warned that only 'armed guards' could enforce the nine-inch limit on lobsters (Canada, Parliament, 1910a: 3).

18 This trend distinguishes the histories of capital accumulation on Canada's West and East coasts (Muszynski 1987). While canning technology facilitated centralization of BC capitalist production, the diversity and dispersal of fisheries and technologies continually undermined efforts to centralize East Coast production (see chapter 3, below).

19 The McLean Commission (Canada 1927a: 12) optimistically predicted that the live trade would have a significant conservation effect, since only large lobsters, nine inches and over, were suitable for this market and the smaller lobsters could be left to mature. However, by 1939, the average catch was only 62 per cent of pre-1914 levels, while the numbers of licences had increased and effort had intensified. The dominion government therefore began limiting seasons and minimum sizes (Nova Scotia 1944: 65).

20 Merchants in Newfoundland made little effort to assume control over the processing of salt fish. Production remained in the hands of the family fishery, and few attempts were made to mechanize the curing process. The most valuable export cure was a light salted product that required labour-intensive sun-drying. Merchants allowed the family to assume production costs, since its labour was largely unpaid. Royal commission reports in 1935 and 1937 opposed changes in this traditional method (Alexander 1980; E. Antler 1981: 150–9).

Under the infamous 'tal qual' port-market system, family-produced salt fish commanded a uniform price, irrespective of quality or grade. Fish could be purchased as low grade and sold as a higher grade (E. Antler 1981: 242). this system also allowed merchants to continue the age-old practice of dispersing market risks over the family fishery: 'They could adjust their prices to the lowest safe expected returns' (205–6). Industrialism was therefore stillborn. No salt fish company produced more than 5,000 quintals of salt fish even as late as 1937 (E. Antler 1981: 152).

21 The schooner used a selective, fixed-longline technology, releasing small dories at sea to fish around the mother ship – a successful but very dangerous method (see Winsor 1987). The schooner could not preserve a catch except through use of salt. Sailing vessels, while having low operating costs, were subject to the vagaries of the weather, especially during winter.

22 Insight into the anarchy that prevailed in the industry can be gleaned from the *Proceedings* of the McLean Commission in 1927. Testimony by Capt. Roland Knickle in Lunenburg, for instance, was essentially a rebuttal of that given by A.H. Whitman of Robin, Jones and Whitman in Halifax: 'The fishermen of Lunenburg have been selling fish ... to these gentlemen for a number of years, and ... they have never by word or deed done anything in a practical manner to encourage the fishermen of Lunenburg to either split, salt, cure or make their fish in a better manner ... [Lunenburg fishers] have not changed their methods because men of Mr. Whitman's calibre have always been willing to purchase their fish no matter in what condition it was ... I do contend that [the fishers] ... are no more guilty than the fish buyers in Halifax who for 30 years have been plodding along in the same old rut' (Canada 1927b: 2848–9, 2,854).

23 According to Antler (1981: 228–9), this system sometimes left merchants 'owing money to the foreign agent.'

24 The testimony of Capt. Henry Winters before the McLean Commission revealed the workings of the truck system as he experienced it (Canada 1927b: 2869–75).

25 Deplorable health and safety conditions on schooners have recently been studied (Winsor 1987) and reveal the depth of clientism in the industry.

26 For example, the dominion minister of fisheries wrote to the Halifax Board of Trade on 5 June 1905 to oppose trawling within the three-mile limit (Canada 1927a: 90).

Chapter 3

1 The terms 'large-scale,' 'competitive-scale,' and 'small-scale' differen-

tiate the types of capital that emerged after 1945 (see chapter 4). They refer to both scale of operations – 'centralization' – and the social organization of capital – 'concentration.'

2 Revisions to the Bank Act and establishment of the Industrial Development Bank in 1944 improved prospects for capitalization. However, Bates argued that the state should be prepared to advance equity capital (Nova Scotia 1944: 129).

3 This section is based, in part, on revised case study material presented in Barrett (1984).

4 The US market for Canadian fish burgeoned after the war (Watt 1963: 62). To meet US shortfall, imports from Canada increased by 130 per cent between 1945 and 1956, and frozen fillets by 80 per cent (Watt 1963: 42).

5 Regier and McCracken (1976: 9) describe Soviet pulse-fishing, 'directed intensively at particular resources until stocks are reduced to very low levels.'

6 Such increased effort, and improved technology, brought Gulf redfish stocks to a crisis in one year (Regier and McCracken 1976: 28).

7 Paradoxically, the Atlantic Development Board had forecast a period of rational growth for the herring fishery which it predicted would reach 800,000 metric tonnes by 1975 (Canada 1968: 37).

8 The company's financial statements reveal that NSP responded to down-turns – either biological or economic – not only by expanding into new fish production but also through diversifying investment holdings (Barrett 1984: 91–3).

9 NSP's decline and fall in the early 1980s and its subsequent restructuring occurred largely in response to high interest rates and excessive debt (see chapters 4 and 5 below, as well as Canada, Task Force 1982: 81–127; Barrett and Davis 1984; Sinclair 1985a; Williams 1987b). NSP's much-trumpeted turnaround between 1984 and 1988 was accomplished largely through divesting six smaller plants and moving into fresh fish. The company claimed that expansion of pre-cooked seafood and non-fish food products would improve its position still further (Kimber 1986b). The mid-1980s boom in fish prices probably underlay the temporary improvement in company fortunes, which seems to have ended with the post-1987 recession.

10 Lunenburg, Guysborough, and Cape Breton counties have been the dominant centres of offshore vessel and plant unionization. Eastern and northern Nova Scotia and Cape Breton have been centres for inshore fisher's struggles for unionization since the 1960s. These trends

probably reflect the dominant position of corporate capital since the war and are perhaps the only unambiguous expressions of the 'proletarianization' thesis posited by Clement (1986) (see chapter 14, below).

11 A government report in 1969 noted the spatial differentiation between regions where NSP had a strong market presence in the costal-zone lobster fishery and southwest Nova Scotia, where it did not. In Prince Edward Island, for example, 3,043 of the 3,566 licensees fished lobster for all or part of the season in 1965, but their income was only 85 per cent of the provincial average, and 58 per cent of the Canadian average. Off-season dependence on forestry, agriculture, and transfer payments was highest among lobster fishers in Prince Edward Island and northern Nova Scotia, where approximately 30 per cent of total income came from these other sources (Canada 1979: 32, 33).

12 Figures were calculated from comparison of data on fish processors provided in Nova Scotia, Department of Trade and Industry (1950) and R.G. Dun and Co. (1922).

13 Figures were calculated from data on fish processors in Nova Scotia, Department of Trade and Industry (1950) and Canada, Environment Canada (1976b).

14 Watt (1963: 20) notes that these factors were pivotal for the development of the fresh and frozen industry in New England in the 1930s: 'By 1940 70% of Boston's fresh and frozen fish shipments were by [refrigerated] truck.' See also White (1956: 16).

15 Clement (1986: 48–9) ignored this trend by relying on aggregated post-war Canadian fish marketing trends. 'Whereas in 1945 ... fresh fish accounted for 38 per cent of value, thirty years later this produce had declined to 21 per cent; the proportion of the market held by frozen fish canned fish, meanwhile fell from 30 per cent to 14 per cent of the market.'

16 The Dominion Bureau of Statistics registered 71 fish curing establishments in Nova Scotia in 1933 and 91 in 1944 (Canada, Dominion Bureau of Statistics, 1934: 46).

17 Kearney (1984b: 167) notes that, in a way that paralleled the lobster industry, 'The development of a technique for canning small juvenile herring and the construction of the first "sardine" canneries resulted in a great expansion of the weir [tidal trap] fishery in the 1880s. The weirs were the most suitable fishing technology for capturing juvenile herring in their shallow water "nursery" areas located along the shore of the Bay of Fundy. For the next 80 years, the weirs dominated the

herring fisheries with total catches averaging about 25,000 metric
tonnes (m.t.) per year.'

18 Watt (1963: 37) recalls that its early transactions were called ' "the old
    $40 loans" and "the $100 loans." '

19 Watt (1963: 36) notes that, as Bates had predicted, this was done par-
    tially to accommodate continuing opposition to draggers.

20 Credit unions established by the co-operative movement in Nova Sco-
    tia and Newfoundland served the same function. However, their long-
    term impact on the coastal-zone fishery has been much less pervasive.

21 The mobility and versatility of the longliner increased fishers' inde-
    pendence from processors. See Watt (1963: 70).

22 Brox (1972) discusses the 'nuclear role of cash' in the structure of
    household plural activity in rural Newfoundland. Social security al-
    lowed household members to reduce the time spent in an exploitative
    cash economy, freeing time for subsistence-related activities.

23 This process had already transformed the forest industry, a backbone
    – especially in eastern and northern Nova Scotia – of the traditional
    rural economy (McMahon 1987).

24 The intent of limited entry is quite another matter. As chapter 8
    points out, dualist economists charged that the subsidy program for
    vessel construction prolonged use of archaic technique and small-scale
    production. For some fishers, limited entry was supposed to eliminate
    these producers. R. Matthews (1983: 194–215) has argued that the
    populist thrust of post-1976 federal fisheries policy cannot be under-
    estimated, given the philosophical orientation of the then minister, Ro-
    meo LeBlanc (also R. Matthews 1988).

25 Increasing corporate concentration and monoculture in these indus-
    tries have gone hand in hand with a cost-price squeeze, producer debt,
    and dependence on capital (McMahon 1987; Murphy 1987). The anal-
    ogy with the fishery is usually seen more in corporate control of out-
    puts (Williams 1977, 1978, 1979, 1987c). Clement (1986: 67),
    however, sees little difference.

26 Innovations in communication technology have made coastal-zone fish-
    ers particularly independent in many aspects of port market relations.
    Women have emerged as vital participants in the harvesting enterprise
    (Stiles 1972; chapter 9, below).

27 Fishers' debt has increased dramatically since 1977 and has added a
    new 'small boat' dimension to the technological imperative. However,
    long-term capital debt is probably not a useful measure of 'depend-

ence.' Where processing capital is involved through backward integration of the harvesting sector, it clearly is. However, accumulation of long-term debt for most fishing enterprises may simply affect their net worth. Whether the debt is excessive in cost-accounting terms is an index of vulnerability, not of proletarianization.

28 As is shown in chapter 4, the new small-capital sector was artisanate, with close social connections to independent commodity production.

Chapter 4

Parts of this chapter are based on an initial project report of the plant-manager survey that was released in 1985 (see Apostle et al. 1985).

1 The roots of the crisis were related to factors both internal and external to the fishing industry. The endemic causes are discussed below. Foreign competition, a declining exchange rate, increasing interest rates, and quotas on preferential fish stocks were precipitating factors cited by the Kirby Task Force in the deepening crisis of 1981 (Canada, Task Force, 1982: 21).

2 The detailed analysis and conclusions of this study are summarized, below, in chapter 5.

3 In terms of the scale factor discussed below, smaller plants are more likely to be owner-managed. For small, competitive, and large plants, 46.4 per cent, 31.4 per cent, and 11.5 per cent, respectively, had owner-managers. Managers of small and competitive plants were more likely to have fathers who work or worked in the fishing industry.

4 Although formally licensed as fish plant operators, 'free lance' outworkers are paid 'piece' wages. They do not regard themselves as fish processors. All of the nine outworkers interviewed had a sum of 3 or lower on the 15-point scale-of-operations index (see chapter 10).

5 The answers to these questions are, as one might expect, quite highly associated. The correlation coefficient was 0.79 between number of workers and replacement value of fixed capital; 0.71 between number of workers and the volume of sales; and 0.85 between volume of sales and replacement value of fixed capital.

6 'Competitive' indicates the nature of this sector and is in keeping with Quijano's (1974) analysis of industrial fragmentation in Peru.

7 Table 6 is based on employment size criterion only; Table 8, on a composite scale based on employment, value of fixed capital, and sales. These differences are most pronounced in the higher proportion of competitive and large plants in the sample distribution.

8 Prior research (e.g. A. Davis et al. 1983) indicated that traditional distinctions between inshore and offshore based on vessel-size classifications were inadequate, since economic conditions often drive smaller boats further offshore and draw larger boats closer inshore. Here, inshore boats are those working within the 12-mile coastal zone; offshore boats, those fishing outside this area.

9 In a third pattern, competitive plants, which combined supply from other plants and inshore boats, got about half of their supply from each source, whereas small plants again depended more on other plants.

10 A survey of 451 fishers selling to these plants suggests that credit has assumed a more reciprocal function than it did under the truck system. Chapter 7 argues that in southwest Nova Scotia provision of inputs or services tends to reflect the relative strength of fishers in the port market.

11 Most fishers and fish processors seem to be price-takers. The crucial factors affecting one's position relate to the variety of species/product market segments in which one participates and the number of alternative outlets in each case. Generally speaking, the fewer the options, the weaker one's position.

Chapter 6

This chapter originated in a paper presented at the International Working Seminar on Social Research and Public Policy Formation in the Fisheries, held at the Institute of Fisheries, University of Trømso, Trømso, Norway, in June 1986. An earlier version was also published in *Marine Policy* 11 (1987): 29–44.

1 Terms such as 'traders' and 'brokers' vary in definition depending on locality. The distinctions made here are the ones most commonly used by the participants themselves.

2 This section is based partially on Mazany (1986), which contains detailed quantitative information on the patterns for fresh, frozen, and shellfish trade.

3 In terms of total exports to all countries, fish exports amounted to only about 8 per cent. See Canada, Statistics Canada (1986).

4 Unfortunately, available data were not disaggregated further.

5 The following discussion is based on quarterly unpublished data from Statistics Canada for 1974–83. Because US landings declined by approximately 24 per cent from 1983 to 1985, some seasonality has dis-

appeared, since seasonal price fluctuations were not as large in these latter years.

6 As we saw in chapter 2, production of salt fish is the oldest industry in Atlantic Canada and antedates permanent European settlement by nearly 100 years. North Atlantic cod was caught, split, salted, and dried there and sold, for over 400 years, either to the Mediterranean countries or the West Indies and Brazil. The world market for salt fish has always been polarized between high and low quality and traditional ethnic and geographical segments.

7 While 74 per cent of 'ordinary' Nova Scotia cure exports go to Puerto Rico, only 17 per cent of 'semi-dry' Newfoundland exports do so. The Canadian Saltfish Corp. stated in its 1984–5 annual report (p. 10) that Puerto Rico accounted for one-third of its total production. This figure probably refers to volume and also includes the full range of cures.

8 Many of the more permanent relationships are based on direct or indirect ownership links. For example, one lobster dealer jointly owns pound facilities in the Maritimes with his two main suppliers. Lobster – more perishable and of higher value – tends to generate even more movement than groundfish.

9 While plant managers in Nova Scotia have an average of eleven years of education (Apostle et al. 1985: 3), fish dealers tend to have first university degrees or nearly completed ones. Eleven of the dealers interviewed had finished degrees, and two others were within a year of completion. The fact that most of this formal training is in business suggests that Canadian processors may be at some disadvantage in their personal interactions with American dealers.

10 Personal communication from Statistics Canada.

11 These commissions begin at 5 per cent and go higher. One company which has irritated some Nova Scotia processors with late payments explains that it is sometimes so inundated by supply that it postpones bookkeeping activities to take care of marketing. If pressed, it will sometimes make payments based on 'round figures' and settle accounts later.

12 Boston accounts for less than 7 per cent of New England's groundfish landings. Because of its perishability, lobster is not sold at the Boston auction. The New Bedford auction is regarded as the trend-setter in scallop prices (Peterson 1985). The authors were permitted to visit the New Bedford auction during their August 1985 visit to Massachusetts but were refused entry to the Boston auction, probably because of the

relatively low summer groundfish landings by Boston boats, the question of countervailing tariffs on Canadian fish, and negative feelings about the George's Bank settlement.

13 NSP's major fresh fish buyer rejects scale comparisons, saying that its hired managers can produce as high quality as the owner-managed independents. Contrary to a prevailing view in the industry, this dealer believes that drag or gillnet fish is superior to hook-and-line and trap fish.

14 Number of employees for the companies interview ranged from 5 to 75, with the median being 16.

15 From a segmentation perspective, use of computers indicates likely involvement in the more central segments of the economy (Apostle, Clairmont, and Osberg 1985)

16 The owner of one of these companies also has some doubts about the utility of computers. As an example, he reports that he recently sent an employee to Logan Airport in Boston to pick up an incoming shipment immediately after being told by airport officials that the shipment was in. At the airport, his employee misquoted the nine-digit shipment number by one digit and was told that it had not arrived yet. The firm received another telephone call from the airport two hours later, saying that the shipment was in. When it asked how long the goods had been there, it was told that they had come in $3\frac{1}{2}$ hours earlier.

17 Differences in labour recruitment and organization parallel these other variations in business structure. Fresh fish and lobster dealers located around the Boston pier accepted high turnover as a fact of life and cultivated a few loyal employees as their 'core' work-force. They either hired casual labour as needed or called in a mobile crew of pier workers for larger jobs. The other companies, located in the Boston suburbs, tried to meet existing wage and salary levels but faced across-the-board turnover. Even in Maine, dealers were complaining that 'Boston wage levels' were appearing, making it difficult to attract workers. One processor offered a good benefit package but found that workers, particularly younger ones, would prefer no benefits in return for higher wages.

18 For Newfoundland lobster captured in spring, fresh water from melting ice creates problems. There are also market-driven downward pressures on prices for relatively large lobsters.

19 Canned fish has some established market niches among low-income

ethnic groups as a source of high protein. As one manager puts it, 'Sardines have more calcium per ounce than milk, more iron per ounce than spinach, and more protein per ounce than steak.' Because these customers are older, some marketing effort is being directed to other audiences, like women, who find the calcium content useful in preventing osteoporosis.

20 The survey of fish processors in Nova Scotia (see chapter 4) revealed that 49 per cent are engaged to some degree in salt fish production – 15 per cent on a small scale, 20 per cent on a competitive scale, and 14 per cent on a large scale.

21 This represents a significant market segment that NSP has not been involved in for some time.

22 By contrast, median annual sales for large Nova Scotia fish processors, with median employment of 121 workers, were only $4 million.

23 During the week of the interview, the Korean wholesaler had his warehouse closed by the local health authority.

24 The US industry also receives government assistance – about half that received by the Canadian industry, according to USITC staff. Of course, such comparisons will depend on whether or not general regional development programs are included as assistance.

25 This view was expressed mostly by processors who did not compete directly with the large companies and by brokers/distributors.

26 See USITC (1986) and US Department of Commerce (1985).

27 Much of the discussion below is based on the work of Wilson (1980, 1986).

28 Atlantic fish is now freighted to California and is also sold on a regular basis by large chain stores in the US mid-west and southwest, as well as the southeast and northeast.

29 One processor indicated that NSP's greatest asset was its ability to supply fish year-round.

30 Wilson (1986: Appendix).

Chapter 7

This chapter is a substantially revised version of a paper published in the *Canadian Journal of Sociology*, 14 (1989): 1–23.

1 While the whole inshore sector may seem irrational and anachronistic, Steinberg, in keeping with the 1980s thrust of fishery policy (see chapter 4), acknowledges that welfare economists do point to the need to consider the 'greatest good for the greatest number' (1984: 35).

2 Clement and Sinclair differ somewhat in their emphasis. For example, Sinclair (1985b: 28) criticizes the sociological accounts for their 'unduly linear model of social history.'

3 These groups may be 'price-takers' in any given case or may be fundamentally 'dependent' on a variety of service capitals (as is everyone in the input-output chain of modern corporate capitalism), but such 'dependence' helps explain the patterns and variations in the fishery.

4 While elements of the labour process distinguish various types of enterprises, the large variations among gear/species-specific technologies in crew size, division of labour, and authority structure make them inadequate criteria for differentiating along class fraction lines.

5 Social ties, especially kin-based, between captain and crew, along with other particularistic relations, are common in small-scale rural industry (Sinclair 1985b: 96, 97).

6 The role of the state is not insignificant. Since the late 1970s, it is much easier for fishers to get a licence to open a fish plant than it is for a plant operator to get fishing licences.

7 However, lobster boats selling to more than one buyer differed little from those that sold to just one buyer.

8 The weak association of the other seven service measures does not render them unimportant. Positive responses on five of these seven items were above 20 per cent and therefore indicate significant individual importance.

9 By contrast, 57 per cent of inshore groundfish draggers located in Shelburne County (the area of next larger concentration) were not tied to fish processors for these services. This trend is discussed further below.

10 Conversely, the independence of the smaller-scale (wood and fibreglass) fleet reflected internalization of costs through self-exploitation. The clear exception to the scale hypothesis overall was the autonomy of inshore scallop draggers (see below).

11 A fourth input item – ice – did not have any significant association with the other three. While 61 per cent of respondents indicated that they receive ice from their buyers, it is not generally a service provided free of charge to all who want it.

12 While most intermediate and large draggers do not need bait, supplies tend to be purchased wholesale or retail (often on a contractually defined basis where the crew is unionized).

13 This pattern is statistically significant only if one groups all longliners irrespective of scale or use of other gear.

14 Thanks to Fred Winsor for this observation.

Chapter 8

Research was carried out in the summer of 1984 by Anthony Davis and Leonard Kasdan. Much of this chapter is extracted from A. Davis and Kasdan (forthcoming). Janice Raymond, Economics Branch, Department of Fisheries and Oceans, Halifax office, provided valuable assistance and information. The manuscript was originally prepared by Donna Edwards; Frances Baker of Antigonish provided professional assistance in the redrafting stages. Many people in the Digby Neck, Long Island, and Brier Island area shared their experiences, thoughts, and concerns with the authors. This study is dedicated to them.

1 These figures are calculated from data presented in Mitchell and Frick (1970: 36).

2 For examples of the gear conflict argument see Canada, Task Force (1982: 31–43).

3 For verification of this, examine Charron (1977).

4 Small draggers were not developed first in the Digby Neck area. Around the same time they are reported as appearing in the Yarmouth area and around Grand Manan Island. Moreover, smaller draggers had been employed by American fishers several years earlier.

5 Quantified information and quotations, unless otherwise specified, were obtained from *summaries* of the annual Narrative Reports submitted by a fisheries officer located in the Digby Neck and Islands area (fisheries district 37). A fisheries officer was given a list of subject areas in which the researchers were particularly interested and prepared the summaries. The information provided has proved indispensable in understanding the fishery.

6 By now, this hull type was being built in boat shops throughout Digby and Yarmouth counties, not just at Cape St Mary's.

7 Such practices are common to fishing peoples and have been widely noted by researchers. For a review of this literature, see J.M. Acheson (1981).

8 On property claims of this sort and their impact on coastal-zone fishing, see, for example, J.M. Acheson (1975), Anderson (1979a), and A. Davis (1984). Notably, this documentation challenges the federal assumption that marine resources are common property and that fishers in common-property circumstances tend to overexploit resources; see H.S. Gordon (1954), C.W. Clark (1981), A. Scott and Neher (1982),

and Canada, Task Force (1982). A growing body of information regarding informal, local systems of access control and management suggests that the imperatives to maximize returns that drive productive activity better explain overexploitation. Such activity will have a different impact on resources than will that focused on livelihood goals.

9 The opposition of line fishers to drag net fishing has a long history in Atlantic Canada. During the 1920s industry-wide protest led to a dominion royal commission, which recommended freezing the size of the trawling fleet, which the government did. Drag net fishing remained strictly regulated until the early 1940s (Canada 1927a). The same types of complaints were made by weir fishers concerning herring seiners and by hook-and-line fishers regarding pollock seining.

10 Pollock was being seined during the 1960s, and increased landings are not arguably associated with drag fishing. However, greater exploitation of pollock is connected directly with the use of more technically sophisticated and costly fishing technologies. Indeed, one district 37 fisheries officer reported in 1962 that hook-and-line fishers were angry at the government for permitting seiners to fish pollock.

11 While the hull design has remained more or less the same since the 1960s, many aspects of fish dragging technology have changed or, at least, become even more technically sophisticated. The pilot house of the 1980s small dragger resembles a cockpit, with a vast array of primary and back-up electronic instrumentation. Fully rigged and equipped for fishing, a new 60-foot wooden-hull vessel costs in excess of $500,000; steel-hulled models sell for over $1 million.

12 See Barrett (1984) and MacDonald (1984) for a comprehensive description and analysis of this process.

13 Several informants report that it was common to record haddock as pollock because of the extremely low landings permissible under federal quota regulations. This would explain the jump in pollock landings reported through this period.

14 Biologists have been warning DFO of a pending crisis. New, very low haddock quotas were announced in the Scotia-Fundy region as a countermeasure. However, recent opposition by fishers and fish buyers has reopened the question.

15 Actual weights can be derived by applying these percentages to landings data (Table 6). For statistical purposes, DFO classifies as inshore catches all fish landed in vessels under 25 gross tonnes; a number of boats in this category actually fish offshore grounds. Conversely, some vessels over 25 gross tonnes with 'offshore' catches regularly fish in-

shore grounds. However, the vast majority of line fishing is captured by the inshore category. On inshore/offshore categories see A. Davis et al. (1983).

16 Dependence on one fishery, especially a closely regulated one such as lobster fishing, can create difficulties for both fishers and federal officers. While catches remain high, few problems will surface, but once they begin dropping, increasing numbers of people will poach and work traps in excess of permissible number.

17 Determined by dividing the totaled landed weights for 1952 and 1983 (Table 33) by the total number of people fishing in 1952 and 1983 (Table 37).

Chapter 9

Help was received from the following people: George Hallett, John F. Kearney, Cindy Lamson, staff members at various offices of the federal Department of Fisheries and Oceans in southwest Nova Scotia and of Canada Customs, and Shirley Buckler, co-ordinator, Word Processing Centre, Saint Mary's University.

1 'Gangen' is an old word for the leaders or short lengths of fishing line fastened at intervals of one fathom along the main or groundline of a trawl as fished from a longliner. Groundfish are caught on baited hooks tied to the free ends of these gangens. Gangens are uniform in length and usually of finer line than that used for the main line. Originally 'gangeing' meant to wrap with wire. Professor George Hallett of the English Department of Saint Mary's University drew the author's attention to its definition in G.M. Story et al (eds.), *Dictionary of Newfoundland English* (Toronto 1982).

2 This breakdown varies somewhat from other classifications. For example, the task force called 20-to-35-foot vessels 'inshore,' 35-to-100-foot vessels 'nearshore,' and vessels over 100 feet 'offshore' (Canada, Task Force, 1982: 14, 15). This provides a good estimation of where, for reasons of safety, boats should fish, but it does not correspond with practice. Correlating boat length with gross tonnage is difficult because of differences in beam. However, all mid-range and many nearshore vessels were registered as being at least 26 tons.

3 Hooks used for fishing halibut are larger and different in shape. On longline gear see A. Davis (1984).

4 Many other fishers use the time between longlining, gillnetting, or her-

ring trapping and lobstering to take a vacation or go hunting for two or three weeks. No business-related patterns were apparent in hunting partnerships (Wadel 1973).

5 For example, Wadel (1973: 8) notes the use of multi-purpose boats in Newfoundland.

6 Frequently, in reference to both domestic and American crews, fishers stated: 'They've got to make a living too!'

7 Another small company buys herring from other processors only for pickling and does not compete with the others.

8 However, many of the longliner fishers with larger vessels (47 to 64 feet) were fishing for halibut off Cape Breton in the summer and were not available for interview. Sales of halibut are somewhat 'exempt' from the usual obligations in any case. Mid-range longliner fishers are given much more leeway as to where they sell their halibut.

9 During the fieldwork period, gillnetters received less than 20 cents per pound for their pollock; cod prices ranged from 26 cents for 'scrod' to 37 cents for 'markets' and 'steakers' got 43 cents.

10 The first-mentioned captain fishes for his family's company and receives the current maximum price being paid to the other two local gillnetters – which varies little because pollock is mostly salted for the stable Iberian trade.

11 When one such longliner fisher lost his boat to fire during the research, the captain of the trap crew ripped his cheque in two when the man tried to pay his bill. The cheque was for $450.

12 However, great care must be taken to ensure that fresh water from heavy rainfalls does not result in a die-off. Fishermen watch carefully for this 'black water' lying on the surface of the salt water. In addition, fishers must ensure that the lobsters are not stolen.

13 Indeed, the performance of successful family fishing firms was a general topic of wharf-side conversations among fishers.

14 This solidarity can be compared with that reported from Buckie, Scotland, by Thompson (1983: 248, 249), and with M.E. Smith (1977: 8) and Epple (1977: 187). Also, see J.M. Acheson (1981: 197) on social distance between fishers and others and the comparison with agriculturists.

15 On one occasion I entered a Gangen Harbour shop where photographic supplies and confectioneries are sold. Three young fishers were just going back to their boat with smoking supplies for a voyage. As I entered I received a blast of 'air freshener.' I asked the salesper-

son: 'What are you spraying for, bugs?' The salesperson replied: 'No, fishermen; didn't you see them just leaving here?'

16 Fishers at Ganger Harbour refer to lower-value cusk and hake as 'shack.' 'Shacking' is the activity of catching these species. One fisher explained that if a fisher is so unfortunate as to land many 'shack,' then he would have to live in a 'shack.'

17 See Thompson (1983: 165, 249) on the attitudes and strategy of fishers who own more than one boat – i.e. 'shore capitalism.' In the transition from smaller to larger ventures, the captain's boats will probably be mastered by sons or other kith and kin.

18 One of the clearest illustrations of the competitive dynamic in operation is that the owner of just such a fleet, one local plant owner, began buying halibut in order to maintain his supply of cod – and this is an exclusively salt fish operation.

19 The crew was anxious until enough fish were caught to ensure a profit. Members started to grumble and exhibit signs of exhaustion and a desire to return to shore as the hold was near being filled to capacity.

20 In five years of fishing, this vessel has caught more fish on only one occasion.

21 See Andersen and Wadel (1972: 146, 147), and see J.M. Acheson (1981: 299) on the role of wives when their husbands are at sea.

22 However, the other trawler captain-shippers may be able to decipher her message. Keeping the price secret from those who do not ship seems to be caused by concern to minimize envy. However, deception about 'luck' seems to stem in part from the desire to protect access to obstacle-free bottom. With automatic pilots and trip-plotters, knowledge of routes clear of wrecks and other obstacles is considered private, while the fish are considered public property. Captains prefer not to have others towing near or across their favourite routes. Compare this with Andersen and Wadel (1972: 122–4, 136, 161, 166), A. Davis (1984: 153, 154), and Stiles (1972: 48; 49).

23 Once there, crew members leave in automobiles driven by their wives. While one vehicle and wife/driver could have picked up the whole crew, no special feature of local geography necessitates the individual 'taxiing' of members. I suspect that this represents a form of re-entry ritual. Compare this with J.M. Acheson (1981: 288).

24 In this case, the buyer was the skipper's cousin.

25 To obtain an even clearer picture of this kin-based infrastructure, one

should add the religious component of the social matrix to that of kin and kith, as has been done in a Scottish study by Nadel (1986).

Chapter 10

Parts of this chapter are based on preliminary hypotheses emerging from the project's survey of plant managers and appeared in a paper published in the *Canadian Review of Sociology and Anthropology*, 24 (1987): 178–212.

1 Surplus labour is central to the dependency analysis of Third World underdevelopment as well (see Quijano 1974; Amin 1976; Bienefeld and Godfrey 1978; Godfrey 1983).

2 Perspectives range from orthodox theories of modernization (Careless 1969; T.W. Acheson 1977; Atlantic Provinces Economic Council 1977) to dependency theory (Veltmeyer 1978; Sacouman 1980). See Barrett (1980) for a review of this literature.

3 An extensive literature examines issues ranging from out-migration to the social-psychological effects of chronic unemployment (Levitt 1960; Retson and L'Ecuyer 1963; Horowitz 1968; D.N. McDonald 1968; Pepin 1968; Wadel 1973; Browne and Wien 1978; Veltmeyer 1979; Postner 1980; Sacouman 1980; Hill 1983). The relevance of Marx's (1967a: 640–8) concepts of relative surplus population to the local situation is the focus of Barrett (1983) and of Veltmeyer (1979).

4 Cutters generally take several months to acquire their basic skills and may take further time to become proficient. Virtually all other plant jobs, save for trades and technical work, can be learned within a week. One manager stated: 'With first-year cutters, the high rollover creates incredible waste. You can see how much yield you're getting by looking at the offal. First-year cutters, I'm sure, would have a 20 per cent loss in yield by comparison to an experienced cutter. I haven't seen any studies to this effect, but I've seen cutters who have been cutting for a year, and those who have been cutting all their lives.'

5 The proportion of supervisors employed by different types of plants tends to be constant, at between 3 and 4 per cent of the total workforce.

6 Out-workers usually have a small outbuilding in their backyard which is inspected annually by the Department of Fisheries and Oceans and are granted a licence to process fish. There are about twenty such operations in the province, with a median value of fixed capital of about

$10,000. Occasionally managers provide loans or second-hand equipment. The workers typically handle up to 18,000 pounds of pickled cod, hake, pollock, or cusk per week, either under contract or on an overload basis. Nine out-workers interviewed employed a total of eighteen workers (see discussion below).

7 Corporate capital has divested the oldest plants, modernized wetfish trawler storage, and entered the higher-unit-value fresh-fish market. Production of fresh fish has further reduced processing in the plants and increased unemployment. The experiment with a freezer factory trawler in 1986 had the same purpose.

8 Approximately 21 per cent of workers in competitive plants and 30.8 per cent of those in small plants worked under a production quota.

9 These trends are somewhat more pronounced in competitive plants than in large ones.

10 Unionized plants employed 39 per cent of the total work-force covered by the plant survey; the 1986 survey of workers included 45.5 per cent unionized workers.

11 One large plant operator was subcontracting the most highly skilled and highly paid part of his fish production – cutting – in order to penalize his unionized work-force. The manager of a nearby competitive plant described the politics behind the move: 'I think they are trying to force the union out with the system they are using now. Given what they are offering other plants to cut for them, they could do it as cheaply at this plant. They're putting the fish out to other plants to have it cut, then they're bringing it in, trimming it, and freezing it. To cut the fish, they only use water to bring the fish in on the tables to the cutters. There is no electricity involved, the plant is sitting there, the tables are sitting there. They have workers on the other end receiving the fillets who can receive the whole fish just as easily. So it's no extra cost to receive those fish and to cut them. Their employee benefits aren't going to be that much greater because they're only eliminating about twelve cutters, and they're still using their trimming work-force, their packing work-force, and their freezing work-force. So it's just a method of saying to the people, "look, you wanted a union, you live with the union". I think this is the method behind a lot of things that have happened at the [plant] in the last two or three years. And it's to a point now where the people are blaming the union. You talk to a lot of the people who are not working there today, or the people who are getting only four hours a week who perhaps worked on a year-round basis at the plant five years ago, and they'll

say, "Look, don't talk union to me. If it wasn't for that union we'd be still working." '

12 A manager in one competitive plant said that workers are hired on a thirty-day basis and usually let go after that period: 'Maybe a month or two later they call them back again ... They couldn't keep up with their performance [quota].'

13 A subsample of six fishers' co-operatives indicated that fishers' wives sometimes served a similar function. Managers would deploy them strategically for labour control, since their interests were seen to coincide with those of management.

14 In 1984, in two types of competitive plants – independent and feeder/ subcontractors for large plants – 49 per cent of workers made less than $6.00 per hour in the former, compared to 57 per cent in the latter.

15 This includes Clark's Harbour but excludes Shelburne and Lockeport.

16 This includes Bedford, Halifax, and Dartmouth.

17 This figure excludes Cape Breton County. Because of the unique problems associated with deindustrialization and structural unemployment in that county, it has been excluded from this analysis. Total unemployment runs over five percentage points above the provincial average, and 31 per cent of all low-income families in the industrial zone are located there.

18 While seasonal unemployment in 1986 was highest in the rural competitive zone and lowest in the rural non-competitive zone, the amount of time out of work for unemployed workers was reversed: less than twenty-four weeks for 68 per cent in the former zone, and twenty-four weeks or more for 73 per cent in the latter.

Chapter 11

Claudia Kingston translated the original draft of this chapter.

1 Ville Française, known in English as the French Town or the French Shore, was named by the neighbouring English settlers to designate Acadian settlements on the eastern shore of St Mary's Bay.

2 In 1951, the population of Clare was 8,409. According to the 1986 census, its population was 9,675 (*Vanguard* 24 March 1987).

3 Lobster fishing started to expand around 1870 with the advent of traps which improved catches (see chapter 2).

4 Women constitute more than 60 per cent of plant employees and more than 80 per cent of workers on the production lines.

5 The law allows the fisher to enter the fish processing sector, but not the reverse: in 1987, processors could not obtain fishing licences unless they were in business before the relevant regulations were enacted.

6 A survey of eleven large and small companies with different operations (groundfish, scallops, and herring) formed part of the fieldwork.

7 In this category are found, among others, cod, haddock, halibut, plaice, pollock, redfish (or ocean perch), and turbot.

8 For example, handling and transportation costs for fish unloaded in Cape Breton and processed in Clare can increase by from 15 cents to 20 cents a pound the selling price for fillets. For a company that produces two million pounds of fillets per year and must absorb part of these additional costs, the amounts involved are sometimes considerable.

9 Herring roe must be frozen within twenty-four hours of extraction.

10 All except one sell 100 per cent of their production to Bay Seafoods. The figure for the fourth plant is over 60 per cent.

11 This enterprise was started in 1980 to process squid. After the market for this species collapsed, the company had to change its orientation. It is currently testing a new product using herring that has been frozen or salted after extraction of the roe.

12 According to an informant, Bay Seafoods intends to enlarge its fish-meal plant in order to increase processing capacity and to absorb herring waste from most of the plants in the region. However, the company's management would not confirm this.

13 The processing consists of sorting (according to size of the muscle), packaging, and freezing.

14 Scallop fishing generally begins in April or May and ends in October.

15 Southwestern Nova Scotia enjoys a ferry service between Yarmouth and Maine; another ferry service links Digby to Saint John.

16 In a few cases, however, former managerial employees have gone into fish processing and started up a new company. They must sell their product elsewhere than to their former employer.

17 There was much uncertainty surrounding this new plant, which was scheduled to start operations toward the end of July. Construction was almost complete, but developers were quiet about operations to be carried out, number of people required, and the like.

18 Ammonia odours are emitted by refrigeration systems.

19 Teams of three or four people, standing very close to each other, feed herring into the machine. The noise level is high, the pace is accelerated, and the work is messy.

20 Employees are usually informed by telephone either the night before or early the same morning that there is work for them.
21 Proximity is generally considered first if the quantities of fish to process are minimal and require only a few hours of work.
22 Certain large fish processing plants have rotating work teams and operate twenty-four hours a day. This is generally not the case in the Clare region's plants.
23 Depending on experience and seniority, some cutters receive up to $8.00 per hour.
24 In some firms, it is forty-four hours and includes Saturday morning.
25 This amount can double on Sunday. People usually work in teams of two. With average speed and fish quality, they can fill a box in approximately one-half hour.
26 A person must work a minimum of fifteen hours per week for the week to be insurable. In other words, the week must be accounted for in the minimum number of work weeks necessary to qualify for UI.
27 At Bay Seafoods, hiring personnel is the responsibility of each plant's supervisor. He or she must see to the training of new recruits and ensure that the rhythm and quality of production are maintained. Training is done on the job, generally by other workers, who receive no particular privilege or treatment for doing so.
28 In fact – and this was confirmed by several informants – multinationals considered settling in Clare, but both times the largest local employers used their influence to exert political pressure to outdistance these potential competitors in wages and recruitment. These multinationals operated in sectors other than fish processing. According to the informants, Michelin was one of them.
29 This is becoming more obvious with growing stratification of the work-force, the structure of the work process, and implementation of more productive and efficient methods. Bay Seafoods has, at least once, used an outside expert to try and improve line productivity.
30 Of the women interviewed, those aged forty and over have an average of grade 8 education, while those in their twenties have completed only high school.
31 Scallops are currently shipped fresh and unprocessed.
32 By the mid-1980s there had been a rise in fish prices on the market, and firms began exploitation of new products such as herring roe.
33 For most of these women, it is their first paid job. It is also an occasion to meet other women outside family and household contexts.
34 While older women, and those entering the job market mainly to get

out of the home, adapt very well to four or five months of paid work during the year, younger ones hope to work year-round and are afraid that the increasing number of plants will lower working time for each of them. Smaller families (those with two or three children), more structured day care (i.e. outside the family network), and acquisition of numerous household appliances make domestic chores easier. The majority of homes visited had most of the up-to-date electric household appliances, including, in many cases, a microwave oven.

35 In Clare, and in many areas of southwest Nova Scotia, the situation of fishers and the processing industry differs from that described by McCay (1987) for Fogo Island in Newfoundland.

36 From ten to fifteen women generally participate in these 'craft nights.'

37 This custom is becoming more and more difficult to maintain, since lots bordering a main road are rarer, and their prices have risen considerably.

38 This is often the case with women working on the herring-cutting machines.

39 The Acadians of Clare call themselves Catholic whether they practise the religion or not.

40 Bay Seafoods has done so for Université Ste-Anne at Church Point.

41 Asked whether she had ever thought of going to work in Yarmouth (about 35 miles from where she lives) one informant replied: 'Well no, I would never go that far away!'

Chapter 12

1 This section is based on an earlier project report (Apostle et al. 1985).

2 The issues that most centrally define these four dimensions are satisfaction with 'living conditions on board,' 'opportunity to be your own boss,' 'your earnings,' and 'time away from home.' Other questions related to the first, and most important, dimension concern cleanliness, trip length, crowding, and job safety.

3 Binkley (1988) points out the extent to which the captains of offshore vessels are having their autonomy eroded by the enterprise allocation system and the drive for quality control.

4 Personal satisfaction with the fishery does not, however, imply that captains would necessarily advise their sons to go into the fishery. In fact, a slight majority said that they would not do so. The main reasons are licensing problems, lack of a good future, and low return for effort.

5 These proportions are high even by Maritime standards. By comparison with three other Nova Scotia and Maritime samples, the captains' figures are relatively high.

6 These homes have a median value of $44,817.

7 These proportions are even greater than those for similar samples of fishers (including crew) from southwest Nova Scotia or from small communities throughout the Maritimes (Apostle, Kasdan, and Hanson 1985: 265).

8 The proportions for a more broad-based sample of marginal workers in the Maritimes are 62.7 per cent and 52.5 per cent, respectively. Where plant workers own their homes, their median values are about $10,000 less than those for captains.

9 Again, these patterns closely mirror those found among marginal workers throughout the Maritimes. There are no major gender differences on these questions.

10 Somewhat less than half (46.2 per cent) of the captain's wives were working for pay, and another 5.5 per cent were looking for paid work. The three major sources of employment were the fishery (35.6 per cent), private services (28.1 per cent), and public services (25.7 per cent); the most frequently reported occupations in these three sectors were plant workers, secretaries, and post-secondary school teachers, respectively. The median 1985 income for captain's wives in the paid work-force was $9,000, and the median contribution to total household income was 29 per cent.

11 The internal stratification identified here is also reflected in the wage and income differences discussed above (chapters 7 and 10). Plant workers' families appear fairly typical of rural towns and areas elsewhere. A substantial majority (66.1 per cent) span only two generations, and by far the larger proportion (82.4 per cent of such units, 54.5 per cent of the total) have both father and mother or husband and wife present. A further 16.8 per cent comprise only a wife and husband, and only 7.5 per cent include members of three generations. The mean number of children for all such two-generation units is a modest 1.94.

12 This invariance across class, ethnic, geographic, and economic (socialist/capitalist) boundaries suggests that gender considerations operate autonomously in this area.

13 Seccombe (1987: 11) reports that 'husbands of employed women do about twice as much work as their counterparts living with homemakers.'

14 These kinds of responses (never/some of the time/always) are used because it is generally agreed that frequency of performance measures
domestic work better than time expended on particular tasks. Joint reports on task performance would facilitate more unbiased measures
(Thiessen 1987: 4). While men are more likely to inflate self-estimates
of contributions, blue-collar respondents apparently supply reasonably
honest answers (Seccombe 1987: 6).

15 Note that the husband's and wife's reports of the contribution by the
man with paid employment are quite consistent, while husbands systematically underestimate the share of the woman with paid employment in tasks identified as women's work. Thus, underestimation is
clear for house cleaning, food preparation, and child care, but not for
house repair.

16 By contrast to Maine, Nova Scotia communities studied are much less
likely to have tourists or retirees from outside the community living
there. J.M. Acheson (1988: 29) notes that organizational involvement
in Maine is differentiated, as locals 'tend to be affiliated with fraternal,
military, and church organizations; the newcomers are far more involved with cultural organizations and service clubs.'

Chapter 13

1 Recent studies have illustrated the dynamics of the processing sector in
coastal communities. See, for example, McFarland 1980; Connelly and
MacDonald 1983; Porter 1983, 1985a, 1985b; Davis 1984; Ilcan
1985a; Sinclair 1985b; Clement 1986; A. Davis and Thiessen 1986;
Willett 1986; McCay 1987; and Thiessen and A. Davis 1988.

2 According to Canada, Employment and Immigration Canada (1987),
there were 175 union locals in Cape Breton, with combined membership of 23,280 in 1985.

3 Feminist proponents in the debate on domestic labour veered too far
in the opposite direction with their valorization of domestic labour in
the capitalist economy; see Vogel (1983), Armstrong, Armstrong, et al.
(1985), and Beechey (1987) on the theoretical history of women's labour issues.

4 A recent literature attempts to clarify the effects of global processes of
economic restructuring on labour markets and household economics.
See, for example, Mingione (1983), Pahl (1984, 1988), Smith, Wallerstein, and Evers (1984), and Redclift and Mingione (1985).

5 See Neis et al. (1988) on the impact of Taylorism in the Newfoundland setting. McFarland (1980), Connelly and M. MacDonald (1985),
MacDonald and Connelly (1986), and Lamson (1986) have also explored the effects of technological changes on fish plant workers in the
Atlantic region.

6 Market conditions and quota allocations have since been unfavourable
to inshore-midshore Cape Breton fishers. Also, a Halifax-based multinational corporation purchased the major processing plant in Northfield. Long layoffs occurred during the spring and summer of 1988.
Since the plant had been reconstructed with assistance from the Department of Regional and Industrial Expansion, the union sought the
intervention of provincial and federal fisheries ministers. The new
company claimed that industry conditions necessitated manipulation of
its substantial province-wide resources in both sectors of the fishery
(*Halifax Mail-Star* June–August 1988) and argued for enterprise allocation to be extended to the midshore range. Fish plant production is
now defined elsewhere, and corporate control of this industry in
Northfield is complete.

7 For the purposes of this research, the concept of 'community' has at
least four distinct connotations. Spatially, it refers to the straggling agglomeration of adjacent but separately established and named localities
initially developed around the various pitheads. Politically, the study
area embraces several types of jurisdictions: federal, provincial, municipal, and regional 'units' are variously grouped for administrative purposes, regardless of more localized aspirations.

Economically, this research covers the area of typical participation in
the labour markets and services developed at the centre of the mining
district. Mobility between job markets in localities within the industrial
area occurs only when jobs are listed through Canada employment
centres, a practice not followed by those employers who prefer patronage networks. It would appear that younger people are the most mobile section of the labour force. An aggregate of population has
therefore been constructed as the potentially largest grouping compris-
ing the community of interest.

Socially, perspectives are oriented first toward a much smaller spatial
scale, typically the neighbourhood, and then toward municipality and
region. Residents of adjacent neighbourhoods can hold strong opinions
about different social attributes characterizing other neighbourhoods.
As suggested above, historical developments underlie these strong at-

tachments/divisions. People usually present their place of birth by neighbourhood first. Neighbourhoods are usually oriented toward pit-heads long since closed.

8 In 1987, the VCR movie-rental outlet seemed a viable entrepreneurial strategy in downtown Northfield. Numerous small businesses advertised rentals for a mere $1–$2 fee. Several larger stores in the downtown core did a thriving business during their extended hours.

9 Theories of culture informed by political economy are relevant here. For example, see Sider (1980, 1986): 'The core of culture is the form and manner in which people perceive, define, articulate, and express their mutual relations. In class societies this form of social perception and this mode of behaviour mediate between the relatively egalitarian aspects of work and daily life and the collectively self-determined aspects of reproducing the domain over time, on the one hand, and on the other, the primarily unequal domain of the appropriate of the product and the reproduction of appropriation' (1986: 120). For more on the difficulties of conceptualizing culture in a historical mode, see Paine (1988) and Pool (1988).

10 Spencer (1988) has identified activity in the region's informal sector as it serves middle-class consumers. Households in the present sample were less likely to see and/or exchange goods and services beyond kin and neighbourhood networks.

11 What is presented here also contrasts with Pahl and Wallace (1985) on the Isle of Sheppey, Kent, where forms of privatization and dependent domesticity accompany deindustrialization and unemployment generates more unemployment. The employed are more likely to be active in informal exchanges and self-provisioning. Life in Sheppey is grim. The authors suggest two ways of dealing with the despair: domestic self-sufficiency or alcohol consumption. In Northfield, there seems a tendency toward the former, although the latter is not absent. Pahl and Wallace's method might help one elaborate on these differences.

12 Current state policy in the area is directed to small-business development, but, of course, resilient informal-sector practices operate with fewer risks of total indebtedness than formal-sector operations (Harder 1988).

13 Women do understand the economic advantage of many wage-labour jobs defined as men's work: having a miner's wage is a 'lucky' situation for the whole household. And women are not subject to the control of men in all aspects of their lives. The purpose here is rather to comment on the division of labour for domestic work.

14 Styles of plant management were altered to accommodate processes of capital intensification and accumulation caused by economic restructuring and state policy.
15 Pay scales are laid out in the company's 1987 policy handbook: trimmers, $6.32; general labourers, $6.58; runners, $7.08; cutters, $7.47; mechanics, $8; refrigeration engineers, $9; welders, $11.77.

Chapter 14

1 Long-recognized cyclical market fluctuations in the fishing industry are not adequately explained (Canada, Task Force, 1982: 38).
2 This close supervision is facilitated by family connections, especially in small plants, and often involves 'self-exploitation' (see chapter 1).
3 Small plants are less likely to join associations, but plant type has no effect on the perceived strength of such associations.
4 Such conflict is not the outcome of conscious government policy, and such incidents are not typical of government-processor relations, but they do reveal the limits of tolerance in the system.
5 Although the situation is not representative of the province, this processor admits that it does not even try to match the wages offered by the local NSP plant but argues that NSP pays artificially high wages made possible by unfair support from government.
6 Small-scale fishers employ up to three hands and usually fish on day trips in boats of up to 45 feet. Intermediate-scale units have from four to ten hands and involve boats of 45 to 75 feet, including purse seiners, longliners, small draggers, and scallopers, with a mixture of company and fisher ownership. Large-scale units average 150 feet, with crews of about eighteen. The trawlers and scallop draggers are all company-owned and go on trips lasting from ten to twenty days (Clement 1984: 18).
7 Case studies from Canada's West Coast and East Coast fisheries suggest the need to modify the idea that fishers are incapable of challenging or reversing unfavourable trends. See Guppy (1986) on BC commercial fishing and R. Matthews (1983) and Sinclair (1984, 1985a, 1985b) on fishers in northwest Newfoundland.
8 The differences between captains, or fishers generally, and other occupational groups reflect not just lower socioeconomic status. Although controls for several socioeconomic factors (particularly education) reduced intergroup differences, none came close to eliminating it. As with the predominantly electoral activities measured in national

studies, political involvement by captains and fishers between elections in Nova Scotia is confined to a minority.

9 Of the captains surveyed, 21.6 per cent reported that the unemployment program affects their ability to fish, and over three-quarters said that the effects were positive. A minority (37.7 per cent) said that they would like changes in UI regulations: in qualifying periods and in eligibility requirements. Lobster captains and fishers who use nets, trap and line gear were particularly positive about the effects of UI, and most opposed changes. One captain stated: 'Small operators must be able to survive in the winter months. We would not be able to fish otherwise.'

10 Set traps were being hauled in order to verify that fishers were adhering to trap limits (*Halifax Chronicle-Herald*, 13 May 1983).

11 There is subregional resentment in some responses. Fishers in Guysborough and Cape Breton indicated dissatisfaction with fishers in the more affluent southwest (where the boat burnings took place). One lobster fisher in Auld's Cove said that 'fishermen on the South Shore have a longer season and more traps. They should be satisfied.'

12 Captains of processor-owned vessels tend to learn of new or changed policies and regulations from DFO or fisheries officers.

13 Only one of the twenty-seven longliner captains belongs to any organization.

14 See Apostle, Kasdan, and Hanson (1985: 263, 266) on the relation between conservative Protestantism in the Shelburne area and usually high job satisfaction in the fishery. The analysis of religion from a political economy perspective requires attention. On the impact of fundamentalism on entrepreneurship see Long (1977).

15 Baptists, Pentecostals, and Lutherans are categorized here as conservative Protestants; Anglicans, United Church members, and Presbyterians, as liberal Protestants. These six groups account for 28.9 per cent, 2.2 per cent, 0.2 per cent, 19.3 per cent, 14.2 per cent, and 2.3 per cent, respectively, of the captains' sample. Roman Catholics were 26.5 per cent, non-believers 4.0 per cent, and others 2.5 per cent.

16 Fishing captains tend to identify income, the way one spends money, and occupation as central elements in a money-based conception of community position. In marked contrast to national samples, in which approximately half of Canadian respondents do not think of themselves as belonging to a social class, the great majority of captains (84.1 per cent) said that they do not think of themselves as belonging to any social class; 13.9 per cent replied that they did belong to a particular

class; and 2 per cent said that they did not know. One captain said that he 'does not put himself above or below anyone else. Everyone is equal.' Others stated that they do not believe in 'class distinctions' or a 'class system.' In 1974, 51.2 per cent of interviewees in the CNES said that they did not belong to a social class, 44.5 per cent said that they did, and 4.2 per cent replied that they did not know. In 1979, 53.7 per cent replied in the affirmative, 53.7 per cent in the negative, and 4.2 per cent did not know. When forced to choose a subjective class designation, the captains (61.0 per cent), more than most national samples (usually less than half), picked the working class. Direct producers are particularly inclined to do so.

17 Thompson (1983: 4, 6) makes the same general point for British fisheries: any 'straightforward Marxist interpretation' – based on the notion that 'the progressive commercialization of the fishing economy' will reduce fishers to 'class-conscious wage labourers' – is no longer 'plausible.' Rather, fishing 'as an occupation does not automatically push men toward a single, simple view of life. On the contrary, it pulls in some very contradictory directions.'

18 Even plants that are part of the largest companies come close to those associated with smaller companies in their tendencies to have weak unions, the freedom to lay people off, low wages, and high proportions of women workers.

19 The questions on political attitude are conventional political-efficacy items. They have been used in the United States since 1952 and have been repeated in Canada and western Europe.

20 Again, these differences are moderately diminished by control for socioeconomic factions but even then remain significant.

21 Non-involvement is consistent with sporadic work stoppages or occasional spontaneous protests. Qualitative data from the survey of fish plant workers indicate that most stoppages or walkouts involved women's concerns. In one large plant, someone 'wanted a leave of absence to care for her husband. We walked out to support her.' In another instance, a female worker emphasized that men 'did not participate in the one-hour strike' that the women staged.

Conclusion

1 While large-scale capital has divested or closed a number of plants in the post-1988 crisis, freezing technology has not as yet become sufficiently decentralized to allow large-scale capital to pursue frozen-fish

production on this basis. However, a crisis often stimulates technological change, and so these circumstances are likely to change. It seems reasonable to predict – in the absence of state intervention to prolong centralized mass-assembly technique – increasing decentralization of production facilities. The state is in fact already proposing studies on the efficiency of intermediate technology in the context of regulation of trawler catch and effort (Canada, Fisheries and Oceans, 1989).

2 Some workers in large plants have recently sought to restrict the export of unprocessed fish. Dewey Waybret represents thirty-two plant operators as president of the South West Nova Scotia Fish Packers Association. He commented: ' "The problem originated in Cape Breton, with the offshore companies and the union ... We want to be flexible and move our fish when the price is high." Waybret said there is no problem with employees of the small inshore plants getting sufficient work to collect unemployment insurance. "We keep our employees going," said Waybret' (Sou'Wester 1 May 1990, p. 5).

3 Brox (1972), Faris (1972), R. Matthews (1976), and E. Antler and Faris (1979) make a persuasive case that regional development policies in Newfoundland have overlooked the formal and informal sources of community vitality that account for continuing attachment to rural communities.

4 There has been organized community resistance to plant closures, particularly in Lockeport, Canso, and Digby. The Canadian Auto Workers, which represents the Lockeport and Canso workers and about 2,200 fish plant workers throughout the province, pressed the provincial and federal governments, as well as NSP, to keep the Lockeport and Canso plants open and to institute more comprehensive planning. DFO subsequently announced a $584-million, five-year package of assistance for Atlantic Canada's fishery (Halifax Chronicle-Herald 8 May 1990, p. 1). Many industry representatives responded, predictably, that the funding does little for inshore fishers, plant workers, or communities (Halifax Chronicle-Herald 8 May 1990, p. 2; Atlantic Fisherman June 1990, p. 3).

5 The fishing industry illustrates the more general problems facing the mass, collective trade-union movement in Britain and Italy. As Lane (1982) and F. Murray (1983) pointed out, unions facing flexible decentralization are handicapped by rigid structures, outmoded forms of communication, and inappropriate ideologies, and they seem unable to decentralize and to address the needs of workers in small rural settings.

6 Inevitably, there are relevant topics that a project like this cannot or does not cover. Continuing difficulties in the fishery, and the deepening of the recession in Western economies, have made aquaculture a more attractive alternative for those involved in the Nova Scotia fishery (Day 1989). The most obvious questions concern control by government and large capital over aquaculture and aquaculture's role in reshaping property rights in the fishery. Similarly, application of scientific knowledge to fish production will alter the Atlantic Canadian fishery. Scientific funding agencies are now supporting a major East Coast bio-technology project designed to use new scientific innovations, particularly in fisheries genetics, to enhance fish production. The fact that the three major corporate actors are sponsors of this project raises the standard question: knowledge for whom? Further, the general topic of fisheries management and enforcement deserves more attention (McMullan, Perrier, and Okihiro 1988; Bannister 1989).

While the existing system is far from the worst available, this study has documented enough structural biases and outright blunders to make further investigation of management issues and broader state policies imperative.

7 This is obviously not the complete solution, especially in terms of species not caught with hook-and-line technology and for plants and regions affected by this catch.

Appendix

Michael Ornstein, associate director, Survey Research Centre, Institute for Social Research, York University, provided helpful comments on the survey design. He bears no responsibility for the sampling decisions made.

1 The sample was constructed from lists provided by the local office of the Department of Fisheries and Oceans, as well as lists compiled internally from previous research projects conducted by the research program's principal investigators.

2 Richard Apostle conducted forty-five interviews, primarily in the Yarmouth, Digby, and Annapolis areas. Gene Barrett, who completed forty-two interviews, concentrated on the South Shore, the Shelburne area, and Cape Breton. Anthony Davis did eight interviews, most on Digby Neck, while Leonard Kasdan did three there. One interview was done in Shelburne County by Suzan Ilcan, an MA graduate student from Dalhousie University who was doing her thesis fieldwork in the

community where the plant was located. In addition, Anthony Davis did several interviews with key firms on the South Shore that were not part of the original sample in order to augment the qualitative understanding of the industry. These South Shore interviews are not included in the data analysis presented above.

3 The selection of only fifty plants was necessitated by the limited resources available. Of the fifty plants chosen, nine had no captains selling to them. Four of these nine plants involved pieceworkers, four were small plants, and one was of competitive scale. This means that the lists actually came from forty-one plants. In addition, seven plants refused to give lists of captains. Of these plants, three were small, two were competitive, and two were large. Six plants refused to give lists of workers. Three of these plants were small, one was competitive, and two were large.

4 Eleven interviews were added between 9 June and 3 November 1987 in one geographic area that would otherwise have gone unrepresented in the study.

5 A slightly higher completion rate for captains was prevented by two factors. First, fishers resumed major fishing activities in April and May. Second, two other university-based research projects were surveying the same population. A higher completion rate for plant workers was inhibited by the mobility of people doing plant work. Many of the incompletes were students who could not be located.

References

Acheson, J.M. 1975. 'The lobster fishery.' *Human Ecology* 3: 183–207
– 1981. 'Anthropology of fishing.' *Annual Review of Anthropology* 10: 275–316
– 1988. *The Lobster Gangs of Maine.* Hanover, NH: University Press of New England
Acheson, T.W. 1977. 'The Maritimes and Empire Canada.' In D.J. Bercuson (ed.), *Canada and the Burden of Unity*, pp. 87–114. Toronto: Macmillan
Adams, W., and J.W. Brock. 1986. *The Bigness Complex.* New York: Pantheon
Agra Europe. 1981. *EEC Fisheries: Problems and Prospects for a Common Policy.* Agra Europe Special Report No. 11. London: Agra Europe
Alavi, H. 1973. 'Peasant classes and primordial loyalties.' *Journal of Peasant Studies* 1 (1): 23–62
Alchian, A., and H. Demsetz. 1972. 'Production, information costs, and economic organization.' *American Economic Review* 62: 777–95
Aldrich, H., and J. Weiss. 1981. 'Differentiation within the U.S. capitalist class.' *American Sociological Review* 57: 59–72
Alexander, D. 1980. 'Newfoundland's traditional economy and development to 1934.' In J. Hiller and P. Neary (eds.), *Newfoundland in the Nineteenth and Twentieth Centuries: Essays in Interpretation*, pp. 17–39. Toronto: University of Toronto Press
– 1981. *The Decay of Trade.* St John's: Memorial University of Newfoundland
Allison, C. 1988. 'Women fishermen in the Pacific Northwest.' In J. Nadel-Klein and D.L. Davis (eds.), *To Work and to Weep: Women in Fishing Economies*, pp. 230–60. St John's: Institute of Social and Economic Research, Memorial University of Newfoundland

Amin, S. 1976. *Unequal Development: Social Formations at the Periphery of the Capitalist System*. Hassocks: Harvester

Andersen, R. 1974. 'North Atlantic fishing adaptations: origins and directions.' In G. Pontecorvo (ed.), *Fisheries Conflicts in the North Atlantic: Problems of Management and Jurisdiction*, pp. 15–33. Cambridge, Mass.: Ballinger

– 1979a. 'Public and private access management in Newfoundland fishing.' In R. Andersen (ed.), *North Atlantic Maritime Cultures*, pp. 229–336. The Hague: Mouton

– 1979b. *North Atlantic Maritime Cultures: Anthropological Essays on Changing Adaptations*. The Hague: Mouton

Andersen, R., and C. Wadel, eds. 1972. *North Atlantic Fishermen: Anthropological Essays on Modern Fishing*. St John's: Institute of Social and Economic Research, Memorial University of Newfoundland

Anderson, D. 1982. 'Small industry in developing countries: a discussion of issues.' *World Development* 10: 913–48

Antler, E. 1977a. 'Maritime mode of production, domestic mode of production, or labour process: an examination of the Newfoundland inshore fishery.' A paper presented before the Northeastern Anthropological Association Symposium on 'Fishermen and Mariners,' March

– 1977b. 'Women's work in Newfoundland fishing families.' *Atlantis* 2: 106–13

– 1981. 'Fisherman, fisherwoman, rural proletariat: capitalist commodity production in the Newfoundland fishery.' PhD dissertation, University of Connecticut

Antler, E., and J. Faris. 1979. 'Adaptations to changes in technology and government policy: a Newfoundland example (Cat Harbour).' In R. Andersen (ed.), *North Atlantic Maritime Cultures*, pp. 129–54. The Hague: Mouton

Antler, S. 1979. 'The capitalist underdevelopment of nineteenth century Newfoundland.' In R.J. Brym and R.J. Sacouman (eds.), *Underdevelopment and Social Movements in Atlantic Canada*, pp. 179–202. Toronto: New Hogtown Press

Apostle, R. 1990. 'Review of uncommon property.' *Canadian Review of Sociology and Anthropology* 27: 247–9

Apostle, R., and G. Barrett. 1989. 'Class, protest and organization in the Nova Scotia fishery.' In Reg Bryon (ed.), *Public Policy and the Periphery: Problems and Prospects in Marginal Regions*, pp. 236–48. Halifax: Queen's Printer

Apostle, R., G. Barrett, A. Davis, and L. Kasdan. 1984. 'Land and sea: the structure of fish processing in Nova Scotia.' Gorsebrook Research Institute, Working Paper No. 1-0284. Halifax: Saint Mary's University

– 1985. 'Land and sea: the structure of fish processing in Nova Scotia. A preliminary report.' Gorsebrook Research Institute, Project Report Series N. 1-85. Halifax: Saint Mary's University

Apostle, R., D. Clairmont, and M. MacDonald. 1978. *Morphology Survey I: Methodology Report.* Halifax: Institute of Public Affairs. Dalhousie University

Apostle, R., D. Clairmont, and L. Osberg. 1980. 'The General Segmentation Survey: methodology report.' Halifax: Institute of Public Affairs, Dalhousie University

– 1985. 'Segmentation and labour force strategies.' *Canadian Journal of Sociology* 10: 253–75

– 1986. 'Economic segmentation and politics.' *American Journal of Sociology* 91: 905–31

Apostle, R., L. Kasdan, and A. Hanson. 1984. 'Political efficacy and political activity among fishermen in southwest Nova Scotia.' *Journal of Canadian Studies* 19: 157–65

– 1985. 'Work satisfaction and community attachment among fishermen in southwest Nova Scotia.' *Canadian Journal of Fisheries and Aquatic Sciences* 42: 256–67

Armstrong, P., H. Armstrong, et al. 1985. *Feminist Marxism or Marxist Feminism: A Debate.* Toronto: Garamond

Atlantic Provinces Economic Council (APEC). 1968. *Atlantic Provinces Fishery.* Pamphlet No. 12. Halifax: APEC

– 1977. *Atlantic Canada Today.* Halifax: APEC

– 1987. *Atlantic Canada Today.* Halifax: Formac

Averitt, R.T. 1968. *The Dual Economy: The Dynamics of American Industry Structure.* New York: Norton

B.C. Packers Ltd. 1977. *Annual Report*

Bain, T. 1956. *Barriers to New Competition.* Cambridge: Harvard University Press

Baker, O. 1979. *Who Benefits from Fishing Vessel Construction Subsidies?* Halifax: Fisheries and Oceans Canada

Balcom, B.A. 1984. *The Cod Fishery of Isle Royale, 1713–58.* Studies in Archaeology, Architecture and History, Parks Canada. Ottawa: Supply and Services

Bannister, R.K. 1989. ' "Orthodoxy and the theory of fishery management': the policy and practice of fishery management theory past and present.' MA thesis, Saint Mary's University, Halifax

Bannock, G. 1981. *The Economics of Small Firms: Return from the Wilderness.* Oxford: Blackwell

Baran, P.A., and P.M. Sweezy. 1966. *Monopoly Capital.* New York: Monthly Review Press

Barrett, L.G. 1976. 'Development and underdevelopment and the rise of trade unionism in the fishing industry of Nova Scotia, 1900–1950.' MA thesis, Dalhousie University

– 1979. 'Underdevelopment and social movements in the Nova Scotia fishing industry to 1938.' In R.J. Brym and R.J. Sacouman (eds.) *Underdevelopment and Social Movements in Atlantic Canada*, pp. 127–60. Toronto: New Hogtown Press

– 1980. 'Perspectives on dependency and underdevelopment in the Atlantic Region.' *Canadian Review of Sociology and Anthropology* 17: 273–86

– 1983. 'Uneven development, rent, and the social organization of capital. A study of the fishing industry of Nova Scotia, Canada.' DPhil dissertation, University of Sussex, Brighton

– 1984. 'Capital and the state in Atlantic Canada: the structural context of fishery policy between 1939 and 1977.' In C. Lamson and A. Hanson (eds.), *Atlantic Fisheries and Coastal Communities: Fisheries Decision-Making Case Studies*, pp. 77–104. Halifax: Dalhousie Ocean Studies Programme, Dalhousie University

Barrett, L.G., and R. Apostle. 1987. 'Labour surplus and local labour markets in the Nova Scotia fish processing industry.' *Canadian Review of Sociology and Anthropology* 24: 178–212

Barrett, L.G., and A. Davis. 1984. ' "Floundering in troubled waters": the political economy of the Atlantic fishery and the Task Force on Atlantic Fisheries.' *Journal of Canadian Studies* 19: 125–37

Batstone, E.V. 1975. 'Deference and the ethos of small town capitalism.' In M. Bulmer (ed.), *Working Class Images of Society*, pp. 116–30. London: Routledge and Kegan Paul

Bechhofer, F., and B. Elliott 1985. 'The petite bourgeoisie in late capitalism.' *Annual Review of Sociology* 11: 181–230

Beechey, V. 1987. *Unequal Work.* London: Verso

Bell, W.P. 1930. 'Why fuss about the trawler?' *Maclean's Magazine* 15 April, 8, 72, 75–6

– 1961. *The 'Foreign Protestants' and the Settlement of Nova Scotia.* Toronto: University of Toronto Press

Benedict, B. 1968. 'Family firms and economic development.' *Southwestern Journal of Anthropology* 24 (1): 1–19

Berger, S., and M. Piore. 1980. *Dualism and Discontinuity in Industrial Societies.* New York: Cambridge University Press

Berkes, F. (ed.), 1989. *Common Property Resources: Ecology and Community-Based Sustainable Development.* London: Belhaven

Beverton, R.J.H. 1953. 'Some observations on the principles of fishery regulation.' *Journal du Conseil permanent international pour l'exploration de la mer* 19: 56–68

Bielby, W., and T. Baron. 1983. 'Organizations, technology, and worker attachment to the firm.' *Research in Social Stratification and Mobility* 2: 77–113

Bienefeld, M., and M. Godfrey. 1978. 'Surplus labour and underdevelopment.' Discussion Paper No. 138. Brighton: Institute of Development Studies, University of Sussex

Biggar, H.P. 1963. *The Early Trading Companies of New France.* New York: Argonaut Press

Binkley, M. and V. Thiessen. 1988. ' "Ten days a grass widow, – forty-eight hours a wife": sexual division of labour in trawlermen's households,' *Culture* 8: 39–50

Blair, J. 1972. *Economic Concentration.* New York: Harcourt, Brace, Jovanovich

Blauner, R. 1969. 'Internal colonialism and ghetto revolt.' *Social Problems* 16 (4): 393–408

Bonacich, E. 1972. 'A theory of ethnic antagonism: the split labour market.' *American Sociological Review* 37: 547–59

Booth, D. 1975. 'Andre Gunder Frank: an introduction and appreciation.' In I. Oxaal, T. Barnett, and D. Booth (eds.), *Beyond the Sociology of Development*, pp. 50–85. London: Routledge and Kegan Paul

– 1985. 'Marxism and development sociology: interpreting the impasse.' *World Development* 13 (7): 761–87

Bowles, R. 1982. *Little Communities and Big Industries: Studies in the Social Impact of Canadian Resource Extraction.* Toronto: Butterworths

Braverman, H. 1974. *Labour and Monopoly Capital: The Degradation of Work in the Twentieth Century.* New York: Monthly Review Press

Breton, Y.D. 1977. 'The influence of modernisation on the modes of production in coastal fishing: an example from Venezuela.' In M.E. Smith (ed.), *Those Who Live from the Sea*, pp. 125–37. St Paul, Minn.: West Publishing

Brewer, A. 1980. *Marxist Theories of Imperialism: A Critical Survey.* London: Routledge and Kegan Paul

Bridges, W. 1982. 'The sexual segregation of occupations: theories of labor stratification in industry.' *American Journal of Sociology* 88: 270–95

Bromley, R. 1985a. 'Small may be beautiful, but it takes more than beauty to ensure success.' In R. Bromley (ed.), *Planning for Small Enterprises in Third World Cities*, pp. 321–41. New York: Pergamon
– ed. 1985b. *Planning for Small Enterprises in Third World Cities*. New York: Pergamon
Brookes, A.A. 1976. 'Out-migration from the Maritime provinces, 1860–1900: some preliminary considerations.' *Acadiensis* 5: 26–56
Brown, D.M. 1978. 'From Yankee to Nova Scotian: Simeon Perkins of Liverpool, Nova Scotia 1762–1796.' MA thesis, Queen's University
Browne, J., and F. Wien. 1978. 'Blacks in the economic structure of southwest Nova Scotia.' In V. D'Oyley (ed.), *Black Presence in Multiethnic Canada*, pp. 99–130. Toronto: Ontario Institute for Studies in Education
Brox, O. 1972. *Newfoundland Fishermen in the Age of Industry: A Sociology of Economic Dualism*. St John's: Institute of Social and Economic Research, Memorial University of Newfoundland
Brusco, S. 1982. 'The Emilian model: productive decentralisation and social integration.' *Cambridge Journal of Economics* 6: 167–84
Brusco, S., and C. Sabel. 1981. 'Artisan production and economic growth.' In F. Wilkinson (ed.), *The Dynamics of Labour Market Segmentation*, pp. 99–113. London: Academic Press
Brym, R.J., and B. Neis. 1979. 'Regional factors in the formation of the Fishermen's Protective Union of Newfoundland.' In R.J. Brym and R.J. Sacouman (eds.), *Underdevelopment and Social Movements in Atlantic Canada*, pp. 203–18. Toronto: New Hogtown Press
Burawoy, M. 1985. *The Politics of Production: Factory Regimes under Capitalism*. London: Verso
Calhoun, S. 1983. *The Lockeport Lockout: An Untold Story in Nova Scotia's Labour History*. Kentville: Kentville Publishing
Cameron, S.D. 1977. *The Education of Everett Richardson: The Nova Scotia Fishermen's Strike, 1970–71*. Toronto: McClelland and Stewart
Canada. 1927a. Royal Commission Investigating the Fisheries of the Maritime Provinces and the Magdalen Islands (McLean Commission), *Report*
– 1927b. Royal Commission Investing the Fisheries of the Maritime Provinces and the Magdalen Islands, Transcript of Evidence, Vol. 8 (PANS MH 6 F Series No. 4)
– 1937. Royal Commission on Price Spreads. Minutes of Proceedings and Evidence, No. 1, 30 October 1934 (typescript)
– 1959. Royal Commission on Price Spreads in the Food Industry. *Report*, Vol. II

Canada. Atlantic Development Board. 1969. *Fisheries in the Atlantic Provinces*. Ottawa: Queen's Printer

Canada. Department of Fisheries. 1937–8. *Annual Report*

Canada. Dominion Bureau of Statistics. 1934. *Fisheries Statistics of Canada, 1933*. Ottawa: King's Printer

– 1946. *Fishery Statistics of Canada, 1944*. Ottawa: King's Printer

Canada. Economic Council of Canada. 1977. *Living Together*. Ottawa: Supply and Services

– 1980. *Newfoundland from Dependency to Self-Reliance*. Ottawa: Supply and Services

Canada. Employment and Immigration Canada. 1987. *Labour Market Analysis: CEC Glace Bay*. Sydney: Employment and Immigration Canada

Canada. Environment Canada. Fisheries Intelligence Branch. 1970. 'List of fish processors, Nova Scotia.' Unpublished data

– 1973. 'List of fish processors, Nova Scotia.' Unpublished data

– 1976a. 'Enumeration of fishermen and fishing craft, 1976.' Fisheries and Marine Service. Unpublished data

– 1976b. 'List of fish processors, Nova Scotia.' Fisheries Statistics and Computer Services Division. Fisheries and Marine Service. Unpublished data

– 1976c. *Policy for Canada's Commercial Fisheries*. Ottawa: Supply and Services

– 1977. *Annual Statistical Review of Canadian Fisheries 1955–1976*. IX, Intelligence Service Division, Marketing Service Branch, Fisheries and Marine Services. Ottawa: Supply and Services

Canada. Fisheries and Oceans. 1978. 'Vessel ownership by size and gear utilization, 1976–78.' Unpublished data.

– 1983. 'List of plants by county.' Unpublished data

– 1984. 'Selected fishery statistics. District 37.' Unpublished data

– 1988. *Canadian Fisheries Statistical Highlights*. Ottawa: Supply and Services

– 1989. *Report of the Scotia-Fundy Groundfish Task Force* (Haché Report). Ottawa: Supply and Services

Canada. Parliament. 1876. Sessional Papers. Report of the Commissioner of Fisheries

– 1910a. Sessional Papers. No. 22a, Report of the Special Commissioner and Inspector of Fisheries on Lobster Industry

– 1910b. Sessional Papers. No. 22, Special Appended Report II. R.N. Venning, 'The Marine Fisheries Committee and the Lobster industry.'

– 1910c. Sessional Papers. No. 22a, Lobster Fishery, Evidence, 1909

– 1939. House of Commons. *Debates*, IV

Canada. Public Accounts. 1987. Vol. 2, Part 2, Section 7. *Transfer Payments:*

Regional and Industrial Contributions under the Industrial Regional Development Act

Canada. Statistics Canada. 1983a. 1981 Census of Canada. SP12FEDO7-211. Ottawa: Supply and Services

– 1983b. 1981 Census of Canada. *Population, Occupied Private Dwellings, Private Households and Census and Economic Families in Private Households. Selected Social and Economic Characteristics. Nova Scotia.* Ottawa: Supply and Services

– 1985. *Exports.* Bulletin 65–202, May

– 1986. *Exports by Commodity.* Bulletin 65-003

Canada. Task Force on the Atlantic Fisheries. 1982. *Navigating Troubled Waters: A New Policy for the Atlantic Fisheries* (Kirby Report). Ottawa: Supply and Services

Canadian Saltfish Corporation. 1984–5. *Annual Report*

Cape Breton Metropolitan Planning Commission. 1982. *Town of Glace Bay Municipal Planning Strategy Background Report.* Sydney: Commission

Cardoso, F.H. 1971. 'Commentario sobre los conceitos de superpobulacao relativa e marginalidade.' *Revista Latinoamericana de Ciencias Sociales* 1–2: 57–76

Careless, J.M.S. 1969. 'Aspects of metropolitanism in Atlantic Canada.' In M. Wade (ed.), *Regionalism in the Canadian Community,* pp. 117–29. Toronto: University of Toronto Press

Carter, R. 1983. *'Something's Fishy': Public Policy and Private Corporations in the Newfoundland Fishery.* St John's: Oxfam

Casanova, P.G. 1965. 'Internal colonialism and national development.' In Studies in Comparative International Development 1/4: 27–37

Chang, M. 1975. 'Newfoundland in transition: the Newfoundland trade and Robert Newman and Company, 1780–1805.' MA thesis, Memorial University of Newfoundland

Charron, T.P. 1977. 'Costs and earnings study of selected fishing enterprises.' Environment Canada, Fisheries Services Directorate. Ottawa: Queen's Printer

Clairmont, D.H., M. MacDonald, and F.C. Wien. 1980. 'A segmentation approach to poverty and low-wage work in the Maritimes.' In J. Harp and J. Hofley (eds.), *Structural Inequality in Canada,* pp. 285–315. Toronto: Prentice Hall

Clark, A.H. 1968. *Acadia: The Geography of Early Nova Scotia to 1760.* Madison: University of Wisconsin Press

Clark, C.W. 1973. 'The economics of over-exploitation.' *Science* 181: 630–4

– 1981. 'Bioeconomics of the ocean.' *Bioscience* 31: 231–7

Clement, W. 1983. *Class, Power and Property*, Toronto: Methuen
- 1984. 'Canada's coastal fisheries: formation of unions, cooperatives, and associations.' *Journal of Canadian Studies* 19: 5–33
- 1986. *The Struggle to Organize: Resistance in Canada's Fishery*. Toronto: McClelland and Stewart
Clow, M.J. 1984. 'Politics and uneven capitalist development: the Maritime challenge to the study of Canadian political economy.' *Studies in Political Economy* 14: 117–40
Cockburn, C. 1983. *Brothers: Male Dominance and Technological Change*. London: Pluto
- 1986. 'The relations of technology, what implication for theories of sex and class?' In R. Crompton and M. Mann (eds.), *Gender and Stratification*, pp. 74–85. Cambridge: Polity
Connelly, P., and M. MacDonald. 1983. 'Women's work: domestic and wage labour in a Nova Scotia community.' *Studies in Political Economy* 10: 45–72
- 1985. 'A leaner meaner industry: a case study of "restructuring" in the Nova Scotia fishery.' Paper presented at the Workshop on the Political Implications of Industrial Restructuring: Primary Producers in Atlantic Canada, Queen's University
Copans, J., and B. Bernier. 1986. 'Présentation.' *Anthropologie et sociétés* 10: 1–11
Copes, P. 1972. *The Resettlement of Fishing Communities in Newfoundland*. Ottawa: Canadian Council on Rural Development
Corbridge, S. 1990. 'Post-Marxism and development studies: beyond the impasse.' *World Development* 18 (5): 623–39
Cowling, K. 1982. *Monopoly Capitalism*. New York: John Wiley
Crutchfield, J.A. 1975. 'An economic view of maximum sustainable yield.' In P.M. Roedel (ed.), *Optimum Sustainable Yield as a Concept in Fisheries Management*. American Fisheries Society Special Publication No. 9. Washington
Curran, J. 1981. 'Class imagery, work environment and community: some further findings and a brief comment.' *British Journal of Sociology* 32 (1): 111–26
Curran, J., and R.J. Burrows. 1986. 'The sociology of petit capitalism: a trend report.' *Sociology* 20 (2): 265–79
Curran, J., and J. Stanworth. 1979a. 'Worker involvement and social relations in the small firm.' *Sociological Review* 27 (2): 317–42
- 1979b. 'Self-selection and the small firm worker – a critique and an alternative view.' *Sociology* 13 (3): 427–44

Dahrendorf, R. 1975. *The New Liberty.* London: Routledge and Kegan Paul

Davis, A. 1984. 'Property rights and access management in the small boat fishery: a case study from southeast Nova Scotia.' In C. Lamson and A. Hanson (eds.), *Atlantic Fisheries and Coastal Communities: Fisheries Decision-Making Case Studies*, pp. 133–64. Halifax: Dalhousie Ocean Studies Programme, Dalhousie University

– 1985. 'You're your own boss: an economic anthropology of small boat fishing in Port Lameron Harbour, southwest Nova Scotia.' PhD thesis, University of Toronto

Davis, A., A.J. Hanson, L. Kasdan, and R. Apostle. 1983. 'Utilization of offshore banks by the small boat fisheries in southwest Nova Scotia.' Research Report, Institute for Resource and Environmental Studies, Dalhousie University, Halifax

Davis, A., and L. Kasdan. 1984. 'Bankrupt government policies and belligerent fishermen responses: dependency and conflict in the southwest Nova Scotia small boat fisheries.' *Journal of Canadian Studies* 19: 108–24

– (forthcoming). *Dire Straits: The Dilemmas of a Fishery, The Case of Digby Neck and the Islands*, Social and Economic Studies Publication No. 38, Institute of Social and Economic Research, Memorial University of Newfoundland, St John's

Davis, A., and V. Thiessen. 1986. 'Making sense of the dollars-income distribution among Atlantic Canadian fishermen and public policy.' *Marine Policy* 10: 201–14

– 1988. 'Public policy and social control in the Atlantic fisheries.' *Canadian Public Policy – Analyse de politiques* 14: 66–77

Davis, D.L. 1986. 'Saint, shrew and strumpet: a crosscultural analysis of the image of fisherwives.' Vermillion, SD: Anthropology Program, University of South Dakota

Day, A. 1989. *Aquaculture in the Maritimes.* Halifax: Nimbus

deRoche, C.P. 1985. *The Village, the Vortex: Adaptation to Regionalism and Development in a Complex Society.* Occasional Papers in Anthropology, No. 12. Halifax: Saint Mary's University

– 1987. 'Workworlds and worldviews: an interpretation of socioeconomic strategies among Cape Breton Acadians.' In C.P. deRoche and J.E. deRoche (eds.), *'Rock in a Stream': Living with the Political Economy of Underdevelopment in Cape Breton.* ISER Research and Policy Paper No. 7. St John's: Memorial University of Newfoundland

Deveau, J.A., ed. 1983. *Clare ou la Ville Française.* Tomes I et II. *Les historiens d'âge d'or de la Baie Ste-Marie.* Pointe-de-l'Eglise, NS: Université Ste-Anne

Dewey, D.J. 1974. 'The new learning: one man's view.' In H.J. Goldschmid, J.M. Mann, and J.F. Weston (eds.), *Industrial Concentration: The New Learning*, pp. 1–14. Boston: Little, Brown

DeWolfe, A.G. 1974. *The Lobster Fishery of the Maritime Provinces: Economic Effects of Regulation*. Ottawa: Department of Environment

Dirlam, J.B., and A.E. Kahn. 1954. *Fair Competition: The Law and Economics of Antitrust Policy*. Ithaca, NY: Cornell University Press

Dore, R. 1986. *Flexible Rigidities*. London: Athlone

Doucet, J.R. 1965. 'L'industrie forestière chez les Acadiens de la Baie Ste-Marie de 1860 à 1930.' MA thesis, Collège Sainte-Anne, Church Point, NS

Drache, D. 1976. 'Rediscovering Canadian political economy.' *Journal of Canadian Studies* 11: 3–18

Dun, R.G., and Co. 1922. *The Mercantile Agency Reference Book*. Toronto: Hunter-Rose

Dun and Bradstreet. 1976. 'List of companies engaged in the fishing industry.' DMI Management Control Report; unpublished data

Durrenberger, E.P., and G. Palsson. 1987. 'Ownership at sea: fishing territories and access to sea resources.' *American Ethnologist* 14 (3): 508–22

Edwards, R. 1975. 'States in corporate stability and the risks of corporate failure.' *Journal of Economic History* 35: 428–57

Elam, M.J. 1990. 'Puzzling out the post-Fordist debate: technology, markets and institutions.' *Economic and Industrial Democracy* 11 (1): 9–37

Ellis, C. 1986. *Fisher Folk: Two Communities on Chesapeake Bay*. Lexington: University Press of Kentucky

Enterprise Cape Breton. 1986. *Community Profile: Cape Breton*. Sydney: Enterprise Cape Breton

Epple, G.M. 1977. 'Technological change in a Grenada; W.I. fishery, 1950–1970.' In M.E. Smith (ed.), *Those Who Live from the Sea: A Study in Maritime Anthropology*, pp. 173–93. St Paul, Minn.: West Publishing

Faris, J.C. 1972. *Cat Harbour: A Newfoundland Fishing Settlement*. St John's: Institute of Social and Economic Research, Memorial University of Newfoundland

Fisheries Council of Canada. 1967. *Annual Review*. Montreal: Fisheries Council of Canada

Foote, R. 1979. *The Case of Port Hawkesbury: Rapid Industrialization and Social Unrest in a Nova Scotia Community*. Toronto: Peter Martin

Foster-Carter, A. 1978. 'The modes of production controversy.' *New Left Review* 107: 47–77

Fothergill, S., and G. Gudgin. 1982. *Unequal Growth: Regional Employment Change in the U.K.* London: Heinemann

Fox, B. 1988. 'Conceptualizing "Patriarchy."' *Canadian Review of Sociology and Anthropology* 25: 163–82

Frank, A.G. 1969. 'Sociology of development and the underdevelopment of sociology.' In A.G. Frank (ed.), *Latin America: Underdevelopment or Revolution*, pp. 21–94. New York: Monthly Review Press

Frank, D. 1974. 'Coal masters and coal miners: the 1922 strike and the roots of class conflict in the Cape Breton coal industry.' MA thesis, Dalhousie University, Halifax

– 1976. 'Class conflict in the coal industry: Cape Breton 1922.' In G.S. Kealey and P. Warrian (eds.), *Essays in Canadian Working Class History*, pp. 161–84. Toronto: McClelland and Stewart

– 1979. 'The Cape Breton coal miners 1917–1926.' PhD dissertation, Dalhousie University, Halifax

– 1988. 'The trial: The King vs. J.B. McLachlan.' *New Maritimes* 6: 3–7

Galbraith, J.K. 1956. *American Capitalism*. Rev. ed. Boston: Houghton Mifflin

– 1967. *The New Industrial State*. Boston: Houghton Mifflin

Gerry, C. 1979. 'The crisis of the self-employed: petty production and capitalist production in Dakar.' In C. Cruise O'Brien (ed.), *The Political Economy of Underdevelopment: Dependence in Senegal*, pp. 126–55. London: Gage

Gershuny, J. 1978. *After Industrial Society: The Emerging Self-servicing Economy*. London: Macmillan

Godfrey, M. 1983. 'Surplus labour as a source of foreign exchange?' *World Development* 11: 945–56

Goffee, R., and R. Scase. 1982. ' "Fraternalism" and "paternalism" as employer strategies in small firms.' In G. Day et al. (eds.), *Diversity and Decomposition in the Labour Market*, pp. 107–24. Aldershot: Gower

– 1985. 'Proprietorial control in family firms: some functions of "quasi-organic" management systems.' *Journal of Management Studies* 22 (1): 53–68

Goldschmid, H.J., J.M. Mann, and J.F. Weston, eds. 1974. *Industrial Concentration: The New Learning*. Boston: Little, Brown

Goldthorpe, J.H. 1984. 'The end of convergence: corporatist and dualist tendencies in modern Western societies.' In J.H. Goldthorpe (ed.), *Order and Conflict in Western European Capitalism*, pp. 315–43. Oxford: Oxford University Press

Gordon, D.M., R. Edwards, and M. Reich. 1982. *Segmented Work, Divided Workers*. New York: Cambridge University Press

Gordon, H.S. 1951. 'The trawler question in the United Kingdom and Canada.' *Dalhousie Review* 31: 177–27

– 1954. 'The economic theory of a common property resource: the fishery.' *Journal of Political Economy* 62: 124–42

Granovetter, M. 1984. 'Small is bountiful: labour markets and establishment size.' *American Sociological Review* 49: 323–34

Guppy, Neil. 1986. 'Property rights and changing class formations in the B.C. commercial fishing industry.' *Studies in Political Economy* 19: 59–81

Haberler, G. 1976. *The Challenge to the Free Market Economy.* Washington: American Enterprise Institute

Haliburton, T.C. 1973. *History of Nova Scotia*, Vols. I and II. Belleville: Mika

Handcock, G. 1977. 'English migration to Newfoundland.' In J. Mannion (ed.), *The Peopling of Newfoundland*, pp. 15–48. St John's: Memorial University of Newfoundland

Hanson, A.J., and C. Lamson. 1984. 'Fisheries decision-making in Atlantic Canada.' In C. Lamson and A.J. Hanson (eds.), *Atlantic Fisheries and Coastal Communities: Fisheries Decision-Making Case Studies*, pp. 1–14. Halifax: Dalhousie Ocean Studies Programme, Dalhousie University

Harder, S. 1988. 'Now that the coal is gone: the state sells entrepreneurialism in Glace Bay.' Paper presented to the Atlantic Association of Sociologists and Anthropologists, Saint Mary's University, Halifax

Hardin, G. 1968. 'The tragedy of the commons.' *Science* 162: 1243–8

Harris, D.J. 1972. 'The black ghetto as a colony: a theoretical critique and alternative formulation.' *Review of Black Political Economy* 2: 3–33

Harrop, G. 1987. *Clarie: Clarence Gillis, M.P. 1940–1957. A Political Memoir: From the Coal Mines of Cape Breton to the Floor of the House of Commons.* Hantsport: Lancelot

Higgins, B. 1956. 'The "dualistic theory" of underdeveloped areas.' *Economic Development and Cultural Change* 4: 99–115

Hill, R. 1983. *The Meaning of Work and the Reality of Unemployment in the Newfoundland Context.* St John's: Community Services Council of Newfoundland and Labrador

Himelfarb, A. 1976. *The Social Characteristics of One-Industry Towns in Canada: A Background Report.* Toronto: Royal Commission on Corporate Concentration

Hodson, R. 1978. 'Labor in the monopoly, competitive and state sectors of production.' *Politics and Society* 8: 429–80

– 1984. 'Companies, industries, and the measurement of economic segmentation.' *American Sociological Review* 49 (3): 335–48

Holland, S. 1976. *Capital versus the Regions.* London: Macmillan

Horwitz, N. 1968. *Seasonal Unemployment in the Province of Newfoundland.* St

John's: Institute of Social and Economic Research, Memorial University of Newfoundland

Hudson, R., and J. Lewis. 1984. 'Capital accumulation: the industrialisation of southern Europe?' In A.M. Williams (ed.), *Southern Europe Transformed*, pp. 179–207. London: Harper and Row

Hughes, G. 1981. *Two Islands: Miscou and Lameque and Their State of Bondage, 1849–1861*. Fredericton: New Brunswick Museum

Ilcan, S. 1985a. 'The position of women in the Nova Scotia secondary fishing industry: a community-based study.' Working Paper No. 8-85. Halifax: Gorsebrook Research Institute, Saint Mary's University

– 1985b. 'The social organization of the fish processing industry of Nova Scotia: a community-based study.' MA thesis, Dalhousie University, Halifax

– 1986. 'Women and casual work in the Nova Scotian fish processing industry.' *Atlantis* 11: 23–34

Inglis, G. 1985. *More than Just a Union: The Story of the NFFAWU*. St John's Jesperson

Innis, H.A. 1954. *The Cod Fisheries: The History of an International Economy*. Toronto: University of Toronto Press

Kalleberg, A.L., M. Wallace, and R.P. Althauser. 1981. 'Economic segmentation, worker power, and income inequality.' *American Journal of Sociology* 87: 651–83

Kay, G. 1975. *Development and Underdevelopment*. London: Macmillan

Kearney, J. 1984a. *Working Together: A Study of Fishermen's Response to Government Management of the District 4A Lobster Fishery*. Université Sainte-Anne, Pointe-de-l'Eglise, NS, mimeo

– 1984. 'The transformation of the Bay of Fundy herring fisheries, 1976–1978: an experiment in fishermen-government co-management.' In C. Lamson and A.J. Hanson (eds.), *Atlantic Fisheries and Coastal Communities: Fisheries Decision-Making Case Studies*, pp. 165–204. Halifax: Dalhousie Ocean Studies Programme, Dalhousie University

– 1986. 'State intervention and the labour process in the southwest Nova Scotia lobster fishery.' Paper presented before International Working Seminar on Social Research and Public Policy in Fisheries: Norwegian and Canadian Experiences, Institute of Fisheries, University of Tromsø, June

Kimber, S. 1986a. 'Claws.' *Canadian Business*, September: 68, 71, 72

– 1986b. 'Rescue at sea.' *Canadian Business*, October: 68–73, 148

Kirby, M. 1990. 'Good solutions difficult to find for problems of fishing industry.' *Financial Post*, 12 March: 14

Kirby, P. 1969. *Industrialisation in an Open Economy: Nigeria 1945–1966.* London: Cambridge University Press

Kitching, G. 1982. *Development and Underdevelopment in Historical Perspective.* London: Methuen

Kottak, C. 1979. *Cultural Anthropology.* 2nd ed. New York: Random House

LaClau, E. 1971. 'Feudalism and capitalism in Latin America.' *New Left Review* 67: 19–38

Lamson, C. 1986. 'On the line: women and fish plant jobs in Atlantic Canada.' *Relations industrielles / Industrial Relations* 41: 145–55

Lamson, C., and A.J. Hanson, eds. 1984. *Atlantic Fisheries and Coastal Communities: Fisheries Decision-Making Case Studies.* Halifax: Dalhousie Ocean Studies Programme, Dalhousie University

Lamson, C., and J.G. Reade. 1987. *Atlantic Fisheries and Social Science: A Guide to Sources.* Canadian Technical Report of Fisheries and Aquatic Sciences, No. 1549.

Lane, T. 1982. 'The unions: caught on the ebbtide.' *Marxism Today*, September: 6–13

Lawson, T. 1981. 'Paternalism and labour market segmentation theory.' In F. Wilkinson (ed.), *The Dynamics of Labour Market Segmentation*, pp. 47–66. New York: Academic

Lee, W.S. 1953. 'The modern fishing trawler.' *Commercial News* 33(4), April

Levellton, C.R. 1973. *Licensing for Atlantic Coast Fisheries: A Policy Discussion Paper.* Environment Canada, Fisheries and Marine Services

Levitt, K. 1969. *Population Movements in the Atlantic Provinces.* Halifax: Institute of Public Affairs

Lewis, J., and A.M. Williams. 1986. 'Factories, farms and families: the impacts of industrial growth in rural Central Portugal.' *Sociologia Ruralis* 26: 320–44

Lewis, W.A. 1963. 'Economic development with unlimited supplies of labour.' In A.N. Agarwala and S.P. Singh (eds.), *The Economics of Underdevelopment*, pp. 400–49. New Delhi: Oxford University Press

Light, I. 1979. 'Disadvantaged minorities in self-employment.' *International Journal of Contemporary Sociology* 20: 31–45

Long, N. 1975. 'Structural dependency, modes of production and economic brokerage in rural Peru.' In I. Oxaal, T. Barnett, and D. Booth (eds.), *Beyond the Sociology of Development*, pp. 253–82. London: Routledge and Kegan Paul

– 1977. *An Introduction to the Sociology of Rural Development.* London: Routledge and Kegan Paul

Lovering, J. 1990. 'Fordism's unknown successor: a comment on Scott's theory of flexible accumulation and the re-emergence of regional economies.' *International Journal of Urban and Regional Research* 14 (1): 159–74

Lozano, B. 1983. 'Informal sector workers: walking out the system's front door.' *International Journal of Urban and Regional Research* 7: 340–64

Lucas, R.A. 1971. *Minetown, Milltown, Railtown: Life in Canadian Communities of Single Industry.* Toronto: University of Toronto Press

McCay, B.J. 1976. 'Appropriate technology and coastal fishermen of Newfoundland.' PhD thesis, Columbia University, New York

– 1981. 'Development issues in fisheries as agrarian systems.' *Culture and Agriculture* 11: 2–7

– 1987. 'Historical observations on old and new world fisheries.' In B.J. McCay and J.M. Acheson (eds.), *The Question of the Commons: The Culture and Ecology of Communal Resources,* pp. 195–216. Tucson: University of Arizona Press

– 1988. 'Fish guts, hair nets, and unemployment stamps: women and work in co-operative fish plants.' In P.R. Sinclair (ed.), *A Question of Survival: The Fisheries and Newfoundland Society,* pp. 105–31. St John's: Institute of Social and Economic Research, Memorial University of Newfoundland

McCay, B.J., and J.M. Acheson (eds.). 1987. *The Question of the Commons: The Culture and Ecology of Communal Resources.* Tucson: University of Arizona Press

McCracken, F.D., and R.D.S. MacDonald. 1976. 'Science for Canada's Atlantic inshore seas fisheries.' *Journal of the Fisheries Research Board of Canada* 33: 2097–2139

McDonald, D.J. 1968. *Population, Migration, and Economic Development in the Atlantic Provinces.* Fredericton: Atlantic Provinces Economic Council

McDonald, I. 1980. 'W.F. Coaker and the balance of power strategy: the Fishermen's Protective Union in Newfoundland politics.' In J. Hiller and P. Neary (eds.), *Newfoundland in the Nineteenth and Twentieth Centuries: Essays in Interpretation,* pp. 148–80. Toronto: University of Toronto Press

MacDonald, M. and M.P. Connelly. 1986. ' "A Cadillac plant": restructuring the labour process in Nova Scotia fish plants.' Paper presented to the Canadian Political Science Association Annual Meetings, University of Manitoba

MacDonald, R.D.S. 1984. 'Canadian fisheries policy and the development of Atlantic coast groundfisheries management.' In C. Lamson and A. Hanson (eds.), *Atlantic Fisheries and Coastal Communities: Fisheries Decision-Making Case Studies,* pp. 15–71. Halifax: Dalhousie Ocean Studies Programme, Dalhousie University

MacEachern, G. 1987. *George MacEachern: An Autobiography – The Story of a*

Cape Breton Labour Radical, ed. by D. Frank and D. Macgillivray. Sydney: University College of Cape Breton Press

MacEwen, P. 1976. *Miners and Steelworkers: Labour in Cape Breton*. Toronto: Samuel Stevens, Hakkert and Co.

McFarland, J. 1980. 'Changing modes of production in a New Brunswick fish packing town.' *Studies in Political Economy* 4: 99–113

McGee, H. 1978. 'The Micmac Indians: the earliest migrants.' In D.F. Campbell (ed.), *Banked Fires: The Ethnics of Nova Scotia*, pp. 14–40. Port Credit: Scribbler's

Macgillivray, D. 1973. 'Cape Breton in the 1920s: a community besieged.' In B. Tennyson (ed.), *Essays in Cape Breton History*, pp. 49–67. Windsor: Lancelot

– 1980. 'Military aid to the civil power: The Cape Breton experience in the 1920s.' In D. Macgillivray and B. Tennyson (eds.), *Cape Breton Historical Essays*, pp. 95–109. Sydney: University College of Cape Breton Press

MacInnis, D.W. 1978. 'Clerics, fishermen, farmers and workers: the Antigonish movement and identity in eastern Nova Scotia, 1928–1939.' PhD dissertation, McMaster University

– 1983. *A Profile of the Municipality of Clare, Digby County, Nova Scotia*. Halifax: Institute for Resource and Environmental Studies, Dalhousie University

McKay, I. 1983. 'Strikes in the Maritimes, 1901–1914.' *Acadiensis* 13: 3–46

McMahon, F. 1987. 'The New Forest in Nova Scotia.' In G. Burrill and I. McKay (eds.), *People, Resources and Power*, pp. 99–105. Fredericton: Acadiensis

McMullan, J.L., D.C. Perrier, and N. Okihiro. 1988. 'Law, regulation and illegality in Nova Scotia lobster fishery.' Paper presented before the Fifth Congress of the Commission on Folk Law and Legal Pluralism, 12th International Congress of Anthropological and Ethnological Sciences, Zagreb, Yugoslavia, July

MacNutt, W.S. 1965. *The Atlantic Provinces*. Toronto: McClelland and Stewart

Machlup, F. 1952. *The Political Economy of Monopoly*. Baltimore: Johns Hopkins Press

Mandel, E. 1978. *Marxist Economic Theory*. London: Merlin

– 1969. 'Capitalism and regional disparities.' *Socialisme* 17 (April–June). Reprinted by New Hogtown

Mann, M.H. 1966. 'Seller concentration, barriers to entry, and rates of return in thirty industries, 1950–1960.' *Review of Economics and Statistics* 48: 296–307

Mann, S.A., and J.M. Dickinson. 1978. 'Obstacles to the development of a capitalist agriculture.' *Journal of Peasant Studies* 5: 466–81

Marchak, P. 1987. 'Uncommon property.' In P. Marchak, N. Guppy, and J. McMullan (eds.), *Uncommon Property: The Fishing and Fish-Processing Industries in British Columbia*, pp. 3–31. Toronto: Methuen

Marchak, P., N. Guppy, and J. McMullan, eds. 1987. *Uncommon Property: The Fishing and Fish-Processing Industries in British Columbia*. Toronto: Methuen

Marshall, B. 1988. 'Feminist theory and critical theory.' *Canadian Review of Sociology and Anthropology* 25: 208–30

Martin, C. 1974. 'The European impact on the culture of a Northeastern Algonquin tribe: an ecological interpretation.' *William and Mary Quarterly* 31: 3–26

Marx, K. 1967a. *Capital*, Vol. I. New York: International
– 1967b. *Capital*, Vol. III. New York: International

Massey, D. 1977. 'Towards a critique of industrial location theory.' In R. Peet (ed.), *Radical Geography: Alternative View Points on Contemporary Social Issues*, pp. 181–99. London: Methuen
– 1978. 'Regionalism: some current issues.' *Capital and Class* 6: 106–25
– 1979. 'In what sense a regional problem?' *Regional Studies* 14: 233–43
– 1984. *Spatial Divisions of Labour: Social Structures and the Geography of Production*. London: Macmillan

Mattera, P. 1980. 'Small is not beautiful: decentralized production and the underground economy.' *Radical America* 14 (Sept./Oct.): 67–76

Matthews, K. 1968. 'History of the West of England – Newfoundland fishery.' DPhil dissertation, Oxford University

Matthews, R. 1976. *'There's No Better Place than Here': Social Change in Three Newfoundland Communities*. Agincourt, Ont.: Book Society of Canada
– 1983. *The Creation of Regional Dependency*. Toronto: University of Toronto Press
– 1988. 'Federal licencing policies for the Atlantic inshore fishery and their implementation in Newfoundland, 1973–1981.' *Acadiensis* 17: 83–108

Matthews, R., and J. Phyne. 1988. 'Regulating the Newfoundland inshore fishery: traditional values versus state control in the regulation of a common property resource.' *Journal of Canadian Studies* 23: 158–76

Mazany, R.L. 1986. 'Trade flows and market structure: Atlantic Canada and New England trade in fish products.' Paper presented at the Conference on Resource Economics in emerging Free Trade, Orono, Maine

Meillassoux, C. 1972. 'From reproduction to production: a Marxist approach to economic anthropology.' *Economy and Society* 1: 93–105

Mellor, J. 1983. *The Company Store: James Bryson McLachlan and the Cape Breton Coal Miners, 1900–1925.* Halifax: Goodread Biographies

Miller, J., and H. Garrison. 1982. 'Sex roles: the division of labor at home and in the work place.' *Annual Review of Sociology* 8:237–62

Mingione, E. 1978. 'Capitalist crisis, neo-dualism and marginalisation.' *International Journal of Urban and Regional Research* 2: 213–21

– 1983. 'Informalization, restructuring and the survival strategies of the working class.' *International Journal of Urban and Regional Research* 7: 311–39

Mitchell, C.C., and H.C. Frick. 1970. *Government Programs of Assistance for Fishing Craft Construction in Canada: An Economic Appraisal.* Canadian Fisheries Reports No. 14, Economic Branch, Fisheries Service, Department of Fisheries and Forestry. Ottawa: Information Canada

Molinari, B.C. 1977. 'The state and the bourgeoisie in the Peruvian fishmeal industry.' *Latin American Perspectives* 4: 103–21

Morris, L.D. 1985. 'Renegotiation of the domestic division of labour in the context of male redundancy.' In B. Roberts, R. Finnegan, and D. Gallie (eds.), *New Approaches to Economic Life: Economic Restructuring, Unemployment and the Social Division of Labour*, pp. 400–16. Manchester: Manchester University Press

Moser, C.O.N. 1978. 'Informal sector or petty commodity production: dualism or dependence in urban development?' *World Development* 6 (9/10): 1041–64

Mouzelis, N. 1988. 'Sociology of development: reflections on the present crisis.' *Sociology* 22: 23–44

Muise, D. 1980. 'The making of an industrial community: Cape Breton coal towns, 1867–1900.' In D. Macgillivray and B. Tennyson (eds.), *Cape Breton Historical Essays*, pp. 76–94. Sydney: University College of Cape Breton Press

Munro, G. 1979. 'The optimal management of transboundary renewable resources.' *Canadian Journal of Economics* 12: 355–77

– 1980. *A Promise of Abundance.* Ottawa: Economic Council of Canada

Murphy, T. 1987. 'Potato capitalism: McCain and industrial farming in New Brunswick.' In G. Burrill and I. McKay (eds.), *People, Resources and Power*, pp. 19–29. Fredericton: Acadiensis

Murray, F. 1983. 'The decentralisation of production – the decline of the mass-collective worker.' *Capital and Class* 19: 74–99

Murray, R. 1977. 'Value and the theory of rent. Part 1.' *Capital and Class* 3: 100–22

– 1978. 'Value and the theory of rent. Part 2.' *Capital and Class* 4: 11–33

– ed. 1981. *Multinationals beyond the Market: Intra-firm Trade and the Control of Transfer Pricing*. Brighton: Harvester

Muszynski, A. 1987. 'Major processors to 1940 and early labour force: historical notes.' In P. Marchak, N. Guppy, and J. McMullan (eds.), *Uncommon Property*, pp. 46–65. Toronto: Methuen

Myrdal, G. 1957. *Economic Theory and Underdeveloped Regions*. London: Duckworth

Nadel, J. 1986. 'Burning with the Fire of God: Calvinism and community in a Scottish fishing village.' *Ethnology* 25: 49–60

National Sea Products Ltd. 1964–76. *Annual Reports*

Neis, B. 1981. 'Competitive merchants and class struggle in Newfoundland. *Studies in Political Economy* 5: 127–43

– 1987. *The Social Impact of Technological Change in Newfoundland's Deepsea Fishery*. St John's: Labour Canada

Neis, B., and the Fishery Research Group. 1988. 'Taylorism at high tide: the degradation of work at Newfoundland's deepsea fish plants.' *New Maritimes* 6 (April): 3–10

Newby, H. 1977. 'Paternalism and capitalism.' In R. Scase (ed.), *Industrial Society: Class, Cleavage and Control*, pp. 59–73. London: Allen and Unwin

– 1979. *The Deferential Worker*. Madison: University of Wisconsin Press

Newfoundland and Labrador. 1981. *Report of the Royal Commission to Inquire into the Inshore Fishery of Newfoundland and Labrador*. St John's: Commission

Northwest Atlantic Fisheries Organization. 1989. *Scientific Council Summary Document No. N1699*. Dartmouth: NAFO

Norton, B. 1988. 'Epochs and essences: a review of Marxist long-wave and stagnation theories.' *Cambridge Journal of Economics* 12: 203–24

Nova Scotia. 1938. Nova Scotia Economic Council, *Proceedings* of the Nova Scotia Fisheries Conference

– 1944. Royal Commission on Provincial Development and Rehabilitation. Vol. IX. Stewart Bates, Report on the Canadian Atlantic Sea Fishery

Nova Scotia. Department of Development. 1986a. *Industrial Cape Breton Fact Book*. Halifax: Department of Development

– 1986b. *1986 Census of Canada: Statistical Release No. 2. Nova Scotia Population Change*. Halifax: Department of Development

Nova Scotia. Department of Fisheries. 1978. 'Port profile.' Unpublished data, Halifax

Nova Scotia. Department of Municipal Affairs. Community Planning Division. 1977. *Municipality of Clare. Background Studies by the Southwestern Nova Scotia Study Team*. Halifax: Province of Nova Scotia

Nova Scotia. Department of Trade and Industry. 1950. *Nova Scotia Fisheries*

Yearbook 1949–50 and Directory of Fish Packers and Processors. Halifax: Department of Trade and Industry

Oakey, R.P. 1985. 'Innovation and regional growth in small high technology firms: evidence from Britain and the U.S.A.' In D.J. Storey (ed.), *Small Firms in Regional Economic Development*, pp. 135–65. Cambridge: Cambridge University Press

O'Brien, P.J. 1975. 'A critique of Latin American theories of dependence.' In I. Oxaal, T. Barnett, and D. Booth (eds.), *Beyond the Sociology of Development*, pp. 7–27. London: Routledge and Kegan Paul

Ommer, R. 1979. 'From outpost to outport: the Jersey merchant triangle in the nineteenth century.' PhD dissertation, McGill University

– 1981. ' "All the fish of the post": resource property rights and development in a nineteenth-century inshore fishery.' *Acadiensis* 10: 107–23

– 1989. 'The truck system in Gaspé, 1822–77.' *Acadiensis* 19 (1): 91–114

Pahl, R.E. 1984. *Divisions of Labour.* Oxford: Basil Blackwell

– 1985. 'The restructuring of capital, the local political economy and household work strategies.' In D. Gregory and J. Urry (eds.), *Social Relations and Spatial Structures*, pp. 242–66. London: Macmillan

– ed. 1988. *On Work: Historical, Comparative and Theoretical Approaches.* Oxford: Basil Blackwell

Pahl, R.E., and C.D. Wallace. 1985. 'Forms of work and privatisation on the Isle of Sheppey.' In B. Roberts, R. Finnegan, and D. Gallie (eds.), *New Approaches to Economic Life: Economic Restructuring, Unemployment and the Social Division of Labour*, pp. 368–86. Manchester: Manchester University Press

Paine, R. 1988. 'That outport culture.' *Canadian Review of Sociology and Anthropology* 25: 148–56

Patrick, H., and T. Rohlen. 1987. 'Small-scale family enterprises.' In K. Yamamura and Y. Yasuba (eds.), *The Political Economy of Japan*: Vol. I, *The Domestic Transformation*. Stanford: Stanford University Press

Patton, D.J. 1981. *Industrial Development and the Atlantic Fishery: Opportunities for Manufacturing and Skilled Workers in the 1980s.* Toronto: James Lorimer

Pearson, N. 1966. *Town of Glace Bay: Urban Renewal Study.* Halifax: Canadian British Engineering Consultant

Pepin, P.-Y. 1968. *Life and Poverty in the Maritimes.* ARDA Research Report No. RE-3. Ottawa: Queen's Printer

Perroux, F. 1955. 'La notion de pole de croissance.' *Economie appliqué* 1–2

Perulli, P. 1990. 'Industrial flexibility and small firm districts: the Italian case.' *Economic and Industrial Democracy* 11 (3): 337–53

Peterson, S. 1985. 'New Bedford's fish auction.' Institute for Employment Policy, Boston University, mimeo

Phillips, P. 1982. *Regional Disparities*. Toronto: James Lorimer

Piore, M.J., and C.F. Sabel. 1983. 'Italian small business development: lessons for U.S. industrial policy.' In J. Zysman and L. Tyson (eds.), *American Industry in International Competition*, pp. 391–421. Ithaca, NY: Cornell University Press

– 1984. *The Second Industrial Divide*. New York: Basic

Pineo, P., J. Porter, and H. McRoberts. 1977. 'The 1971 census and the socio-economic classification of occupations.' *Canadian Review of Sociology and Anthropology* 14: 91–102

Pool, G. 1988. 'Anthropological uses of history and culture.' *Acadiensis* 18: 226–37

Porter, M. 1983. 'Women and old boats: the sexual division of labour in a Newfoundland outport.' In E. Gamarnikow et al. (eds.), *The Public and Private*, pp. 91–105. London: Heinemann

– 1985a. ' "She was skipper of the shore crew": notes on the history of the sexual division of labour in Newfoundland.' *Labour / Le Travailleur* 15: 105–23

– 1985b. 'The "tangly bunch": outport women of the Avalon Peninsula.' *Newfoundland Studies* 1: 77–90

– 1987. 'Peripheral women: towards a feminist analysis of the Atlantic Region.' *Studies in Political Economy* 23: 41–72

Postner, H.H. 1980. *A New approach to Frictional Unemployment: An Application to Newfoundland and Canada*. Ottawa: Economic Council of Canada

Proskie, J. 1959. *Operations of Modern Fishing Craft, Atlantic Seaboard*. Ottawa: Queen's Printer

Quijano, A.O. 1974. 'The marginal pole of the economy and the marginalised labour force.' *Economy and Society* 3: 393–428

Rainnie, A.F. 1984. 'Combined and uneven development in the clothing industry: the effects of competition on accumulation.' *Capital and Class* 22 (Spring): 142–56

– 1985a. 'Small firms, big problems: the political economy of small businesses.' *Capital and Class* 25 (Spring): 140–68

– 1985b. 'Is small beautiful? Industrial relations in small clothing firms.' *Sociology* 19 (2): 213–21

Raymond, J.L., ed. 1985. *Scotia-Fundy Region Fishing Community Profiles*. Ottawa: Department of Fisheries and Oceans, Economic Branch

Redclift, N., and E. Mingione, eds. 1985. *Beyond Employment: Household, Gender and Subsistence*. London: Basil Blackwell

Redfield, R. 1947. 'The folk society.' *American Journal of Sociology* 52: 292–308

Regier, H.A., and F.D. McCracken. 1976. 'Science for Canada's shelf-sea fisheries.' *Fisheries Research Board Report*, No. 3

Reich, M., D.M. Gordon, and R.C. Edwards. 1973. 'A theory of labor market segmentation.' *American Economic Review* 63: 359–65

Remiggi, F.W. 1979. ' "La lutte du clerge contre le marchand de poisson": a study of power structures on the Gaspé north coast in the nineteenth century.' In L.R. Fischer and E.W. Sager (eds.), *The Enterprising Canadians: Entrepreneurs and Economic Development in Eastern Canada, 1820–1914*, pp. 185–99. St John's: Maritime History Group, Memorial University of Newfoundland

Retson, G.C., and R.R. L'Ecuyer. 1963. *A Study of Rural Problems in Madawaska County, New Brunswick*. Ottawa: Department of Agriculture

Robertson, I., ed. 1988. *Prince Edward Island Land Commission of 1860*. Fredericton: Acadiensis

Robichaud, H.J. 1964. 'Atlantic fisheries in review.' *Atlantic Advocate* 55: 19–20, 22–3

– 1966. 'Atlantic fisheries: blueprint for an orderly revolution.' *Atlantic Advocate* 57: 13, 15–16

– 1967. 'From inshore to offshore.' *Atlantic Advocate* 58: 18–20

Roemer, M. 1970. *Fishing for Growth: Export-Led Development in Peru, 1950–1967*. Cambridge: Harvard University Press

Rosenberry, W. 1988. 'Political economy.' *Annual Review of Anthropology* 17: 161–85

Roy, N., et al. 1981. 'Cost of production in the Newfoundland fish products industry'. Discussion Paper No. 190, Economic Council of Canada. Ottawa: Supply and Services

Rubery, J., and F. Wilkinson. 1981. 'Outwork and segmented labour markets.' In F. Wilkinson (ed.), *The Dynamics of Labour Market Segmentation*, pp. 115–32. London: Academic

Rubin, P.H. 1978. 'The theory of the firm and the structure of the franchise contract.' *Journal of Law and Economics* 21 (1): 223–33

Ryan, S. 1983. 'Fishery to colony: a Newfoundland watershed, 1793–1815.' *Acadiensis* 12: 34–52

Sabel, C., and J. Zeitlin. 1985. 'Historical alternatives to mass production: politics, markets and technology in nineteenth-century industrialisation.' *Past and Present* 108 (Aug.): 133–76

Sacouman, R.J. 1979. 'Underdevelopment and the structural origins of Antigonish Movement co-operatives in eastern Nova Scotia.' In R.J. Brym and R.J. Sacouman (eds.), *Underdevelopment and Social Movements in Atlantic Canada*, pp. 107–26. Toronto: New Hogtown

– 1980. 'Semi-proletarianisation and rural underdevelopment in the Maritimes.' *Canadian Review of Sociology and Anthropology* 17: 232–45

– 1981. 'The "peripheral" Maritimes and Canada-wide Marxist political economy.' *Studies in Political Economy* 6: 135–50

Sacouman, R.J., and D. Grady. 1984. 'The state versus the Woods Harbour 15 + 1043.' *The Maritimes* 2 (Mar.): 4–6

Safa, H.J. 1986. 'Runaway shops and female employment: the search for cheap labor.' In E. Leacock and H.I. Safa (eds.), *Women's Work*, pp. 58–71. South Hadley, Mass.: Bergin and Garvey

Samson, R. 1984. *Fishermen and Merchants in 19th Century Gaspé: The Fishermen-Dealers of William Hyman and Sons*. Ottawa: Parks Canada

Schaefer, M.B. 1957. 'Some considerations of population dynamics and economics in relation to the management of marine fisheries.' *Journal of the Fisheries Research Board of Canada* 14: 669–81

Schmitz, H. 1982. 'Growth constraints on small-scale manufacturing in developing countries: a critical review.' *World Development* 10 (6): 429–50

– 1989. 'Flexible specialization – a new paradigm of small-scale industrialization?' Discussion Paper No. 261, Institute of Development Studies, University of Sussex, Brighton

Schumacher, E.F. 1973. *Small Is Beautiful: Economics as if People Mattered*. New York: Harper and Row

Schumpeter, J.A. 1950. *Capitalism, Socialism and Democracy*. 3rd ed. New York: Harper and Brothers

Scott, A. 1955. 'The fishery: the objectives of sole ownership.' *Journal of Political Economy* 63: 116–24

Scott, A., and P.A. Neher. 1982. *The Public Regulation of Commercial Fisheries in Canada*. Ottawa: Economic Council of Canada

Scott, A.J. 1987. 'The semiconductor industry in south-east Asia: organisation, location and the international division of labour.' *Regional Studies* 21: 143–60

Scott, J.C. 1976. *The Moral Economy of the Peasant*. New Haven: Yale University Press

Seccombe, W. 1987. ' "Helping her out": the participation of husbands in domestic labour when wives go out to work.' Paper presented at the 1987 Annual Meetings of the Canadian Sociology and Anthropology Association, McMaster University, Hamilton

Semmler, W. 1982. 'Theories of competition and monopoly.' *Capital and Class* 18 (Winter): 91–116

Servan-Schreiber, J.J. 1968. *The American Challenge*. New York: Atheneum

Sharp, E. 1976. *A People's History of Prince Edward Island*. Toronto: Steel Rail

Shinohara, M. 1968. 'A survey of the Japanese literature on small industry.' In B.F. Hoselitz (ed.), *The Role of Small Industry in the Process of Economic Growth*, pp. 1–113. The Hague: Mouton

Shrank, W.L., et al. 1980. *The Relative Productivity and Cost-effectiveness of Various Fishing Techniques in the Newfoundland Ground-fishery*. Discussion Paper No. 180, Economic Council of Canada. Ottawa: Supply and Services

Sider, G.M. 1976. 'Christmas mumming and the New Year in outport Newfoundland.' *Past and Present* 71: 102–25

– 1980. 'The ties that bind: culture and agriculture, property and propriety in the Newfoundland village fishery.' *Social History* 5: 1–39

– 1986. *Culture and Class in Anthropology and History: A Newfoundland Illustration*. Cambridge: Cambridge University Press

Simons, H. 1948. *Economic Policy for a Free Society*. Chicago: University of Chicago Press

Sinclair, P. 1983. 'Fishermen divided – the impact of limited entry licensing in northwest Newfoundland.' *Human Organization* 42: 307–13

– 1984. 'Fishermen of northwest Newfoundland – domestic commodity production in advanced capitalism.' *Journal of Canadian Studies* 19: 34–47

– 1985a. 'The state goes fishing: the emergence of public ownership in the Newfoundland fishing industry.' Research and Policy Papers No. 1. St John's: Institute of Social and Economic Research, Memorial University of Newfoundland

– 1985b. *From Traps to Draggers: Domestic Commodity Production in Northwest Newfoundland, 1850–1982*. St John's: Institute of Social and Economic Research, Memorial University of Newfoundland

Singer, H., and R. Jolly. 1973. 'Unemployment in an African setting: lessons of the Employment Strategy Mission to Kenya.' *International Labour Review* 107 (2): 3–115

Smetherman, B.C., and R.M. Smetherman. 1973. 'Peruvian fisheries: conservation and development.' *Economic Development and Cultural Change* 21: 338–51

Smith, J., I. Wallerstein, and H. Evans, eds. 1984. *Households and the World Economy*. Beverly Hills: Sage

Smith, M.E. 1977. *Those Who Live from the Sea: A Study in Maritime Anthropology*. St Paul, Minn.: West Publishing

Smith, W.A. 1963. 'History of the company.' Unpublished manuscript.

Lunenburg Sea Products Collection, Public Archives of Nova Scotia, Halifax, MGS Nos. 646–970

Solinas, G. 1982. 'Labour market segmentation and workers' careers.' *Cambridge Journal of Economics* 6: 331–52

Spencer, G. 1988. 'Informal economic practice as workers' self-activity: a case study in an underdeveloped community.' Paper presented to the Atlantic Association of Sociologists and Anthropologists, Saint Mary's University, Halifax

Steinberg, C. 1984. *Structure and Price Determination in Maritime Port Markets: A Study of Fishermen / Buyer Relations.* Canadian Industry Report of Fisheries and Aquatic Sciences, No. 149. Ottawa: Department of Fisheries and Oceans

Stiles, G. 1972. 'Fishermen, wives and radios: aspects of communication in a Newfoundland fishing community.' In R. Andersen and C. Wadel (eds.), *North Atlantic Fishermen: Anthropological Essays on Modern Fishing*, pp. 35–60. St John's: Institute for Social and Economic Research, Memorial University of Newfoundland

Stolzenberg, R.M. 1978. 'Bringing the boss back in: employer size, employee schooling, and socioeconomic achievement.' *American Sociological Review* 43: 813–28

Storey, D.J. 1982. *Entrepreneurship and the New Firm.* London: Croom Helm
– ed. 1983. *The Small Firm: An International Survey.* London: Croom Helm
– 1985. 'Introduction.' In D.J. Storey (ed.), *Small Firms in Regional Economic Development: Britain, Ireland and the United States,* Cambridge: Cambridge University Press

Storper, M. 1990. 'Industrialization and the regional question in the Third World: lessons of postimperialism; prospects of post-Fordism.' *International Journal of Urban and Regional Research* 14 (3): 423–44

Storper, M., and R. Walker. 1984. 'The spatial division of labor: labor and the location of industries.' In L. Sawers and W.K. Tabb (eds.), *Sunbelt / Snowbelt: Urban Development and Regional Restructuring,* pp. 19–47. New York: Oxford University Press

Surrette, R. 1987. 'Hooked on fish.' *Report on Business Magazine* 4 (Nov.): 52, 54, 56, 58

Tallman, R.D. 1975. 'Peter Mitchell and the genesis of a national fisheries policy.' *Acadiensis* 4: 66–78

Thiessen, V. 1987. 'A correspondence analysis of the division of household labor.' Paper, Department of Sociology and Anthropology, Dalhousie University, Halifax

Thiessen, V., and A. Davis. 1988. 'Recruitment to small boat fishing and public policy in the Atlantic fisheries.' *Canadian Review of Sociology and Anthropology* 25: 603–27

Thompson, P. 1983. *The Nature of Work: An Introduction to Debates on the Labour Process*. London: Macmillan

Thompson, P., with T. Wailey and T. Lummis. 1983. *Living the Fishing*. History Workshop Series. London: Routledge and Kegan Paul

Tolbert, C., P. Horan, and E.M. Beck. 1980. 'The structure of economic segmentation: a dual economy approach.' *American Journal of Sociology* 85: 1095–1116

Tunstall, J. 1968. *Fish: An Antiquated Industry*. London: Fabian Society

United Nations, Food and Agricultural Organization. 1987. *World Nominal Catches*, Vol. 64. Rome: FAO

United States. Department of Commerce. National Marine Fisheries Service. 1985. *Fisheries of the United States*. Washington: Government Printing Office

United States International Trade Commission (USITC). 1985. *Certain Dried Salted Codfish from Canada*. Determination of the Commission in Investigation No. 731-TA-199 (Final). USITC Publication No. 1711. Washington

– 1986. *Certain Fresh Atlantic Groundfish from Canada*. Prehearing Report to the Commission and Parties. Investigation No. 702-TA-257 (Final). Washington

Upton, L.F.S. 1977. 'The extermination of the Boethucks of Newfoundland.' *Canadian Historical Review* 58: 133–53

– 1979. *Micmacs and Colonists: Indian-White Relations in the Maritimes, 1713–1867*. Vancouver: University of British Columbia Press

Uzzell, J.D. 1980. 'Mixed strategies and the informal sector: three faces of reserve labor.' *Human Organisation* 39: 40–9

Veltmeyer, H. 1978. 'Dependency and underdevelopment: some questions and problems.' *Canadian Journal of Political and Social Theory* 2: 55–71

– 1979. 'The capitalist underdevelopment of Atlantic Canada.' In R.J. Brym and R.J. Sacouman (eds.), *Underdevelopment and Social Movements in Atlantic Canada*, pp. 17–36. Toronto: New Hogtown Press

Vogel, L. 1983. *Marxism and the Oppression of Women: Toward a Unitary Theory*. New Brunswick, NJ: Rutgers University Press

Wade, M. 1975. 'After the Grand Derangement: the Acadian return to the Gulf of St. Lawrence and to Nova Scotia.' *American Journal of Canadian Studies* 5: 42–65

Wadel, C. 1969. *Marginal Adaptations and Moderization in Newfoundland*. St John's: Institute of Social and Economic Research, Memorial University of Newfoundland

– 1973. 'Now, Whose Fault Is That?': The Struggle for Self-esteem in the Face of Chronic Unemployment. St John's: Institute of Social and Economic Research, Memorial University of Newfoundland

Watkins, M. 1963. 'A staple theory of economic growth.' Canadian Journal of Economics and Political Science 29: 141–58

– 1980. 'The staple theory revisited.' In J.P. Grayson (ed.), Class, State, Ideology and Change, pp. 373–85. Toronto: Holt, Rinehart and Winston

Watt, J.W. Personal Papers, File No. 94-640-500, in possession of family

– 1963. A Brief Review of the Fisheries of Nova Scotia. Halifax: Department of Trade and Industry

Weiss, L. 1988. Creating Capitalism: The State and Small Business since 1945. New York: Basil Blackwell

Wheelock, I. 1983. 'Competition in the Marxist tradition.' Capital and Class 21 (Winter): 18–47

White, D.J. 1956. The New England Fishing Industry. Cambridge: Harvard University Press

Wilkie, M. 1977. 'Colonials, marginals and immigrants: contributions to a theory of ethnic stratification.' Comparative Studies in Society and History 19 (1): 67–95

Willett, L. 1976. 'The cultural ecology of outport Newfoundland.' BA honours thesis, Dalhousie University, Halifax

– 1980. 'Egbost: the social causes of emigration from Lewis.' MLitt thesis, Aberdeen University

– 1986. 'Gangen Harbour: a study of relations between fishing captains and fish buyers.' Unpublished manuscript, Gorsebrook Research Institute, Saint Mary's University, Halifax

Williams, R. 1977. 'Fish or cut bait.' This Magazine 11 (May–June): 4–7

– 1978. 'Nova Scotia: "Fish at my price or don't fish."' Canadian Dimensions 13: 29–33

– 1979. 'Inshore fishermen, unionisation and the struggle against underdevelopment today.' In R.J. Brym and R.J. Sacouman (eds.), Underdevelopment and Social Movements in Atlantic Canada, pp. 161–78. Toronto: New Hogtown

– 1987a. 'The poor man's Machiavelli: Michael Kirby and the Atlantic fisheries.' In G. Burrill and I. MacKay (eds.), People, Resources, and Power, pp. 67–73. Fredericton: Acadiensis

– 1987b. 'The restructuring that wasn't: the scandal at National Sea.' In G. Burrill and I. MacKay (eds.), People, Resources, and Power, pp. 74–83. Fredericton: Acadiensis

– 1987c. 'Is there life after underdevelopment? An afterword.' In G. Burrill and I. MacKay (eds.), *People, Resources, and Power*, pp. 193–200. Fredericton: Acadiensis

Wilson, J.A. 1980. 'Adaptation to uncertainty and small numbers exchange: the New England fresh fish market.' *Bell Journal of Economics* 11: 491–504

– 1986. 'Free trade in the fisheries of Atlantic Canada and New England: the conditions for mutual gain.' Paper presented at the Conference on Resource Economics in emerging Free Trade, Orono, Maine

Winsor, F. 1987. 'A history of occupational health and safety in Nova Scotia's offshore fishery, 1915–1985.' MA thesis, Atlantic Canada Studies, Saint Mary's University, Halifax

Wolpe, H. 1975. 'The theory of internal colonialism: the South African case.' In I. Oxaal, T. Barnett, and D. Booth (eds.), *Beyond the Sociology of Development*, pp. 227–52. London: Routledge and Kegan Paul

Wong, B. 1987. 'The role of ethnicity in enclave enterprises: a study of the Chinese garment factories in New York City.' *Human Organisation* 46: 120–30

Wynn, G. 1981. *Timber Colony: A Historical Geography of Early Nineteenth Century New Brunswick*. Toronto: University of Toronto Press

Young, J.K. 1962. 'A progressive firm plans further advances.' *Atlantic Fishermen and Shipping Review*, August

Contributors

Pauline Barber, after obtaining an MA in social anthropology from the University of Auckland, New Zealand, emigrated to Canada in 1976. She has taught extensively at universities in the areas of development, gender, and social change. In 1989 she completed her doctoral thesis at the University of Toronto, household economic strategies and deindustrialization in Cape Breton. A comparative study of class, gender, and culture in the southwest Nova Scotian fishery was projected for post-doctoral research, and subsequent research is planned to introduce comparative international perspectives, both from the European periphery and from Australia.

Kevin Barrett received his bachelor of commerce degree from Mount Allison University in 1983. He then went on to earn the designated chartered accountant's degree in 1985. Barrett lives in Liverpool, NS.

Anthony Davis is an associate professor of sociology and anthropology at St Francis Xavier University, Antigonish, NS. Born and raised in Nova Scotia, he has since 1974 been doing research and writing about socioeconomic organization and public policy in the Atlantic Canadian fisheries.

Marie Giasson is a post-doctoral fellow of the Department of Anthropology at Laval University. In addition to teaching Social Anthropology of the Middle East, she has carried out field research on maritime communities in Brazil, Israel, and Canada (southwest Nova Scotia and Quebec's Lower North Shore). Interested in rural development, community studies, and ethnicity, she is involved with international development organizations such as FAO and CIDA and is currently conducting fieldwork on a forestry development project in Honduras.

Leonard Kasdan is professor of social anthropology and resource and environmental studies at Dalhousie University. He has carried out research in the Middle East, Scotland, and Atlantic Canada. His main interests are cultural and social change, particularly as affected by changes in technology and government policies.

Leigh Mazany is an associate professor of economics at Dalhousie University. She has been engaged in fisheries research since 1982 and has published a number of monographs and articles. She has also served as adviser to the Eastern Fishermen's Federation on a study of seal damage and to the Fisheries Council of Canada on countervailing duties.

Lawrence Willett, after graduating from the Maritime Forest Ranger School, worked for twenty years as a wildlife and forestry technician in Nova Scotia and latterly as a waterfowl bander in Labrador and a guide in northwestern Ontario. He graduated from Dalhousie and Aberdeen universities with theses on cultural ecology, emigration and religion, and environmental philosophy. Willett has done fieldwork in Newfoundland and the Outer Hebrides, has lectured in cultural ecology at Saint Mary's, and has worked as a researcher and consultant in archaeology, ethnohistory, socioeconomic studies, wildlife management, and environmental studies.